4th International Workshop & Summer School on Plasma Physics 2010

Journal of Physics: Conference Series Volume 516

Kiten, Bulgaria
5-10 July 2010

ISBN: 978-1-63439-297-6
ISSN: 1742-6588

Printed from e-media with permission by:

Curran Associates, Inc.
57 Morehouse Lane
Red Hook, NY 12571

Some format issues inherent in the e-media version may also appear in this print version.

Printed by Curran Associates, Inc. (2014)

For permission requests, please contact the Institute of Physics
at the address below.

Institute of Physics
Dirac House, Temple Back
Bristol BS1 6BE UK

Phone: 44 1 17 929 7481
Fax: 44 1 17 920 0979

techtracking@iop.org

Additional copies of this publication are available from:

Curran Associates, Inc.
57 Morehouse Lane
Red Hook, NY 12571 USA
Phone: 845-758-0400
Fax: 845-758-2634
Email: curran@proceedings.com
Web: www.proceedings.com

Table Of Contents

PAPERS

PLASMA SOURCES, DIAGNOSTIC AND TECHNOLOGY

PLASMA MODELLING AND FUNDAMENTALS

FUSION PLASMA AND MATERIALS

Preface

Fourth International Workshop & Summer School on Plasma Physics 2010

The Fourth International Workshop & Summer School on Plasma Physics (IWSSPP'10) is organized by St. Kliment Ohridsky University of Sofia, with co-organizers TCPA Foundation, Association EURATOM/IRNRE, The Union of the Physicists in Bulgaria, and the Bulgarian Academy of Sciences. It was held in Kiten, Bulgaria, at the Black Sea Coast, from July 5 to July 10, 2010. The scientific programme covers the topics **Fusion Plasma and Materials; Plasma Modeling and Fundamentals; Plasma Sources, Diagnostics and Technology.**

As the previous issues of this scientific meeting (IWSSPP'05, *J. Phys.: Conf. Series* **44** (2006) and IWSSPP'06, *J. Phys.: Conf. Series* **63** (2007), IWSSPP'08, *J. Phys.: Conf. Series* **207** (2010)), its aim was to stimulate the creation and support of a new generation of young scientists for further development of plasma physics fundamentals and applications, as well as to ensure an interdisciplinary exchange of views and initiate possible collaborations by bringing together scientists from various branches of plasma physics.

This volume of *Journal of Physics: Conference Series* includes 34 papers (invited lectures, contributed talks and posters) devoted to various branches of plasma physics, among them fusion plasma and materials, dc and microwave discharge modelling, transport phenomena in gas discharge plasmas, plasma diagnostics, cross sections and rate constants of elementary processes, material processing, plasma-chemistry and technology. Some of them have been presented by internationally known and recognized specialists in their fields; others are MSc or PhD students' first steps in science. In both cases, we believe they will raise readers' interest.

We would like to thank the members of both the International Advisory Committee and the Local Organizing Committee, the participants who sent their manuscripts and passed through the (sometimes heavy and troublesome) refereeing and editing procedure and our referees for their patience and considerable effort to improve the manuscripts. We would like to express our gratitude to the invited lecturers who were willing to pay the participation fee. In this way, in addition to the intellectual support they provided by means of their excellent lectures, they also supported the school financially.

E. Benova

Group Photo

Peer review statement

All papers published in this volume of *Journal of Physics: Conference Series* have been peer reviewed through processes administered by the proceedings Editors. Reviews were conducted by expert referees to the professional and scientific standards expected of a proceedings journal published by IOP Publishing.

Journal of Physics: Conference Series **516** (2014) 012001 doi:10.1088/1742-6596/516/1/012001

Microwave measurement of electrical fields in different media – principles, methods and instrumentation

Plamen I Dankov
Sofia University "St. Kliment Ohridski", Faculty of Physics, James Bourchier blvd., Sofia 1164, Bulgaria

E-mail: dankov@phys.uni-sofia.bg

Abstract. This paper, presented in the frame of 4th International Workshop and Summer School on Plasma Physics (IWSSPP'2010, Kiten, Bulgaria), is a brief review of the principles, methods and instrumentation of the microwave measurements of electrical fields in different media. The main part of the paper is connected with the description of the basic features of many field sensors and antennas – narrow-, broadband and ultra-wide band, miniaturized, reconfigurable and active sensors, etc. The main features and applicability of these sensors for determination of electric fields in different media is discussed. The last part of the paper presents the basic principles for utilization of electromagnetic 3-D simulators for E-field measurement purposes. Two illustrative examples have been given – the determination of the dielectric anisotropy of multi-layer materials and discussion of the selectivity of hairpin-probe for determination of the electron density in dense gaseous plasmas.

1. Introduction

The determination of the high-frequency electrical (E) fields is not a primary type of measurement in the microwave range like the measurements of power, frequency, spectrum, device's S-parameters and noise. In fact, the local E-field magnitude around a given radiator or in a given electro-magnetic (EM) media could be determined by the same equipment for measurements of power, spectrum and S-parameters – power meters, spectrum analyzers and network analyzers. A necessity to obtain an actual information about the E-field intensity, orientation and distribution appears for three basic groups of measurement: 1) EMC/EMI measurements (electromagnetic compatibility/interference); 2) antenna measurements (mainly for communication purposes) and 3) characterization of different materials and media (crystals, reinforced substrates, liquids, powders, absorbers, thin films, biological tissues, gaseous plasmas, etc) and imaging. The EMC/EMI measurements will not be considered here; they are very important in the communications and for determination of the safety standard for non-ionisating radiation [1]. We will focus our discussions here on the main principles, methods and instrumentation of the E-field measurements; the variety of E-field sensors and their implementation and finally, on the utilization of the modern EM 3-D simulators as an assistant tool for measurement purposes.

2. Electrical-field measurements' principles and instrumentation

2.1. Measurement principles

Let us start with the discussion of figure 1, where the conditional scheme of the E-field measurements is presented. The unknown signal falls over a sensor or an antenna and the equivalent response after a conversion (into dc, RF or IR power; or directly) is received and the detected signal is displayed. This

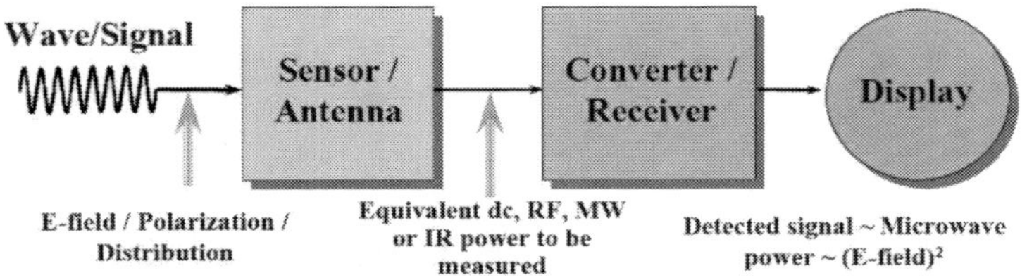

Figure 1. Conditional block scheme for measurement of high-frequency electric fields.

displayed response is usually proportional to the microwave power (or E-field square). The basic power sensors are the well-known termistors, thermocouples and diode detectors [2]. In the next section we will focus our attention to the description of variety of E-field probes and antennas as an effective receiving equipment for determination of the local E field and imaging.

There exist three basic methods for receiving of the unknown signal – see figure 2. The simplest way is the power detection. The detector converts the radio-frequency (RF) power into an equivalent dc or low-frequency (1–100 kHz) current/voltage. The sensitivity is not high – the minimal detecting power is ~ 10^{-7} mW (or –70 dBm) and the dynamic range between the maximal and minimal receiving power is typically 30-40 dB up to 70 dB. More effective is the down-conversion (or mixing), where the unknown signal is mixed with a controlled signal by a local oscillator (LO) and the response is measured at the intermediate frequency (IF) – typically several hundred MHz. The response could be digitalized by analog-to-digital converter (ADC) and then treated by a digital signal processing (DSP). The main benefit of the mixing is that the IF frequency falls into the frequency range, where the noise components is formed mainly by the thermal noise and the low-frequency $1/f$-noise does not appear. Thus, the sensitivity increases – the minimal detecting power is now ~10^{-12} mW (or –120 dBm) and the dynamic range increases up to 100–110 dB. The biggest sensitivity and dynamic range (~140 dB) can be achieved by the sampler – here the LO is a frequency synthesizer, which produces a fixed array ("comb") of LO-frequencies with high enough levels, stable amplitudes and frequencies.

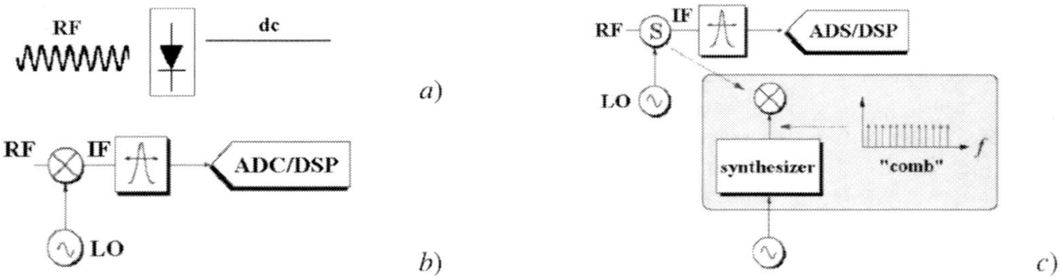

Figure 2. Three basic methods for signal registration: *a*) by a detector; *b*) by a mixer; *c*) by a sampler.

2.2. Main measurement schemes

Figure 3 illustrates three basic schemes for determination of the E fields. When the media under test (MUT) produces own RF fields, the simplest scheme consists of a receiving antenna (probe) A_R and a spectrum analyser SA (preferable highest sensitivity tool) or wattmeter W (cheaper tool with lower sensitivity). This scheme is applicably also for "near-field" scanning/imaging. If the MUT is passive, it could be illuminated by a known RF signal (A_T) and the response signal (A_R, A_T) (transmitted and/or reflected) will be a measure of the MUT dielectric properties. This case, known as a "free-space" method, is applicable using network analyzer NA [3]. The most sensitive way to determine the dielectric properties is the resonance method – now the MUT is a resonator itself (or its filling) and the field response is a good measure of the material properties of this media using NA. Figure 4 illustrates the combination of the necessary equipment (W, SA and NA) – the "dream" of each RF engineer.

Journal of Physics: Conference Series **516** (2014) 012001 doi:10.1088/1742-6596/516/1/012001

Finally, let us discuss the measurement strategy. It is important to note, that the power-flux density $\overline{S}$ is the basic "measurable" parameter in the microwave range. If $\overline{S}$ is known (measured), the necessary

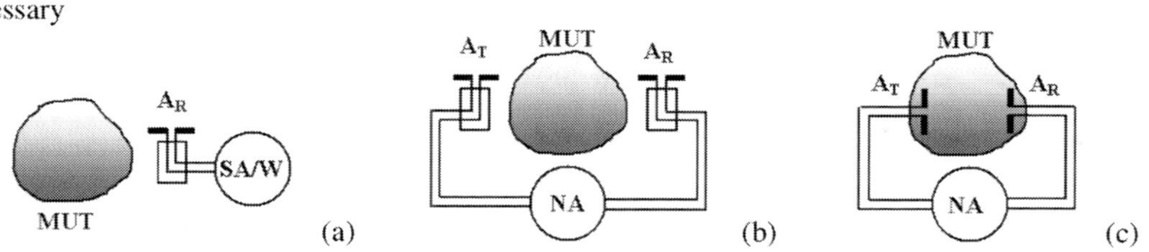

Figure 3. Three basic measurement schemes for determination of the E fields in a media under test (MUT) ($A_{T,R}$ – transmit/receive antennas): (a) characterization of the own transmitted fields in MUT by SA (spectrum analyser) or W (wattmeter). This application is very suitable also for "near-field" scanners; (b) illumination of MUT and measurement of the field response by NA (network analyser) – "free-space" method; (c) MUT as a resonator – characterization of its dielectric properties of MUT

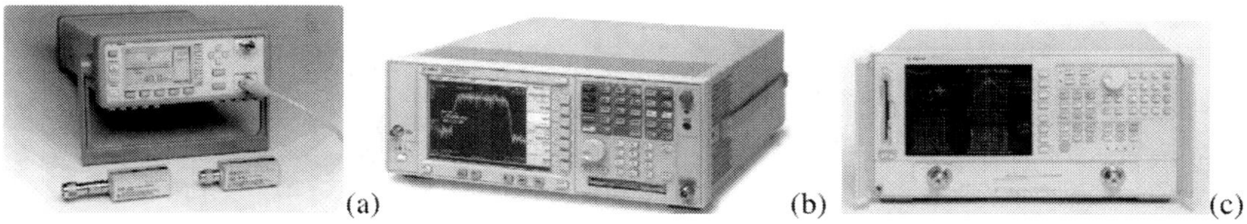

Figure 4. Basic measurement equipment: (a) wattmeter W; (b) spectrum analyser SA; (c) network analyser NA

averaged fields $\overline{E}$ and $\overline{H}$ can be easy expressed through $\overline{S}$ for plane waves using the expressions (1, 2) (see figure 5(a)). A direct measurement of the E field is also possible, when the antenna factor *AF* of the measurement antenna is known – see figure 5(b). The antenna factor is the ratio of incident local E-field intensity E_{local} (upon the antenna) to the measured voltage V_{meas} that is produced at the antenna terminal – a 50-ohms load.

$$\overline{S} = 0.5\,\mathrm{Re}(\mathbf{n}.\mathbf{S}) = 0.5\,\mathrm{Re}(E_p \times H_p^*)_z = \mathrm{Re}(\overline{E}_{rms} \times \overline{H}_{rms}^*)_z \quad (1)$$

$$\overline{E} = 19\sqrt{\overline{S}}\text{, V/m;}\quad \overline{H} = 4.8\times10^4\sqrt{\overline{S}}\text{, A/m;}\quad \overline{S}\text{, W/m}^2\text{;} \quad (2)$$

$$E_{local}\text{, V/m} = AF\text{, 1/m} \times V_{meas}\text{, V} \quad (3)$$

$$AF(f)\text{, 1/m} = 9.73/(\sqrt{g}\,\lambda_0\text{, m}) \quad (4)$$

g – antena gain as a ratio

$$P_R = P_T g_T g_R (\lambda_0/4\pi d)^2 \quad (5)$$

$$d > R_{ffz} = 2D^2/\lambda_0 \quad (6)$$

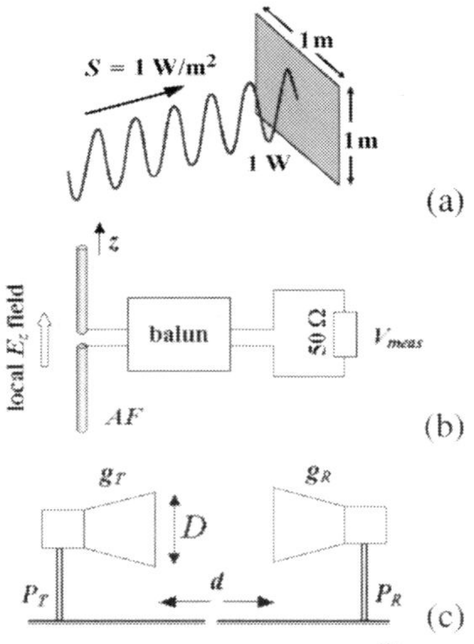

Figure 5. Basic relations between the average power-flux density $\overline{S}$, average E-field intensity $\overline{E}$ and average magnetic (H-) field intensity $\overline{H}$. Denotations: E_p, H_p – peak-amplitude fields, E_{rms}, H_{rms} – "root-mean-square" fields, known also as average or effective ("measurable") fields; E_{local} – local E field; *AF* – antenna factor; g – antenna gain; $P_{T,R}$ – transmitted/received power; λ_0 – wavelength; *d* – distance between the antennas; R_{ffz} – antenna far-field zone radius; *D* – antenna diameter/dipole length

If the test antenna is reference, the frequency-dependent parameter AF is usually given by the manufacturer; if not, the AF can be derived from the antenna gain g using the expression (4). The unknown antenna gain g can be easy obtained by the known two-antenna method – see figure 5(c). Two equal antennas – receiving and transmitting, are placed at a distance d, which is greater than the so-called "far-field-zone" radius R_{ffz} of the antenna – equation (6). If the both antennas are equal, their common gain $g = g_T = g_R$ can be calculated using the equation (5), if the transmitted P_T and received power P_R have been measured. Thus, the presented equations (1–6) can help practically in each situation for E-field measurement.

3. Field sensors

In this section we discuss the variety of field sensors, probes and antennas for receiving of the signal. The discussion includes mainly the problems how to do sensors broadband, miniaturised and tuneable. This is not a full description – only the well-known and some modern solutions have been given.

3.1. Simple dipoles, monopoles and apertures

The simplest way to "see" the local E field in a given place is the wire probe. Figure 6 (a), (b) illustrates the most spread field sensors [4, 5] – dipoles and monopoles from electric (open wires) and magnetic (loops) types. The length of the dipoles is usually chosen $\sim\lambda_0/2$, while the length of the monopole is $\sim\lambda_0/4$, but this is not a rigorous rule (the monopole is half of dipole placed on top of a ground plane). Contrariwise, if the length of the used probe is $D \langle\langle \lambda_0/2$, this is a typical local sensor for detecting the field, nevertheless its small sensitivity. The strongest signal in the feed lines appears, if the E-probe is parallel to the E field direction, or if the H-probe loop plane is perpendicular to the H field. A popular example is illustrated in figure 7 for excitation of the TE_{011} mode in the dielectric resonator – two positions of the E- and H-probes are given, for which this mode could be (or could not be) excited.

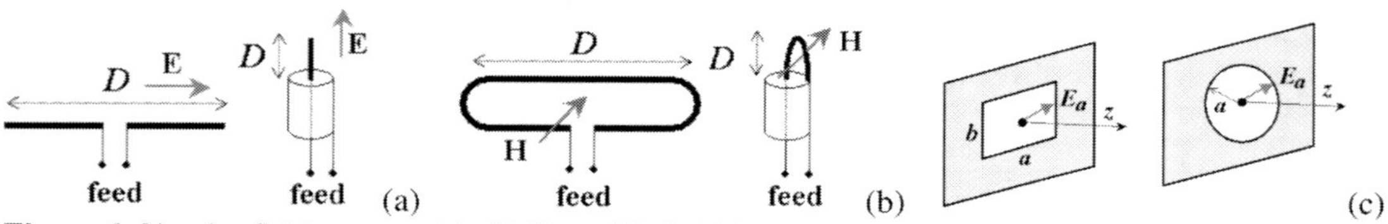

Figure 6. Simples field sensors: (a), (b) E- and H-field dipole and monopole; (c) Huygens' apertures ($a.b << \lambda_0^2$)

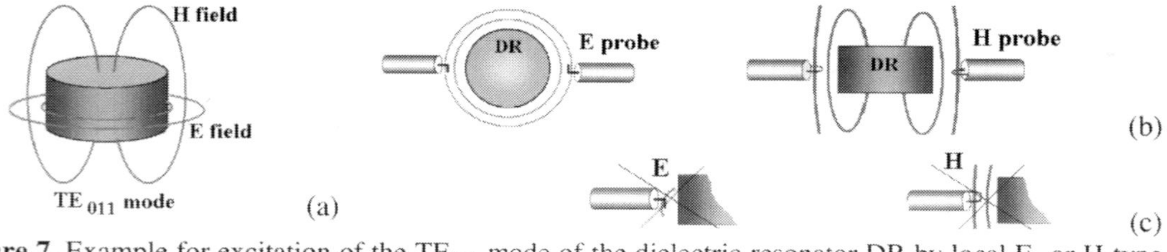

Figure 7. Example for excitation of the TE_{011} mode of the dielectric resonator DR by local E- or H-type probes: (a) mode fields, (b) the TE_{011} mode is excited; (c) the TE_{011} mode is not excited.

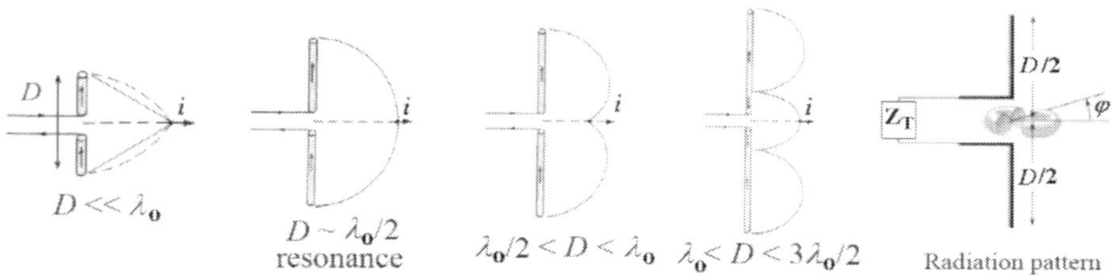

Figure 8. E-field dipole with increasing length and the concept for the radiation pattern introduction [4]

If the length of the dipole wires is $\lambda_0/2$, the probe becomes resonance (see figure 8) and the current response in the centre increases. The gain of similar probe is ~1.64; for example, using expression (4) for resonance dipole at 2.45 GHz the antenna factor AF is 62.06 1/m or 35.85 dB/m. Increasing the dipole length, the distribution of the induced current changes and the concept of the radiation pattern should be introduced (figure 9(a)). The definition of the radiation pattern is the normalized E-field distribution $F(\varphi,\theta) = E(\varphi,\theta)/E_{max}$ according to the azimuthal and elevation angles to the maximal field.

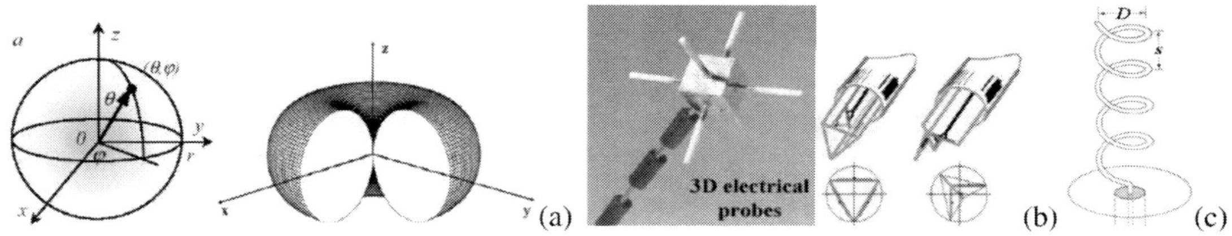

Figure 9. (a) Azimuthal φ and elevation θ angles in the antenna far-field region and example for an omnidirec-tional radiation pattern; (b) 3-D E probes; (c) electrical monopole as a helix for circularly-polarised E field

This distribution should be determined at a distance $r > R_{ffz}$; this condition restricts the applicability of the antenna as close as possible to a given radiator. For example, the considered above 2.45-GHz resonance dipole has a far-field-zone radius R_{ffz} ~ 61.2 mm (or ~$5\lambda_0$). Figure 9(a) presents one of the most spread radiation-pattern diagrams – the so-called "omnidirectional" diagram of the resonance dipole. This diagram means that the dipole has biggest sensitivity perpendicularly on its axis (along z), but can "detect" the E field practically from each directions excepting exactly the direction 0z. An impor-tant parameter is also the antenna beamwidth – the angle width of the radiation diagram at level –3 dB from the maximum. The –3-dB beamwidth is a good measure of the "directivity" of an antenna. An pure isotropic radiation pattern does not exist. Thus, if the task is to measure the total E field in a given point, $E_\Sigma = \sqrt{E_x^2 + E_y^2 + E_z^2}$, 3-D dipoles or monopoles should be used – see examples in figure 9(b).

The regular dipole is suitable for detection of E field with a linear polarization, when the travelling wave keeps one direction of his E-field. For the waves with circular polarization the end of the E-field vector describes a circle. The right-hand circular polarization RHCP is defined for transmitting antenna, when the E-field vector rotates clockwise (note, that due to the reciprocal principle the parameters of one passive antenna in transmitting or receiving regime are fully equal). Figure 9(c) shows an example for monopole with helix wire, suitable for measurement of circularly polarized E fields.

Other types of simple local field sensors are the elementary Huygens' apertures (see figure 6(c)). If the aperture area is small compared to λ_0^2, the radiation pattern is wide, practically without back lobe – figure 10(a). As long the aperture diameter $2a$ increases, the radiation diagram becomes narrower, the antenna gain increases, but side lobes appear – see the example in figure 10(b) for circular horn antenna with radius $a = 3\lambda_0$. For very low frequency the dimension of similar apertures becomes extremely big. Figure 10(c) shows a useful solution – utilization of a pair of spherically curved E- and H-type dipoles, which acts like a quasi-Huygens aperture with very similar radiation diagram like the normal aperture.

The aperture antennas – horns, reflectors, slot antennas, planar patches, lens antennas, etc. [4–5] are wide spread also like high-gain and high-efficiency field sensors, but they will not be discussed here.

3.2. Broad-band, ultra-wideband and frequency-independent sensors

The resonance thin-wire dipoles have narrow (~10 %) normalized bandwidth BW = $\Delta f/f_0$, where $\Delta f = f_2 - f_1$ is the difference between the frequencies $f_{1,2}$ for achieving of –10-dB return losses in the reflected from the antenna signal and f_0 is the central frequency determined by the dipole length. The reason of the narrow-band operation is the high quality factor of the considered $\lambda_0/2$-resonator; this behaviour is due to the strong coherent diffraction between the both wire ends [4]. In fact, the short dipole with length $D << \lambda_0/2$ has much wider BW, but the radiation efficiency (and therefore, the sensor sensitivity) is very weak. One can increase the BW of the resonance dipole by using of more complex wire geometries – see figure 11(a), (b). The main reason for increasing of BW in these structures is the more complicated incoherent diffraction from the curved edges in these cases.

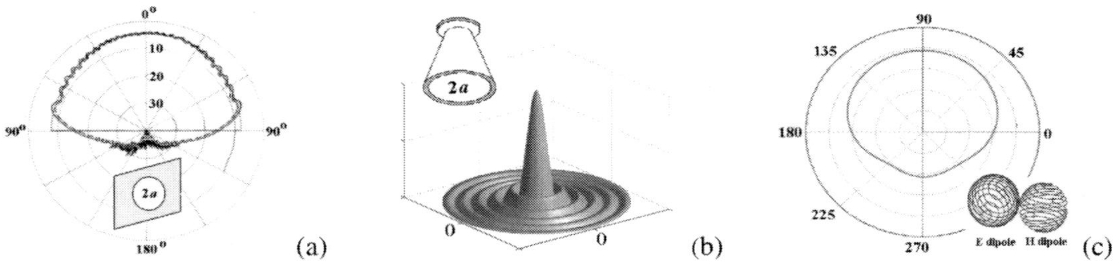

Figure 10. Radiating apertures [6]: (a) radiation pattern of small circular aperture with radius $a = 0.42\lambda_0$; (b) 3-D radiation pattern of circular horn with radius $a = 3\lambda_0$; (c) imitation of a reduced-size quasi-Huygens source at very low frequency by a coupled pair of H and E curved dipoles and view of its radiation pattern.

Thus, the quality factor decreases and the BW become wider (for example the ratio f_2/f_1 becomes up to 2:1 and even 3:1 – figure 11(c)). The upper frequency limit of these constructions is determined by the parasitic excitation of high-order modes that causes pattern narrowing and multiple radiation lobes.

Another more effective way for achieving of ultra-wide band antenna operation is the adopting of frequency-independent (figure 12(a), (b)) or frequency-scaled geometries (figure 12(c), (d)). The log-periodic antenna is defined as a structure whose electrical properties vary periodically with the logarithm of the frequency and a multi-decade operation can be reached. In fact, the frequency independence occurs for a certain low-frequency cut-off (for which the longest tooth length is $\lambda_0/4$).

3.3. Traveling-wave sensors: surface-wave and leaky-wave probes

The considered above dipoles/monopoles are typical resonant field sensors, which operation is based on the formed standing waves in the fed wire conductors. The travelling-wave sensors have non-resonant action, based on an effective propagation of slow or fast waves along the antenna axis [5]. The slow-wave antennas ($v_{ph}/c < 1$) are based on the radiation of surface waves along delaying systems – dielectric rods, helixes, series of metal rods (Yagi-Uda antennas – see figure 12(a)), etc. Since the surface waves radiate only at discontinuities, nonuniformities and curvatures at the open interface, the total pattern of this "end-fire" type of antennas is formed by the interference between the feed lines and the end-terminal patterns. Usually the surface-wave antenna has narrow BW and limited gain (less than 20 dB) – see figure 12(b). A very interesting solution for surface-wave E-field probe is the printed version, so-called quasi-Yagi antenna, which can be applied up to millimetre-wave range.

The leaky-wave antenna is basically a waveguiding structure that possesses a mechanism that permits it to leak power in a tilt direction all along its length during the propagation of fast waves ($v_{ph}/c > 1$).

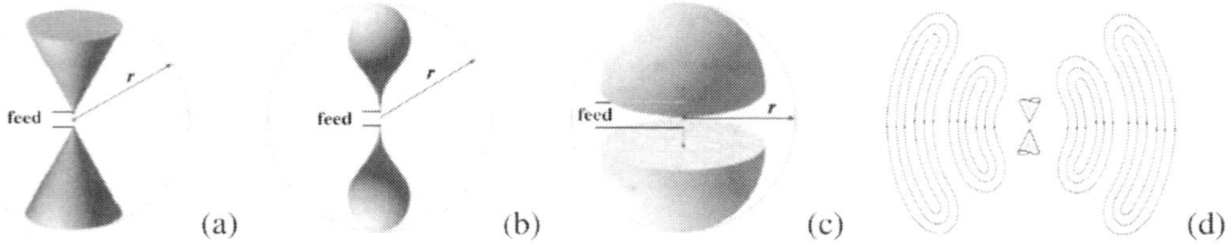

Figure 11. Broad-band dipoles: (a) bi-conical; (b) tapered (intermediate bandwidth BW); (c) hemispherical BW (wide BW); (d) the bi-conical dipole is very suitable for detection of quasi-spherical waves [4].

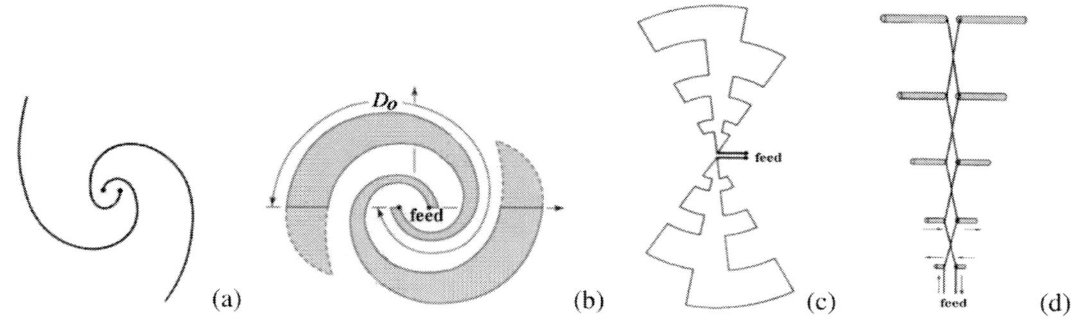

Figure 11. Ultra-wide band and frequency-independent dipoles: (a) spiral dipole; (b) planar spiral dipole; (c) log-periodical antenna; (d) log-periodical dipole with crisscross connection [4].

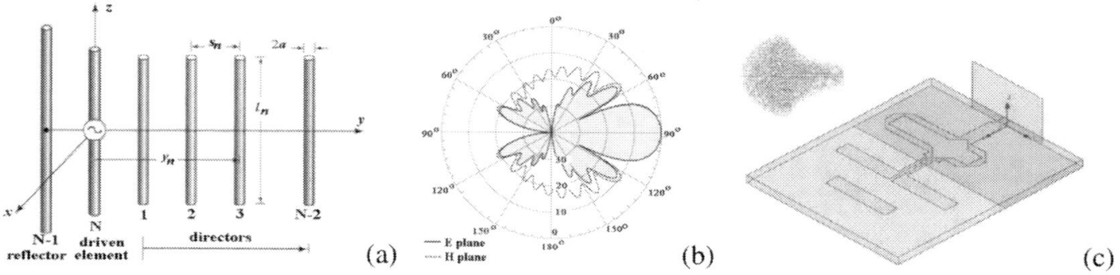

Figure 12. Travelling-wave dipoles: (a) ordinary Yagi-Uda TV antenna; (b) its radiation diagram; (c) planar quasi-Yagi antenna – an effective sensor for detection or excitation of surface waves [4].

To be able to radiate effectively before its end, this antenna should be relatively long (~20 λ_0); the beamwidth of such antenna depends on the leakage constant and the beam direction depends on the frequency. Thus, the leaky-wave antenna can be scanned by varying the frequency. A typical example for leaky-wave antenna is the array of nonresonant slots in rectangular waveguides (not shown here).

3.4. Miniaturized sensors using fractals and metamaterials

The achieving of broadband operation is an important task for the E-field probes, but the other important task is the size miniaturization. The simplest way is the placing of the dipole in a dielectric body; the reducing factor is $1/\sqrt{\varepsilon_r}$; ε_r is the dielectric constant. The other way, especially at low frequencies, is the use of lumped LC elements. But there exist more efficient methods given below.

The utilization of the fractal geometry is one of the so-called natural-based antenna designs. In this case the miniaturization of the antennas is based on employing of the fractals [4, 5]. The original term "fractal" means broken or irregular fragments to describe a family of complex shapes that possess an inherent and self-similarity or self-affinity in their geometrical structures. Figure 13(a) shows how the known straight dipole could be reconstructed to the so-called Koch fractal dipole – it is composed from small but exact copies of itself. The result after the 5 stages is an increasing of the effective dipole length, nevertheless that the length of the straight line is equal in all the stages. Big 2-D sizes' reduction can be achieved with the Hilbert curve. It is an example of a space-filling fractal curve that is self-avoiding (i.e., has no intersection points) – see figure 13(b). Used as GPS antenna the size can be minimized more than 3 times. Figure 13(c) shows the Sierpinski gasket dipole. Similar antenna elements have multiband operation; this is due to the many self-similar objects, which radiate in different frequency ranges. The main disadvantage of the fractal antennas is the decreased radiation efficiency.

The other attractive way for miniaturisation of the antenna dimensions is the using of the metamaterial filling. There are a variety of engineered electromagnetic materials [4], implemented into the antenna design. We will consider here only an example for using of embedded-circuit metasubstrates.

Figure 13. Miniaturised dipoles using fractals: (a) Koch fractal dipole formed by 5 stages; (b) Hilbert-curve dipole used for extremely small GPS antenna inserted into USB devices; (c) Sierpinski's bow-tie dipole

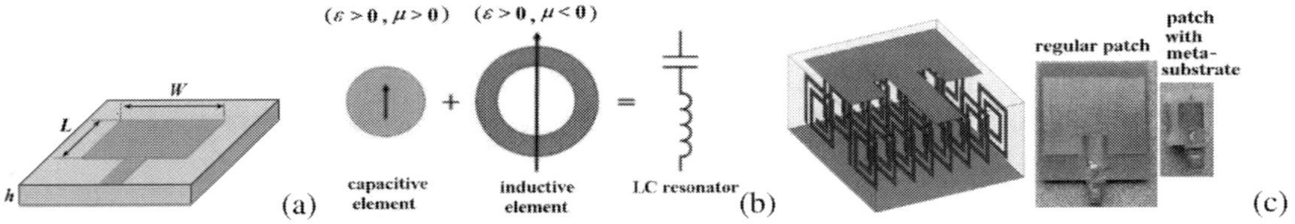

Figure 14. Miniaturised patches: (a) ordinary microstrip patch; (b) concept for using of metamaterials; (c) miniaturized microstrip patch using embedded circuit metasubstrate compared with the original patch

The regular dielectrics and magnetodielectrics used in the antenna design have positive values of the dielectric and magnetic constants ($\varepsilon > 0$, $\mu > 0$). Plasmas ($\varepsilon < 0$, $\mu > 0$) and gyrotropic ferrites ($\varepsilon > 0$, $\mu < 0$) in external magnetic field can acquire negative values of one of the material constants. The materials with double negative material constants ($\varepsilon < 0$, $\mu < 0$) are not found in the nature, but can be physically realizable. How these materials can help for the miniaturization? The operation frequency of the microstrip patch made by regular dielectrics is ~ $1/\sqrt{LC}$, where C is the element capacitance and the L is the small inductance (figure 14(a), (b)). If the filling is a properly designed embedded-circuit meta-substrate, the absolute values of the effective material constants ($|\varepsilon|$ and especially $|\mu|$) can strongly increase, the corresponding element C and especially L increase, and the resonance frequency decrease. This means that for the same frequency the size of the microstrip patch is considerable smaller (more than 4 times) – see figure 14(c). The radiation patterns of the both patches are similar but the BW of the patch with metasubstrate is wider.

3.5. Reconfigurable sensors. Plasma dipoles and patches

Let us finally consider the possibility to tune the operation frequency of the E-field sensors. One very efficient way to do reconfigurable antennas is the utilization of dc-controlled MEMS switches (Micro Electro-Mechanical System). Figure 15(a), (b) gives an example for a planar reconfigurable monopole with 2 MEMS's ensuring 3 different effective wire lengths and operating at 3 different frequencies – figure 15(c). The speed of the modern MEMS switches is high enough and this fact allows a relatively easy online tuning of the antenna frequency range, which is important for communication applications. The plasma antennas also have possibility to be tuneable – figure 16. There exist two main ways to construct reconfigurable plasma antenna. The first one is the changing of the effective dipole length by the control of the plasma discharge length (in this case the plasma column is used instead of the metal wire). The second way is the using of the plasma medium as an electronically controlled metematerial. Figure 16(b) shows a simple construction – regular patch covered by plasma sheath. The operation frequency of the antenna can be changed by control of the plasma frequency – figure 16(c).

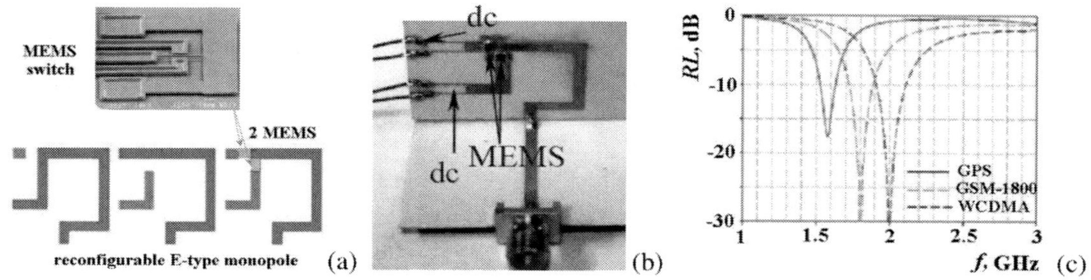

Figure 15. Reconfigurable E-field monopoles: (a) Printed monopole with 2 MEMS for obtaining of three wire lengths; (b) experimental monopole; (c) measured return losses for the three cases

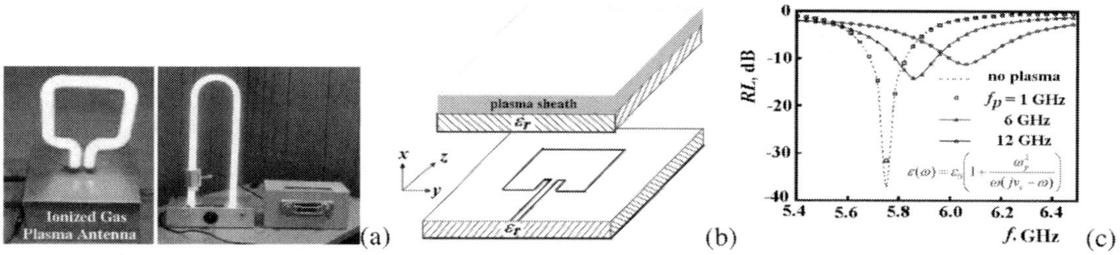

Figure 16. Plasma dipoles and patches: (a) ionized gas plasma loops; (b) microstrip patch with controlled plasma sheath; (c) measured return losses of the microstrip plasma patch in the range 5.4-6.4 GHz

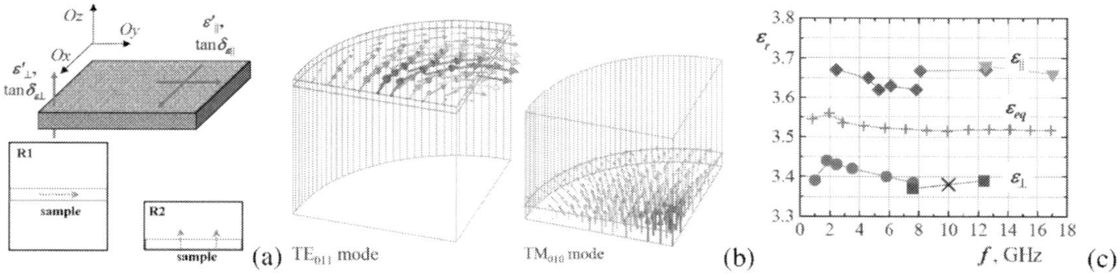

Figure 17. Substrate anisotropy: (a) two-resonator method; (b) E-field distribution of the considered modes; (c) measured parallel, perpendicular and equivalent dielectric constant Ro4003 substrate by two-resonator method

4. Utilisation of 3D electromagnetic simulators for E-field measurement purposes

Finally, we briefly consider a specific, but important problem – the utilization of the EM simulators as assistant tool for microwave measurements, including the E-field measurements. The modern 3-D simulators based of powerful EM methods [4] (FDTD, FEM, MoM, etc.) have great capability for accurate numerical simulations of very complicated EM structures. In fact, the modern style of design of the RF and antenna engineers is exactly based on the construction of extremely detailed structure models with expectations to achieve more accurate design. Similar approach is used with more or less success to do simulations, which support the measurement process in the microwave range. Contrariwise, we propose an opposite approach for application of the 3-D simulators for measurement purposes – construction of stylized 3-D models without unnecessary elements (connectors, excitation loops or probes, tuning elements, etc.), which ensure simplicity, accuracy and computer time and memory efficiency. In order to be more concrete, we present two examples.

4.1. Example 1 – determination of dielectric substrate anisotropy using two-resonator method

The modern RF substrates used in the electronics have well expressed dielectric anisotropy – different dielectric parameters (dielectric constant ε_r and dielectric loss tangent $\tan\delta_\varepsilon$) parallel and normal to the substrate surface – figure 17(a). The anisotropy is not included into the manufacturer catalogues; this fact is very inconvenient for many RF and antenna designers, which want to do more accurate design. Due to this reason we developed the two-resonator method [7] for measurement of the dielectric anisotropy of multilayer materials. This method employs two resonators, which support modes with appropriate orientation of the E fields in order to determine the necessary parameters parallel and perpendicular to the substrate surface. For this purpose we use 3-D simulators to support the measurements. Figure 17(b) shows the simulated E-field distribution for two simple cases – for TE_{011} mode in the first cylindrical resonator R1 (the E field is parallel to the sample surface) and for TM_{010} mode in the second resonator R2 (the E field is normal to the sample surface). The presented pictures illustrate the used style for construction of the 3-D resonator models. They have been built as pure cylindrical and plane surfaces without any coupling details. In order to ensure coincidence between the measured and calculated resonance parameters (frequency and Q factor) of pure (without sample) resonators equivalent parameters have been introduced – equivalent diameters and equivalent wall conductivity. Fast simulations can be realized, if the whole resonators have been splitted and only symmetrical parts of the resonator have been simulated (1/8 R1 and 1/4 R2) using appropriate magnetic-wall boundary conditions – "E-field symmetry" (if the E field is parallel to the surface) or "H-field symmetry" (if the E field is perpendicular to the surface). Finally, the simulations of the resonators with samples easy give the dielectric parameters. This approach ensures both accuracy and fast simulations, and the measurement errors are kept quite low, ±3 % for ε_r and ±7 % for $\tan\delta_\varepsilon$. An example for the measured dielectric constant anisotropy is shown in figure 17(c) for one of the most popular RF substrate Ro4003.

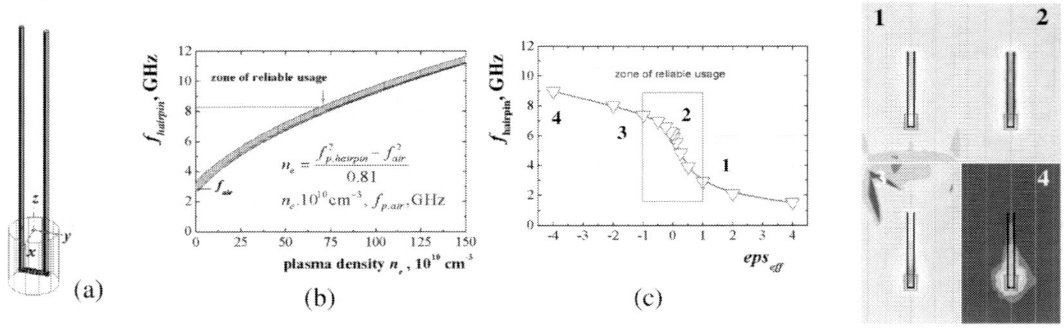

Figure 18. Hairpin resonator as a microwave probe for determination of the plasma density: (a) 3-D model; (b) dependence of the hairpin resonance frequency v/s the plasma density; (c) dependence of the resonance frequency v/s the effective plasma permittivity and simulated E-field distributions in 4 different points

4.2. Example 3 – sensitivity of the hairpin resonance probe for determination of plasma density

The hairpin resonator can be used as a very effective E-field probe for determination of the plasma density – figure 18. The main benefit of this method is that the measurements are based on determination of the resonance frequency f_{haipin} instead of any absolute measurements of E-filed magnitudes, which ensure better sensitivity, accuracy and measurement simplicity. The hairpin probe is a quarter-wave two-wire resonator, which can be relatively easy designed for TEM-mode operation [9]. The open end of the structure has maximum of the E field of the standing wave, which makes this resonance probe enough sensitive to the changes of the dielectric parameters of the surrounding medium – the plasma media in this case. The opposite short end has maximum of the H field, which allows to construct quite stable and reliable coupling between the resonator and the feed coaxial cable using H-type loop probe (figure 6(b)). But how to determine the measurement sensitivity of this resonance probe?

For this purpose we construct a simplified 3-D model of the hairpin resonator with a small ceramic support at the short end without introducing the coupling loop model – figure 18(a). Then we simulate this structure for different effective dielectric constant of the plasma medium (ε_{eff} in the interval from 4 to –4). The results have been presented in figure 18(c). We can see that the measurement sensitivity is higher (the curve slope in bigger) in the interval $\varepsilon_r \subseteq (1, -1)$. Moreover, the E-field distributions in several specific points (1–4) over the plotted curve have been given, which confirm the loss of sensitivity of the hairpin probe for $|\varepsilon_{eff}| > 1$. In these points (3, 4) the maximum of the E filed of the standing wave changes from the open end (when $\varepsilon_{eff} > 0$) to the short end (due to the negative sign of ε_{eff}) and the probe does not more react to the dielectric constant changes.

Acknowledgments
The author thanks to Scientific Research Fund of Sofia University for the necessary support.

5. References

[1] 1995 *Handbook of Electromagnetic Compatibility* ed. Perez R, Academic Press Inc.
[2] Agilent® Application Note 1449-1/2/3/4 *Fundamentals of RF and Microwave Power Measurements*; Agilent® Application Note 150-1 *Spectrum Analysis. Amplitude and Frequency Modulations*; on-line: www.agilent.com
[3] 2005*Free-space measurement seminar* Agilent Technology "; on-line: www.agilent.com
[4] 2008 *Modern Antenna Handbook* ed. Balanis C A, John Wiley & Sons Inc.
[5] 2007 *Antenna Engineering Handbook* ed. Volakis J, McGraw-Hill Co.
[6] Orfanidis S J, *EM Waves and Antennas* on-line: www.ece.rutgers.edu/~orfanidi/ewa
[7] Dankov P I 2006 *IEEE Trans. on MTT* **54** 1534-44
[8] Dankov P I, Stefanov P, Gueorguiev V and Ivanov T 2010 *J. Phys.: Conf. Ser.* **207** 202-9

4th International Workshop & Summer School on Plasma Physics 2010 IOP Publishing
Journal of Physics: Conference Series **516** (2014) 012002 doi:10.1088/1742-6596/516/1/012002

Microwave plasma torches used for hydrogen production

F M Dias, N Bundaleska, J Henriques, E Tatarova, and C M Ferreira

Instituto de Plasmas e Fusão Nuclear, Instituto Superior Técnico, 1049-001 Lisboa, Portugal

E-mail: francisco.dias@ist.utl.pt

Abstract. A microwave plasma torch operating at 2.45 GHz and atmospheric pressure has been used as a medium and a tool for decomposition of alcohol in order to produce molecular hydrogen. Plasma in a gas mixture of argon and ethanol/methanol, with or without water, has been created using a waveguide surfatron launcher and a microwave generator delivering a power in the range 0.2–2.0 kW. Mass, Fourier Transform Infrared, and optical emission spectrometry have been applied as diagnostic tools. The decomposition yield of methanol was nearly 100 % with H_2, CO, CO_2, H_2O, and solid carbon as the main reaction products. The influence of the fraction of Ar flow through the liquid ethanol/methanol on H_2, CO, and CO_2 partial pressures has been investigated, as well as the dependence of the produced H_2 flow on the total flow and power. The optical emission spectrum in the range 250–700 nm has also been detected. There is a decrease of the OH(A-X) band intensity with the increase of methanol in the mixture. The emission of carbon atoms in the near UV range (240–300 nm) exhibits a significant increase as the amount of alcohol in the mixture grows. The obtained results clearly show that this microwave plasma torch at atmospheric pressure provides an efficient plasma environment for hydrogen production.

1. Introduction

In the recent years, great attention has been paid to the new hydrogen based technologies. Hydrogen is considered as a reliable alternative energy source, which liberates a considerable quantity of ecologically pure energy per unit weight (120 kJ/g) without emitting greenhouse gases [1]. Moreover, hydrogen can be viewed as a renewable fuel and it can be easily converted into electricity by fuel cells. Therefore, investigation of new hydrogen sources is of the utmost importance. The idea to focus on alcohols, especially ethanol, is due to the fact that it is a renewable resource in contrast to CH_4, which is usually used for this purpose. Ethanol can be obtained from cellulosic materials, corn or sugar.

Presently, hydrogen is produced by fuel processing technologies converting different hydrocarbons into hydrogen rich stream. The main conventional techniques used to produce hydrogen from hydrocarbon fuels (steam reforming, partial oxidation, and autothermal reforming) have problems like catalyst poisoning, size and weight requirements, compactness, dynamic behaviour (sluggish response), and limitations on hydrogen production from heavy hydrocarbons [1–5]. Among the possible innovative options, plasma systems can provide original solutions to these drawbacks [4–17]. In the plasma, the overall reforming reactions are in fact the same as in conventional reforming, but the energy and the free radicals used for the reforming reaction are provided by the plasma itself. The plasma eliminates the need for a catalyst in the systems. It plays a catalytic role because highly active species such as electrons, ions and radicals may significantly enhance the reaction rate. Plasma systems can be applied to various hydrocarbons including natural gas, gasoline, heavy oils, and bio fuels, as well as biomass. Moreover, the plasma high energy density leads to compactness of the system and fast response times can be achieved due to the electrical operation of the system. Generally, non-thermal plasmas due to their non-equilibrium properties have low power requirements and capacity to induce physical and chemical reactions in gases at relatively low temperature [8]. Different types of non-thermal plasmas have been used for reforming of hydrocarbons: the corona discharge [10]; the dielectric barrier discharge (DBD) [12, 14]; gliding arc technology. The comparison between different arc discharge-based systems shows that the gliding arc technology meets the best performances due to its relative set-up simplicity and high flexibility, which allows operation over a wide range of flow rates and the treatment of a large amount of chemical species [8]. Different DC and AC corona discharges operating at room temperature have been applied to decompose alcohols for hydrogen production [15–17].

In this context, microwave discharge based systems, which are electrodeless, have not yet been thoroughly investigated as plasma reformers. Some results obtained with microwave discharges at reduced pressure are presented in [18]. Microwave plasmas operating at atmospheric pressure have been the subject of increased attention during the last decade as well. It has been shown that these torches provide suitable conditions to dissociate molecules in the abatement systems, to burn out chemical and biological warfare agents, and to atomize materials [19–30]. One of the main merits of microwave plasma torches at atmospheric pressure is that they make it possible to inject large power densities into the plasma and thus achieve high densities of active species of interest. An essential advantage of microwave plasmas over other types of discharges, such as gliding arcs, corona and dielectric barrier discharges, is the absence of electrodes [1–3]. In this way, problems such as electrode erosion, discharge pollution, and additional energy consumption for cooling are avoided. Besides, working at atmospheric pressure would also make the industrial application less complicated. The capability of microwave plasmas generated by surface waves at atmospheric pressure to decompose various alcohols (methanol, ethanol, butanol, and propanol) and to produce hydrogen has been tested in [31]. In particular, surface wave driven discharges offer several advantages over other types of discharges: they can operate with high stability and reproducibility over a large range of operation conditions; they require lower power to achieve similar values of electron density (note that electrons are responsible for the creation of active species and for the first step in the decomposition processes); they produce large plasma volumes with different possible geometries providing high enough residence times. The present investigation aims at finding optimal conditions for the plasma decomposition process of alcohol, i.e., the highest H_2 yield, the appropriate gas flow, input power and reforming substance (methanol/ethanol with or without water). For this purpose, plasma in a gas mixture of argon and methanol/ethanol (CH_3OH, C_2H_5OH) has been created using a waveguide surfatron based set-up [32]. The influence of the Ar "bubbling" flow through the liquid methanol/ethanol on the H_2, CO_2, and CO partial fluxes has been investigated. The dependence of the produced H_2 flow on the total Ar flow and power delivered to the launcher have also been analyzed. In order to investigate the contribution of steam reforming reaction, water vapor has also been introduced into the mixture. Mass spectrometry has been used to detect molecular hydrogen produced by methanol/ethanol decomposition. Fourier Transform Infrared Spectroscopy has been applied to detect different species such as H_2O, CO_2, and CO molecules, which are produced as a result of plasma driven decomposition processes and subsequent association processes. The optical emission spectrum in the range 250–700 nm was investigated to detect the presence of the species of interest in the discharge.

2. Experimental set-up

A microwave plasma torch operating at 2.45 GHz and atmospheric pressure has been used. Plasma in a gas mixture of argon and alcohol (methanol/ethanol), with or without water, has been created using a waveguide surfatron launcher [32] and a microwave generator delivering a power in the range 0.2–2.0 kW. The generator is connected to a waveguide (WR-340) system, which includes an isolator, directional couplers, and a 3-stub tuner. The above system is terminated by a movable short-circuit, which allows the maximization of the electric field at the launcher position. The discharge takes place in a quartz tube (1.5 cm inner and 1.8 cm outer diameters, respectively). The quartz tube is inserted vertically and perpendicularly to the waveguide wider wall. The total gas flow used in the experiments ranges from 500 to 4000 sccm. The argon gas used has a purity of 99.999 %. Alcohol (methanol/ethanol) is introduced into the discharge by bubbling Ar through the alcohol filled tank at room temperature as shown in the figure 1. The bubbling method allows one not to use any additional energy source in the process. The total flow Q_t consists of two parts. One part goes through the tank containing the alcohol to drag the molecules (bubbling). The resulting mixture (Ar + alcohol) flow Q_A is combined with the second part of pure Ar flow Q_{Ar} and the total flow $Q_t = Q_A + Q_{Ar}$ is then introduced into the quartz tube, where the discharge takes place. When water vapor is additionally introduced into the plasma, the pure Ar flow is divided in two different fractional "bubbling" flows. The first one is formed by "bubbling" of Ar through the alcohol tank, Q_A, and the second one is formed by "bubbling" of Ar through a water tank, Q_W. Then, the two fractional flows are combined and the total gas flow $Q_t = Q_A + Q_W$ is introduced into the discharge. The water in the tank was heated up to 70 °C.

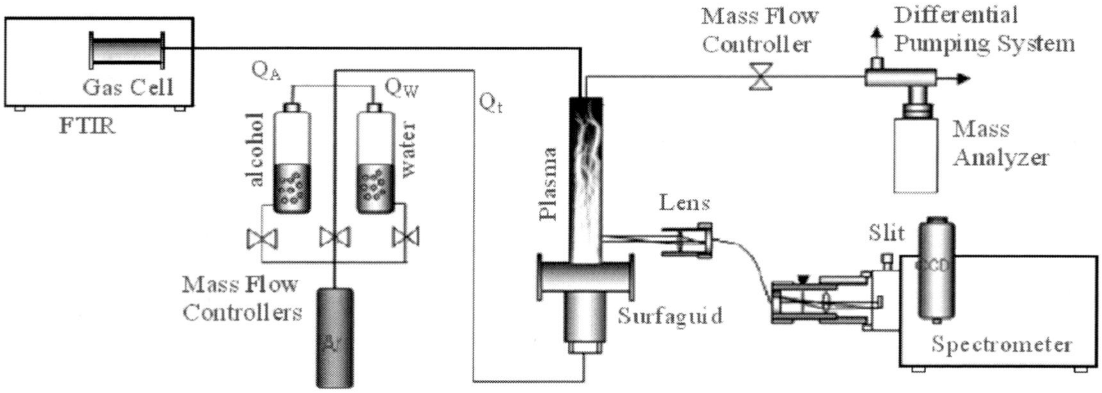

Figure 1. Experimental set-up

Three types of diagnostics have been carried out. Mass Spectrometry and Fourier Transform Infrared Spectroscopy (FTIR) have been used to analyze the exhaust gas stream fluxes. Gas sampled portions of the output gas stream from the plasma reactor were directed to the mass analyzer (Stanford Research System RGA 200) and to the FTIR spectrometer. The absorption spectra of H_2O, CO, CO_2, and C_2H_5OH (ethanol) species were detected by a FTIR Termo Nicolet 5700 spectrometer in the wave number range of 1000–4000 cm^{-1}. At the same time, light emitted by the plasma radiative species was collected perpendicularly to the discharge tube by an optical fibre and directed to the entrance slit of a Jobin-Yvon Spex 1250 spectrometer equipped with a liquid–nitrogen cooled CCD camera. The plasma emission in the 250–850 nm range has been investigated.

3. Experimental results

In order to demonstrate the feasibility and effectiveness of the microwave plasma decomposition processes of methanol and ethanol, the hydrogen yield dependence on the fluxes and the microwave input power has been investigated both in Ar and Ar + water environments. Separate experiments have been performed for methanol and ethanol.

Journal of Physics: Conference Series **516** (2014) 012002 doi:10.1088/1742-6596/516/1/012002

3.1. Decomposition of methanol

3.1.1. Influence of total flow rates and input power. The influence of the Ar "bubbling" flow Q_A, total flow Q_t, and microwave power delivered to the launcher on the molecular hydrogen production have been investigated. The species in the exhaust gas stream have been detected in order to investigate the effectiveness of the argon microwave plasma to break down the bonds of methanol molecules and release hydrogen molecules. There is a clear jump in H_2 partial pressure, as measured with the mass analyzer, when the microwave plasma is turned on, which is a clear evidence of H_2 production from methanol decomposition by the plasma. The time evolution of the H_2 partial pressure detected by the mass analyzer is presented in figure 2. The H_2 partial pressure increases with the Ar "bubbling" flow. The relative fractions of Q_A in the total flow Q_t were 3, 6 and 9 % in the above case. The investigation shows that the variation of Q_A is linearly correlated with the actual methanol flux (see figure 3). The CO partial pressure also shows linear dependence on Q_A. This is expected since H_2 and CO are the methanol dissociation products according to the reaction $CH_3OH \rightarrow 2H_2 + CO$ (R1). Other species detected by the mass analyzer (Ar, N_2, H_2O) do not show changes when the plasma is turned on or turned off, which indicates that they are not related to processes in the plasma. Also, a small black deposit appears on the discharge wall when methanol is introduced in the Ar plasma environment.

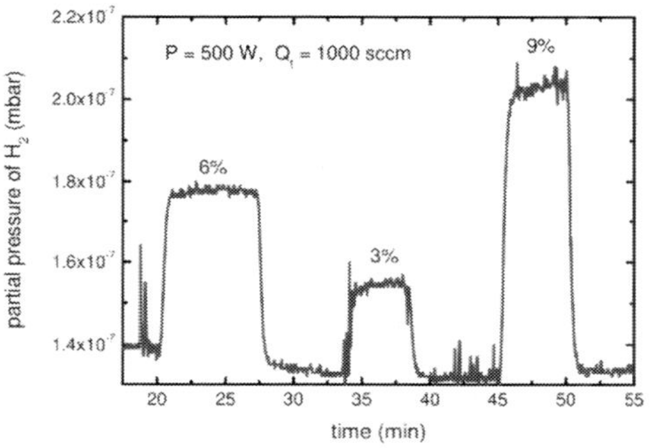

Figure 2. Time evolution of H_2 partial pressures during the plasma decomposition of methanol.

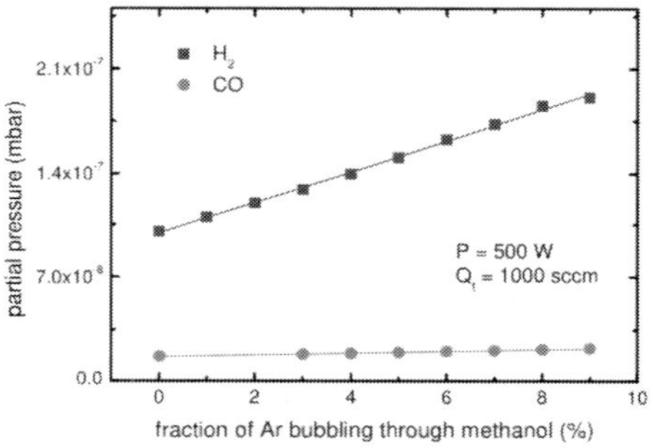

Figure 3. Dependence of the H_2 and CO partial pressures on the fraction of Ar bubbling through methanol.

The molecular hydrogen production dependence on the pure methanol flow for different total flows is shown in figure 4. The solid line shown on this figure represents the theoretical dependence according to the dissociation reaction (R1). As seen, for a total flux Q_t in the range 500–2000 sccm the produced hydrogen partial flow agrees with the predictions from (R1). This means that every CH_3OH molecule yields two H_2 molecules as a result of the plasma stimulated dissociation process (nearly 100 % efficiency of the process).

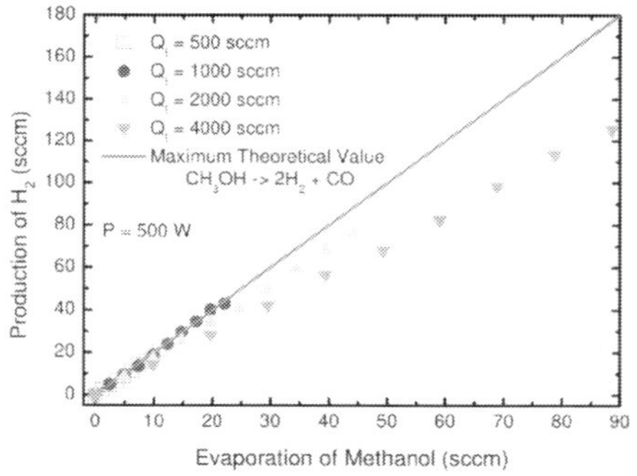

Figure 4. Hydrogen flow vs. pure methanol flow.

The microwave input power is also an important parameter in the plasma decomposition process. The microwave power absorbed per unit volume is linearly correlated with the electron number density. The electrons are responsible for the first step of the decomposition processes since they produce reactive species in the discharge via collisions with the gas atoms and molecules. The dependence of the hydrogen yield on the total microwave power delivered to the launcher in the range 100–700 W is shown in figure 5. Here, the total flow Q_t and the fractional Ar "bubbling" Q_A flow are kept constant. The observed increase in hydrogen yield with the input power can be explained by the fact that more active species are generated at higher microwave power. However, some saturation in hydrogen production is observed for microwave powers greater than 400 W, which corresponds to a power density of about 10 W/cm^3. This suggests that increasing power density above this value induces intensive chemistry involving dissociation of hydrogen molecules followed by association processes of highly reactive H atoms with other radicals to form hydrogen containing species.

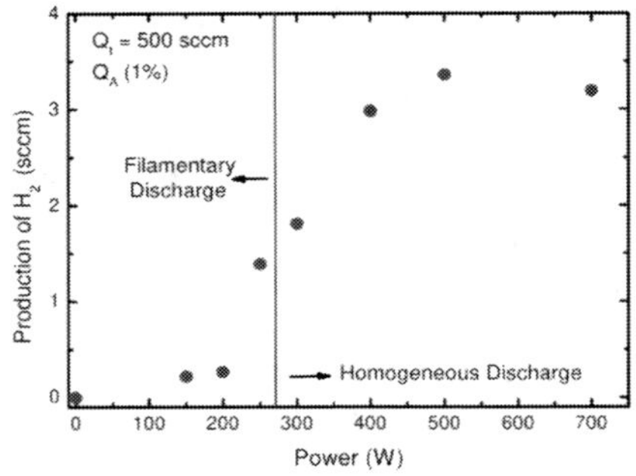

Figure 5. Dependence of H_2 flow on power

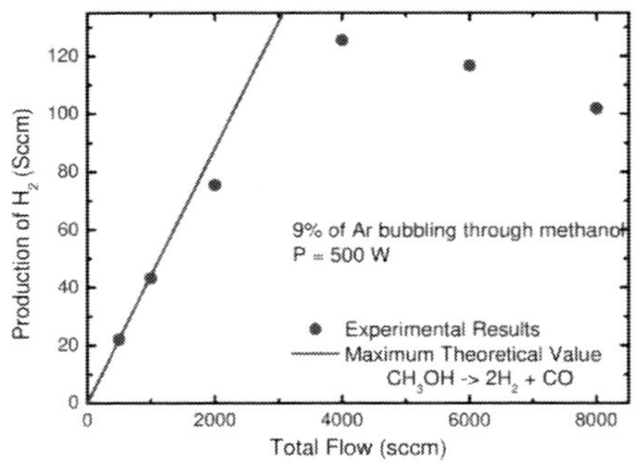

Figure 6. Dependence of H_2 flow on total flow

The influence of the total flow on the H_2 production is presented in figure 6. The fractional Ar "bubbling" Q_A flow is kept constant. As seen, for small total flow the H_2 production reaches the maximum theoretical value, i.e., the introduced methanol is all reformed. As the total flow increases, strong deviations from the maximum theoretical value and even a decrease in H_2 production above 4000 sccm are observed. This can be explained by the decrease of the amount of methanol in the mixture. Moreover, the residence time also decreases at higher gas velocities.

3.1.2. Effect of additional water vapor flow. The effect of a small addition of water vapor on the hydrogen and the carbon oxide/dioxide yields has also been investigated (figure 7). To this end, Ar "bubbling" flow through the water tank, Q_W, was combined with the Ar–methanol flow Q_A.

The water in the tank has been heated to about 70°C for this purpose. The corresponding chemical reaction is $CH_3OH + H_2O \rightarrow 3H_2 + CO_2$ (R2), and an increase in the H_2 and CO_2 yields is observed. It should be noted that the methanol decomposition reaction (R1) is followed by the so called water–gas shift reaction $CO + H_2O \rightarrow H_2 + CO_2$ (R3). The relative contribution of the latter reaction will depend on the quantity of introduced water: if there is enough water, CO should not be present at all. For very small methanol amounts, we observe in figure 7 a changing slope of H_2 production as water is added. The reason is that 3 hydrogen molecules per single methanol molecule are formed instead of 2. So, if the water quantity is larger or equal to that of methanol we can expect a 3:2 increase in H_2 production.

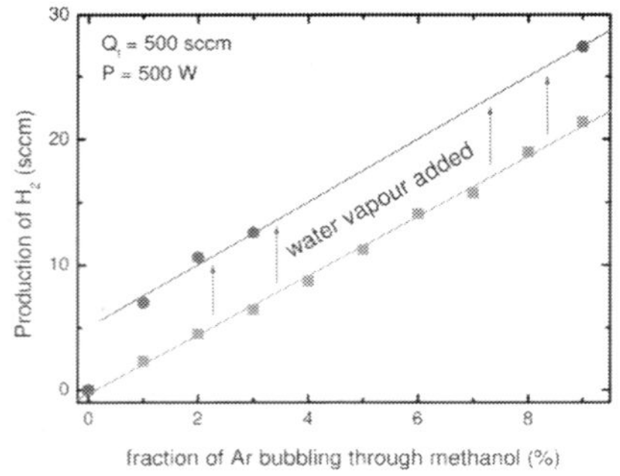

Figure 7. Produced H_2 flow vs. Ar "bubbling" Q_A flow through methanol

3.2. Decomposition of ethanol

3.2.1. Influence of total flow rates. An ethanol solution 96 % ethanol + water has been introduced in the Ar plasma environment by the "bubbling" method described above. Figures 8(a, b) show two FTIR spectra in the 1000–4000 cm^{-1} wave number range as obtained before and after the microwave plasma torch is applied, respectively, for 500 W of applied power, 500 sccm total Ar flow, and two different fractional Ar "bubbling" flows. The spectral lines at wave numbers around 3000 and 1000–1500 cm^{-1} are a signature of the ethanol presence in the exhaust stream before the plasma is turned on. Obviously, ethanol molecules are very efficiently decomposed in the Ar plasma environment and CO_2 and H_2O are observed as the by-products.

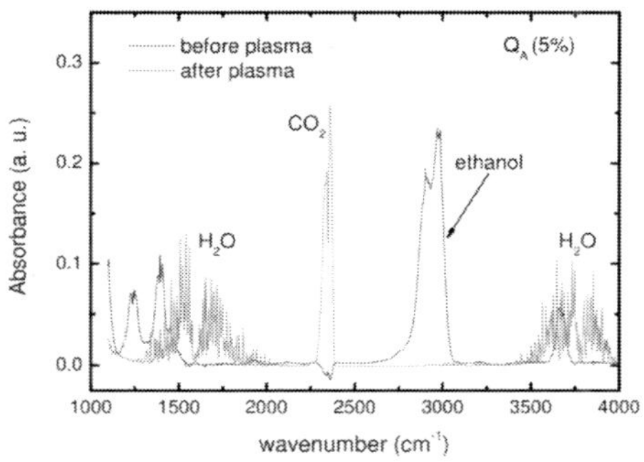

(a)

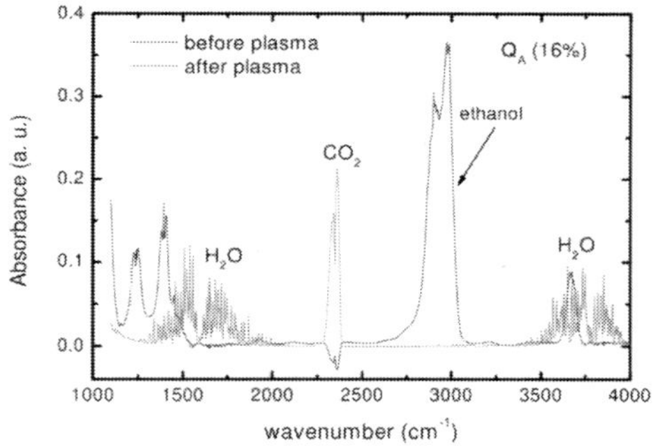

(b)

Figure 8. Infrared absorption spectra of Ar + ethanol mixture before and after plasma processing for two different fractional Ar "bubbling" flows Q_A (a) 5 and (b) 16 %. Total flow Q_t = 500 sccm.

Ethanol dissociates through the reaction $C_2H_5OH \rightarrow CO + 3H_2 + C$ (R4). The presence of CO_2 and water in the exhaust stream can be explained by the conversion of part of CO into CO_2 as a result of the water gas shift reaction (R3). Moreover, the presence of a small quantity of H_2O can initiate the steam reforming reaction $C_2H_5OH + H_2O \rightarrow 2CO + 4H_2$ (R5).

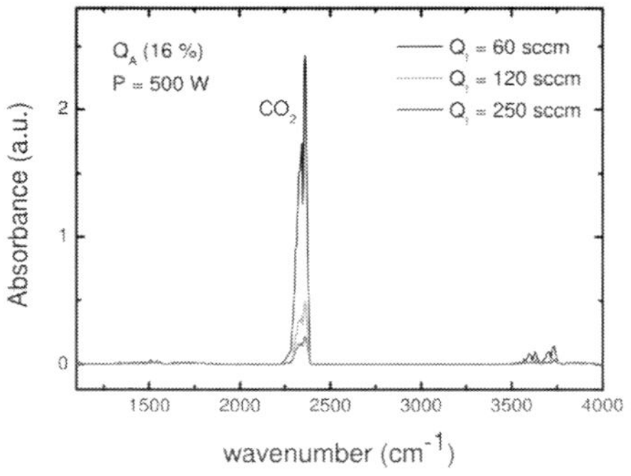

Figure 9. Infrared absorption spectra of Ar + ethanol mixture for different Q_t and constant Q_A.

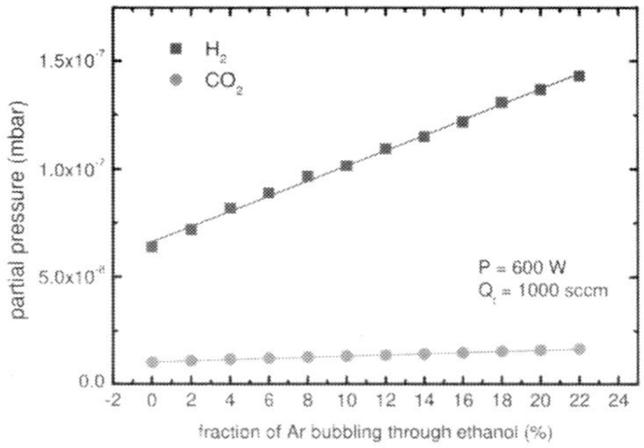

Figure 10. Dependence of H_2 and CO_2 partial pressures on the fraction of Ar bubbling through ethanol.

Indeed, the ethanol solution contains about 4 % of water, therefore steam reforming and water gas shift reactions do occur. The CO_2 absorption clearly decreases when the total Q_t flow increases at constant ethanol percentage in the mixture (figure 9). It should be noted that Q_A or the ethanol percentage is kept constant. Thus, the relative amount of ethanol in the mixture decreases as the total gas flow increases, consequently the amount of CO_2 decreases. As in the methanol case, a black deposit on the quartz wall is formed due to the introduction of ethanol into the plasma. This indicates that some recombination reactions, such as $C + C$ (+ wall) $\rightarrow$ C_2 (+ wall), leading to C_2 formation are in play. Furthermore, association reactions such as $C_2 + C_2$ (wall) $\rightarrow$ graphite produce solid carbon (graphite) on the tube wall, as observed in the experiment. The dynamic behavior of the plasma decomposition system exhibits jumps in the H_2 partial pressure when the fractional Ar "bubbling" flow Q_A (i.e., the ethanol amount) increases (not shown here). The dependence of the H_2 and CO_2 partial pressures on the fraction of Ar bubbling through ethanol is presented in figure 10. As seen, the dependence is linear and the CO_2 partial pressure is nearly one order of magnitude smaller than that of molecular hydrogen.

3.2.2. Effect of additional water vapor. In order to investigate the influence of the steam reforming reaction, an additional quantity of water vapor was introduced by "bubbling" of Ar through the water tank as done before. The influence of the introduced amount of "pure" ethanol or ethanol + water into the plasma on the H_2 production is shown in figure 11. Three different values of the fractional Ar "bubbling" flow, Q_A, passing through the ethanol tank have been used, i.e., 5, 10 and 15 % (Q_W varies from 95 to 85 % of the total flux). In fact, the H_2 yields nearly double when water vapor is added. The infrared absorption spectra of the argon-ethanol plasma exhaust stream with and without water addition are shown in figure 12. As seen from this figure, water addition into the plasma causes a decrease in CO_2 absorption and the appearance of CO absorption. The intensity of H_2O bands dramatically decreases. Moreover, the intensity of the CO line increases with increasing fractional Ar "bubbling" flow Q_A, i.e., when the amount of C_2H_5OH in the plasma increases. These results demonstrate that the steam reforming reaction (R5) is dominant when water vapor is introduced into the plasma. Furthermore, no carbon black deposition was observed in this case, which seems to confirm that the steam reforming reaction is the dominant process of H_2 production here.

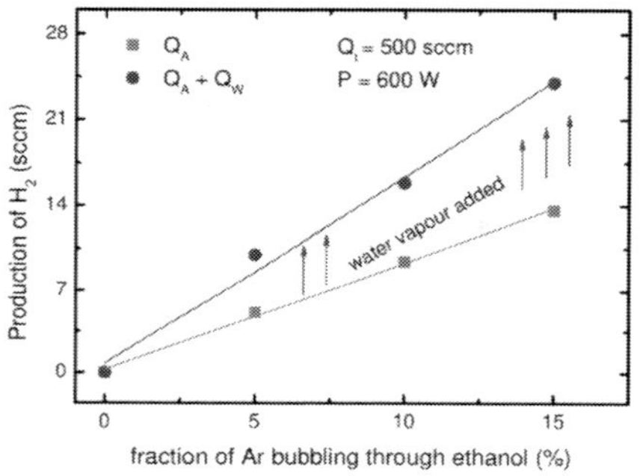

Figure 11. Produced H_2 flow vs. Ar "bubbling" Q_A flow through ethanol

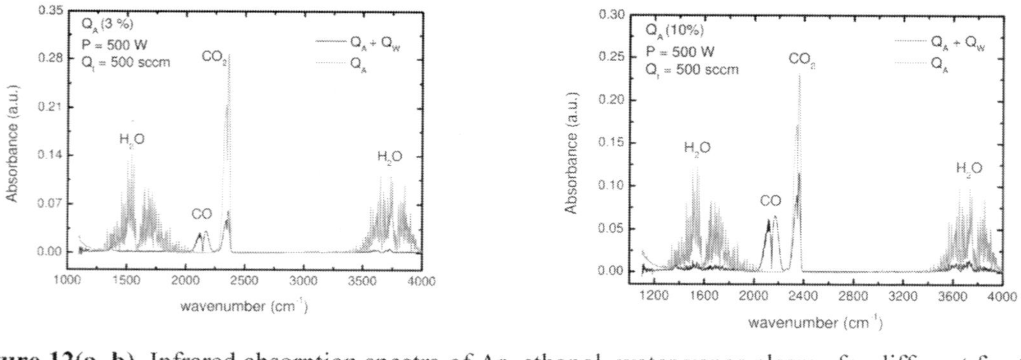

Figure 12(a, b). Infrared absorption spectra of Ar–ethanol–water vapor plasma for different fractional Ar "bubbling" flows Q_A.

4. Characterization of argon – alcohol – water plasma torch

The presence of alcohols and water vapor in the argon plasma torch change significantly the plasma characteristics. Optical emission spectroscopy has been applied to measure the gas temperature and the electron density as key plasma parameters. Usually, the rotational temperature of the OH radical is determined as a measure of the gas temperature. Here, the emission spectrum of the $OH(A^2\Sigma^+, v=0 - X^2\Pi_i, v'=0)$ (Q_1 branch) band in the range 306–315 nm was detected. The measured spectrum is shown in figures 13, 14. The rotational temperature has been determined by using the classical Boltzmann plot method. Assuming equilibrium, the intensity of the emission lines (I_R) of a rotational band can be expressed as [29]:

$$I_R = aA_j H \frac{c}{\lambda} \exp\left[\frac{-E_j hc}{kT_R}\right] \tag{1}$$

where a is a constant, A_j is the transition probability of rotational level j, h is the Planck's constant, c is the speed of light in vacuum, λ is the wavelength of the line, k is the Boltzmann's constant, T_R is the rotational temperature, and E_j is the energy of the emitting state. Using (1), the rotational temperature can readily be calculated from the slope of the Boltzmann plot $\log(I_R\lambda/A_j)$ vs. E_j.

The results were obtained in Ar–methanol and Ar–methanol–water vapor plasmas at constant total flow of 500 sccm and constant microwave power delivered to the launcher. The estimated gas temperature is typically around 3000 K when Q_A changes from 1 to 10 %. The presence of water in the plasma increases the gas temperature to about 3500–4000 K. The influence of the total flow Q_t on the gas temperature has also been investigated. The increase of Q_t from 500 to 4000 sccm contributes to an increase in gas temperature of about 15 %.

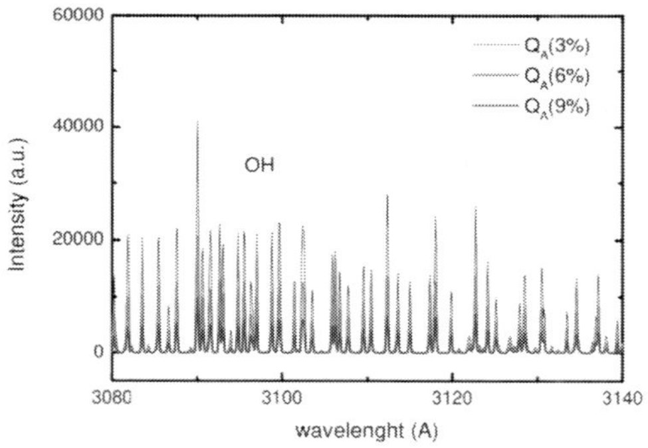

Figure 13. Intensity of OH(A-X) band emitted from Ar–methanol plasma at different Q_A.

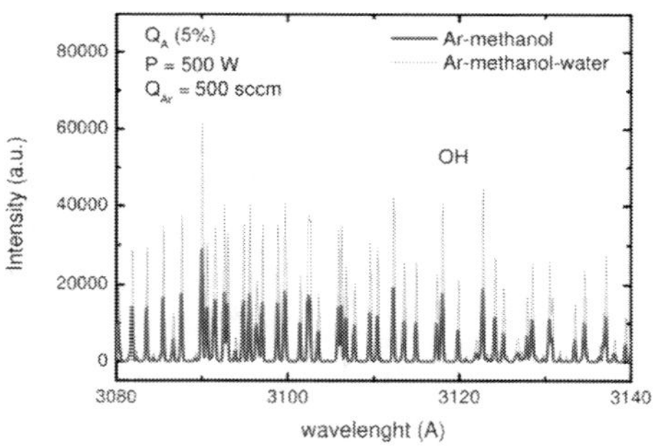

Figure 14. Intensity of OH (A-X) band emitted from Ar–methanol and Ar–methanol–water plasma

Figure 13 shows the spectrum of the OH(A-X) band emitted by the Ar–methanol plasma. As seen, there is a decrease in the OH(A-X) band intensity with the increase of methanol in the mixture. This suggests that the OH molecule dissociates with the increase of methanol. Adding water vapour increases the intensity of OH as expected (figure 14).

The electron density has been determined using the well known Stark broadening method. Since H_β was not detected in the present experiments, the H_α line has been used for this purpose. The measured line profiles have been well fitted by a Voigt function. Estimations show that resonance and natural broadening are negligible under the present conditions. A deconvolution procedure has been used to determine the full width at half-maximum of the Gaussian ($\Delta\lambda_G$) and Lorentzian ($\Delta\lambda_L$) components.

Stark and van der Waals broadening convolute to give a Lorentzian component. The measured gas temperature has been used to estimate van der Waals broadening (~ 0.116 Å) and the latter was deconvoluted from the Lorentzian full width at half maximum to estimate the pure Stark broadening. Assuming that Ar atoms are the main perturbers, the expression for van der Waals broadening of the H_α line given in [30] has been used. The electron density has been calculated by fitting the widths listed by Gigosos *et al* [34] for a reduced mass of 0.976, which corresponds to the present emitter – perturber conditions, i.e., the pair hydrogen–argon.

The values of electron density so determined at 2 cm from the launcher vary in the range 3.3×10^{13} – 5.4×10^{14} cm^{-3}, for a total flux varying from 500 to 2000 sccm, an Ar "bubbling" flux through alcohol kept constant at 4 %, and 500 W microwave power delivered to the launcher. It should be noted that the densities determined in this way correspond to radially averaged values.

5. Conclusions

A microwave plasma torch operating at 2.45 GHz and atmospheric pressure conditions has been used to create a plasma environment to decompose alcohols. Methanol/ethanol and water vapor have been introduced into the discharge by bubbling of Ar through the alcohol and water. Mass spectroscopy and Fourier Transform Infrared Spectroscopy have been applied to detect the molecular hydrogen produced by methanol/ethanol decomposition and H_2O, CO_2 and CO molecules in the exhaust gas stream.

The H_2 partial pressure increases when the Ar "bubbling" flow through methanol increases up to 10%, i.e., with the increase of the methanol amount. The highest H_2 yield in the case of methanol was obtained for relatively low argon flows (Q_t up to 2000 sccm). In fact, nearly 100 % decomposition of methanol was achieved by the Ar–methanol plasma torch operation. The optimal microwave power is about 400 W. Addition of water increases the H_2 production by about 50% but high H_2O flows influence the discharge stability.

The partial hydrogen flux in the exhaust stream of the Ar–ethanol plasma torch was observed to increase when the amount of ethanol (correlated with the Ar "bubbling" flow Q_A through the ethanol tank) increases. H_2O, CO, and CO_2 were detected as by-products of the plasma decomposition processes. Solid carbon as a black deposit on the tube walls was also observed. In Ar–ethanol–water vapor environment the measured H_2 partial pressure is nearly doubled in respect to the Ar–ethanol plasma. The steam reforming reaction $C_2H_5OH + H_2O \rightarrow 2CO + 4H_2$ is the most likely source of H_2 production in this case, which is confirmed by the fact that no solid carbon was observed. In addition, the main plasma characteristics, i.e., gas temperature and electron density, were experimentally determined for the Ar–methanol and Ar–methanol–water vapor plasmas. Gas temperatures around 3000 K and 3500 K – 4000 K were obtained for Ar–methanol and Ar–methanol–water vapor plasmas, respectively. The values of electron density vary in the range 3.3×10^{13} – 5.4×10^{14} cm^{-3}, for a total flux varying from 500 to 2000 sccm.

This work clearly shows that the present microwave plasma torch at atmospheric pressure can be very efficient for H_2 production from methanol/ethanol.

Acknowledgement

This work was supported by the Portuguese *Fundação para a Ciência e a Tecnologia* (FCT), under the research contract PTDC/FIS/108411/2008.

References

[1] Holladay J D, Hu J, King D L and Wang Y 2009 *Cat. Today* **139** 244–60
[2] Petitpas G, Rollier J-D, Darmon A, Gonzalez-Aguilar J, Metkemeijer R and Fulcheri L 2007 *Int. J. Hydrogen Energy* **32** 2848–67
[3] Biniwale R B, Mizuno A and Ichikawa M 2004 *Appl. Catalysis A: General* **276** 169–77
[4] Sekiguchi H and Mori Y 2003 *Thin Solid Films*, Elsevier, Jeju Island, South Korea 44–8

[5] Bromberg L, Cohn D R, Rabinovich A and Alexeev N 1999 *Int. J. Hydrogen Energy* **24** 1131–7
[6] O'Brien C J, Hochgreb S, Rabinovich A, Bromberg L and Cohn D R 1996 *Proceedings of the Intersociety Energy Conversion Engineering Conference, Hydrogen Production via Plasma Reformers* (IEEE, Piscataway, NJ, Washington DC, USA) 1747–52
[7] Czernichowski A, Czernichowski P and Wesolowska K, Fuel 2004 *Cell Science, Engineering and Technology* (American Society of Mechanical Engineers, NY, USA) 669–76
[8] Paulmier T and Fulcheri L 2005 *J. Chem. Eng.* **106** 59–71
[9] Hammer T, Kappes T and Baldauf M 2004 *Cat. Today* **89** 5–14
[10] Mutaf-Yardimci O, Saveliev A V, Fridman A A and Kennedy L A 1998 *Int. J. Hydrogen Energy* **23** 1109–11
[11] Matsui Y, Kawakami S, Takashima K, Katsura S and Mizuno A 2005 *Energy and Fuels* **19** 1561–5
[12] Jiang T, Li Y, Liu C-J, Xu G-H, Eliasson B and Xue B 2002 *Cat. Today* (Elsevier Science B.V., Washington DC) 229–35
[13] Sobacchi M G, Saveliev A V, Fridman A A, Kennedy L A, Ahmed S and Krause T 2002 *Int. J. Hydrogen Energy* **27** 635–42
[14] Sarmiento B, Javier Brey J, Viera I G, Gonzalez-Elipe A R, Cotrino J, Rico V J 2007 *J. Power Sources* **169** 140–3
[15] Aubry O, Met C, Khacef A and Cormier J M 2005 *J. Chem. Eng.* **106** 241–7
[16] Sekine Y, Urasaki K, Kado S, Matsukata M and Kikuchi E 2004 *Energy&Fuels* **18** 455–9
[17] Sekine Y, Matsukata M, Kikuchi E, Kado S and Haga F 2004 (Prepr. Pap. Am. Chem. Soc.) *Div. Fuel Chem.* **49**(2) 914–5
[18] Yanguas-Gil A, Hueso J L, Cotrino J, Caballero A and Gonzalez-Elipe A R 2004 *Appl. Phys. Let.* **85**(18) 4004–6
[19] Dubois D, Merbahi N, Eichwald O, Yosfi M and Benhenni M 2007 *J. Appl. Phys.* **101** 053304
[20] Xinpei Lu and Laroussi Mounir 2005 *J. Appl. Phys.* **98** 023301
[21] Ono Ryo and Tetsuji Oda 2005 *J. Appl. Phys.* **97:** 013302
[22] Chen C K and Phillips J 2002 *J. Phys. D: Appl. Phys.* **35** 998–1009
[23] Santiago I and Calzada M D 2007 *Appl. Spectroscopy* **61** 725
[24] Ricard A, Jaoul C, Gaboriau F, Gherardi N and Villeger S 2004 *Surface&Coating Technology* **287** 188–9
[25] Dato A, Radmilovic V, Lee Z, Phillips J and Frenklach M 2008 *Nano Letters* **8** 2010
[26] Weigle J C, Luhrs C C, Chen C K, Perry W L, Mang J T, Lopez G P and Phillips J 2004 *J. Phys. Chem.* **108** 18601
[27] Gleiman S, Chen C K. Datye A and Phillips J 2002 *J. Mat. Sci.* **37** 3429
[28] Mills R, Ray P, Mayo R, Nansteel M, Dhandapanian B and Phillips J 2005 *J. Plasma Phys.* **71** 877
[29] Phillips J and Chen C K 2008 *Int. J. Hydrogen Energy* **33** 7185
[30] Oliveira C, Souza Corrêa J A, Gomes M P, Sismanoglu B N and Amorim J 2008 *J. Appl. Phys.* **93** 041503
[31] Jimenez M, Yubero C and Calzada M D 2008 *J. Phys. D: Appl. Phys.* **41** 175201
[32] Moisan M and Zakrzewski Z 1991 *J. Phys. D: Appl. Phys.* **24** 1025
[33] Timmermans E A H, Jonkers J, Rodero A, Quintero M C, Sola A, Gamero A, Schram D C and van der Mullen J A M 1999 *Spectrochim. Acta B* **54** 1085

Energetic Hydrogen Atoms in High Frequency Plasmas

E Felizardo[1], E Tatarova[1*], F M Dias[1], C M Ferreira[1] and B Gordiets[2]

[1] Instituto de Plasmas e Fusão Nuclear, Instituto Superior Técnico, 1049-001 Lisboa, Portugal
[2] Lebedev Physical Institute of the Russian Academy of Sciences, Moscow, Russia

E-mail: e.tatarova@ist.utl.pt

Abstract. Generation of energetic hydrogen atoms, with energy in the range 4–8 eV, was detected throughout the volume of a surface wave generated (500 MHz) plasma column in H_2 at pressure p = 0.01 mbar. The H_β, H_γ, H_δ, and H_ε, line profiles were found to be bi-Gaussian towards the plasma column end. The kinetic temperatures corresponding to the Doppler broadening of the H_β, H_γ, H_δ, lines are higher than the rotational temperature of the hydrogen molecular Fulcher-α band and the wall temperature. At pressure p = 0.2 mbar, the kinetic temperature of excited H (n = 4–7) atoms, as determined from the fitting of the spectral lines with a single-Gaussian profile, increases with upper level principal quantum number. The experimental results have been analyzed in the framework of a global self-consistent kinetic model describing this surface wave sustained plasma column.

1. Introduction

The analysis of Balmer line shapes has usually been applied to detect energetic hydrogen atoms in various types of gas discharges [1, 2]. The presence of such atoms results in extended "wings" in the line profiles. The Balmer lines spectra emitted by different types of DC glow and RF discharge plasmas have typical multimode behaviour, i.e., they exhibit widely broadened "wings" and a sharp top [3, 4]. Different processes and mechanisms can be at the origin of such phenomena. Usually two or three components of the line profiles may be clearly distinguished [4]. The central, narrow component originates from low-energy dissociative excitation and direct electron impact excitation processes. Collisions of highly energetic electrons with vibrationally excited H_2 molecules result in dissociative excitation, whose products may have high kinetic energies depending on the molecular state [1]. Dissociative broadening contributes to the overall processes, but it does not explain far-lying "wings" found in some DC and RF discharges, corresponding to fast atoms with kinetic energies of hundreds of electron volts. Since high DC electric fields are present in near-electrode sheaths, excessive broadening may result from acceleration of charges such as H^+, H_3^+ ions, with subsequent conversion into fast neutrals. Two sources of these fast atoms have been suggested, *viz.*, charge exchange collisions between fast ions and the background H_2 gas molecules producing energetic atoms that move towards the electrode and the formation of energetic H atoms at the electrode surface due to fast ion bombardment [4–7]. Microwave and HF discharges are, however, free of such strong DC electric fields and the radiated profiles usually exhibit single mode behaviour [8, 9]. Nevertheless, generation of fast H atoms, with energies in the range 2–4 eV, in microwave plasma has also been reported in [10]. In a series of works, selective hydrogen line broadening has been detected in microwave discharges and its afterglows when there is no significant broadening of noble gas lines or hydrogen molecular lines [8, 11, 12].

This study presents spectroscopic measurements of Balmer lines emitted by a surface wave sustained high frequency H_2 plasma source operating at $\omega/2\pi = 500$ MHz and low pressure conditions (0.01–0.3 mbar). Results on the line shape and the emission intensities of excited H($n = 4$–7) hydrogen atoms, and the emission intensities of the Q-branch of the Fulcher-α band [$d^3\Pi_u$ (v = 0) → $a^3\Sigma_g^+$ (v = 0)] are presented and discussed. Bi-Gaussian shape of the H_β, H_γ, H_δ, H_ε lines detected at a pressure $p = 0.01$ has been observed. Different average energies have been determined from the measured Balmer lines and the rotational distribution of hydrogen molecular lines. The results are analyzed in the framework of a previously developed self-consistent model describing the entire discharge structure [13]. This model couples in a self-consistent way wave electrodynamics, electron and heavy particles kinetics, and gas thermal balance (see [13] for details).

2. Experimental set-up and conditions

A classical surface wave sustained discharge has been used as a plasma source. The discharge is created using a surfatron-based set-up (figure 1) [14]. The microwave power is provided by a 500 MHz generator, whose output power was varied from 40 to 250 W. The discharge takes place inside a Pyrex tube with internal and external radii of 2.25 and 2.5 cm, respectively (figure 1).

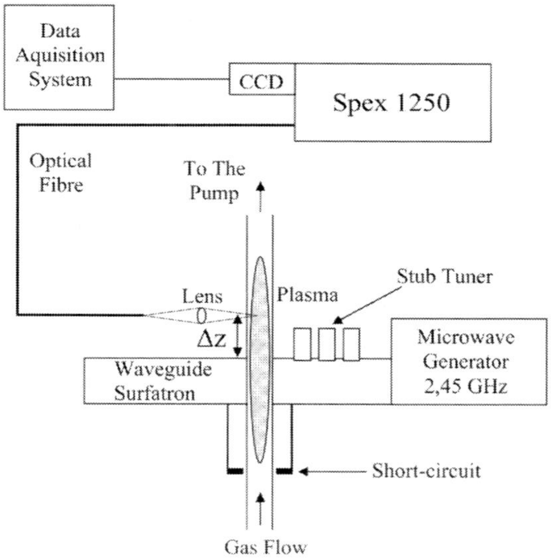

Figure 1. Experimental set-up

The background gas is injected into the discharge tube at flow rates from 0.4 to 20 sccm under laminar gas flow conditions. The electromagnetic surface waves are coupled into a Pyrex tube via the launching gap of a surfatron and travel along the interface between the plasma and the tube. The discharge is sustained by the wave electric field. The wave simultaneously propagates and creates its own propagation structure. The wave power is progressively dissipated by the plasma electrons along the wave path and the absorbed power per unit length, as well as the electron density, decrease gradually towards the plasma column end. Under the present conditions, this corresponds to a decrease in electron density from about (1–1.5)×10^{10} cm^{-3} at the beginning of the plasma column to about (2–3)×N_{cr} ($N_{cr} \approx 3.09\times10^9$ cm^{-3}). Taking into account the total power delivered to the launcher and subtracting the reflected power, one can estimate the average absorbed power per unit volume. It ranges in the limits 0.1–0.2 W/cm^3. Estimations indicate that the effective electric field intensity sustaining the discharge ranges from about 5 V/cm to 10 V/cm.

The optical system used consists of a 1.25 m focal length (visible light) Jobin-Yvon Spex 1250 spectrometer, with a holographic ruled diffraction grating (2400 gmm^{-1}) that provides a nearly flat response between 300 and 800 nm, equipped with a liquid-nitrogen cooled CCD camera. The slit was set at 10 μm in all cases. The light emitted by the plasma is collected perpendicularly to the discharge tube axis by a bundle optical fibre. The measurements correspond to some radially averaged value over the plasma cross-section. The spectral profiles of the H_β, H_γ, H_δ, H_ε, lines, corresponding to the transitions H[(n = 4–7) → (n = 2)] have been measured. The temperature of the gas molecules was determined from measurements the H_2 rotational temperature using the Q-branch of the Fulcher-α band rotational spectrum [$d^3\Pi_u$ (v=0) → $a^3\Sigma_g^+$ (v=0)] in the 600–617 nm wavelength range [8].

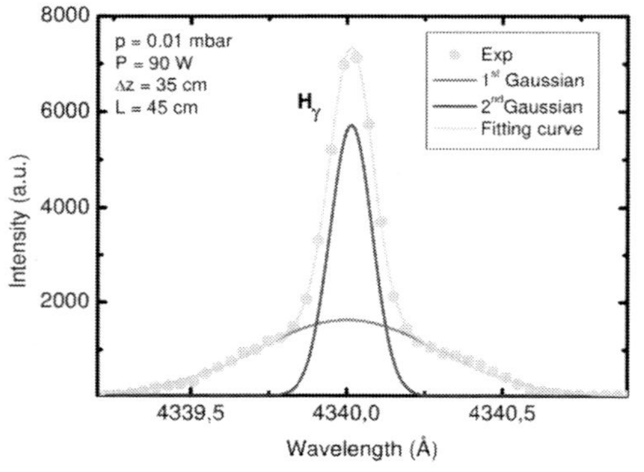

(a)

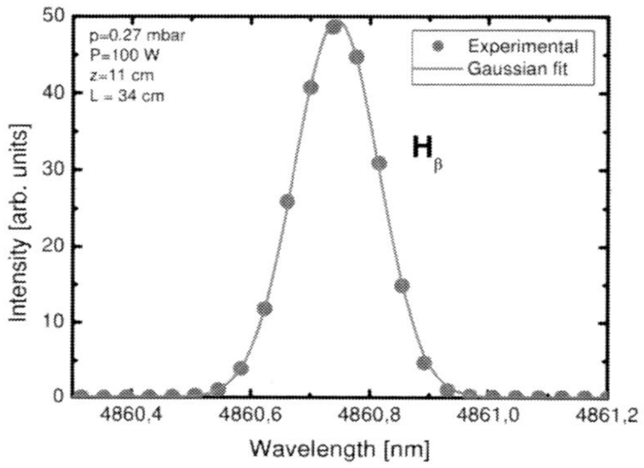

(b)

Figure 2. Measured profiles of Balmer lines fitted with Gaussian profiles: (a) H_γ line; (b) H_β line.

3. Results and discussion

In general, the experimental profile is a convolution of the Doppler, instrumental, natural (lifetime), Stark, van der Waals, and resonance profiles and of fine structure [15]. The instrumental function width was determined to be $\Delta\lambda_{app}$= 0.04 Å. Taking into account the typical electron densities and microwave electric field intensities for the present conditions, Stark, van der Waals and resonance broadening may be ignored. So, we can assume that the experimental profiles are a convolution of the Doppler and instrumental profiles. In order to determine the "pure" Doppler broadening, the fine structure of the Balmer lines has also been taken into account [16]. The measured profiles of the H_β, H_γ, H_δ, H_ε, lines (H_α is critical because of self-absorption) are well fitted by single- or two- Gaussian profiles (much better than by a Voigt profile) as shown in figures 2(a, b). GRAMS/32® and Mat Lab based software have both been used for this fitting. The bi-Gaussian profiles are observed at the lower pressures (figure 2(a)), but when pressure increases up to 0.2 mbar the broad base disappears and the bi-Gaussian profiles become single ones (figure 2(b)). At low pressures, the broad base is observed to develop downstream, as the distance from the launcher increases (figure 3). After such fittings, the average kinetic energies corresponding to Doppler broadening have been determined assuming Maxwellian energy distributions. A cascade approximation [17] has been applied to obtain the average kinetic energy corresponding to H_γ Doppler broadening.

The axial variation of the fractional population of energetic atoms in the level $n = 5$ is shown in figure 4, for a pressure of 0.01 mbar. Figure 5 shows the axial variation of the atomic kinetic temperature as inferred from the analysis of the H_γ line, at 0.2 mbar pressure. Also shown here for comparison is the theoretical prediction from the model of reference [13].
The axial variation of the rotational temperature corresponding to the rotational band of the transition $d^3\Pi_u$ ($v = 0$) $\rightarrow$ $a^3\Sigma_g^+$ ($v = 0$) is shown in figure 6, for the same conditions as in figure 5. Also shown for comparison is the axial variation of gas temperature predicted by the theoretical model as well as that of the wall temperature, obtained from infrared sensitive measurements using an optical thermometer. As seen from figures 5 and 6, the measured rotational temperatures are much lower than the atomic ones, which brings evidence for selective heating of H($n = 5$) atoms.

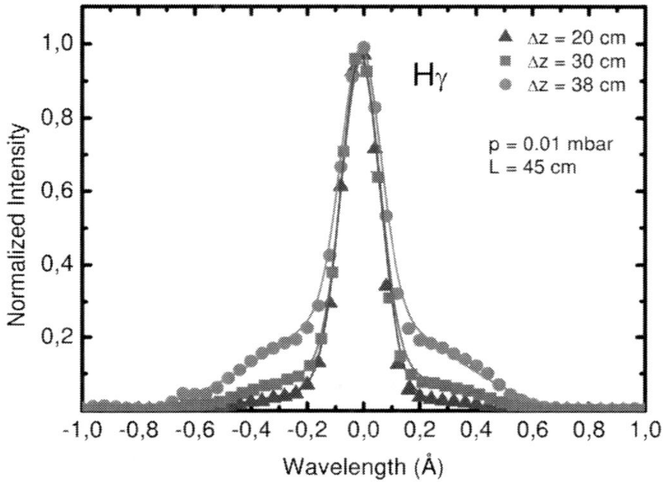

Figure 3. Axial evolution of H_γ line (p = 0,01 mbar; L = 45 cm)

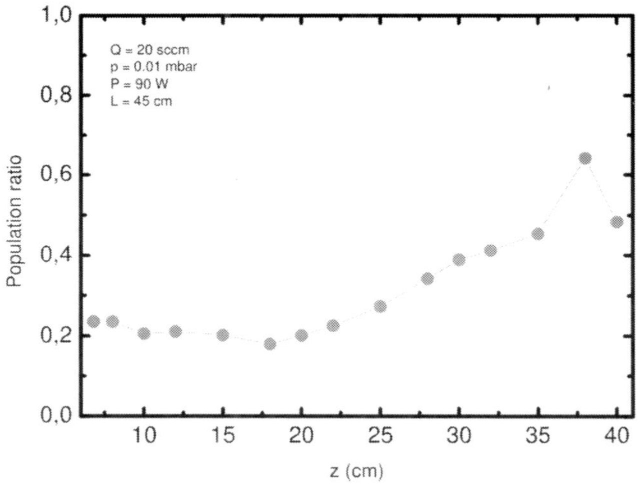

Figure 4. Axial variation of the fractional density of energetic excited atoms (n = 5)

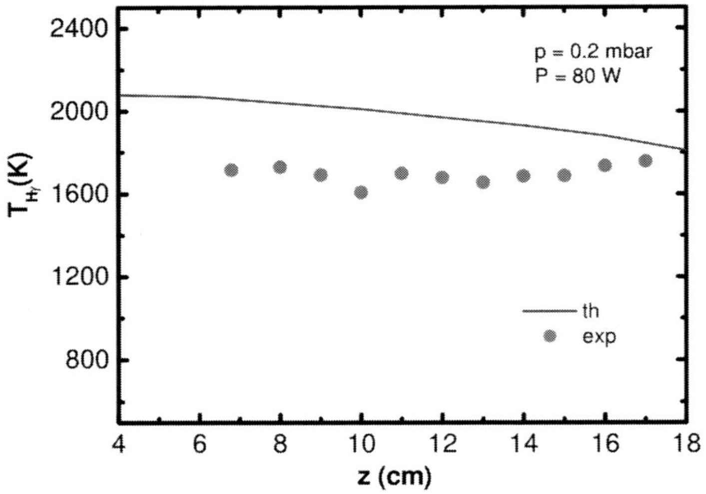

Figure 5. Atomic temperature axial variation

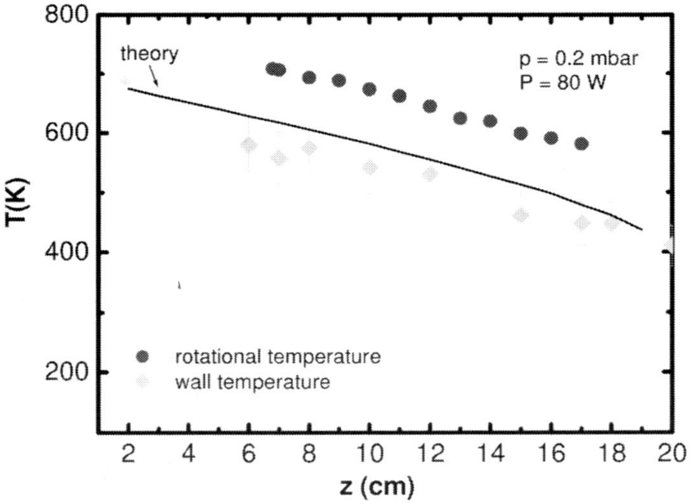

Figure 6. Axial variation of the gas temperature

Data on the Balmer (β, γ, δ, ε) line broadening have been recorded at fixed axial position $\Delta z = 15$ cm from the launcher at pressure $p = 0.2$ mbar. The length of the plasma column is 16 cm under such conditions. The kinetic temperatures determined from these lines are shown in figure 7. A striking result is observed: hydrogen atoms excited in higher levels appear to be hotter than those excited in lower ones. For example, the kinetic temperature of H($n = 7$) is higher than that of H($n = 4$). Let us try to find out the reason for such behaviour. Linear Stark splitting caused by a high frequency field with an average intensity $\langle E\rangle = E_0/\sqrt{2}$ (assuming that the microwave field at 500 MHz acts as a dc field) induces the following broadening ΔE of the Balmer lines $[n \to 2]$ [17]:

$$\Delta E_n(\text{eV}) \approx 7.9\times10^{-9}(n^2-4)\times\langle E\rangle[\text{V/cm}] \tag{1}$$

This broadening will add some additional "splitting temperature" to the Doppler temperature, i.e., some error in determining the "pure" Doppler temperature may occur. However, estimations show that this additional "splitting" temperature varies in only the range 5–50 K for the different levels, for a field $\langle E\rangle$ = 5–10 V/cm. For example, for the level $n = 7$, for which the Stark effect is maximum, the error is less than 50 K, *i.e.*, less than 5 %. For the lower levels the error is still much smaller. Hence, Stark splitting has negligible influence under the present conditions.

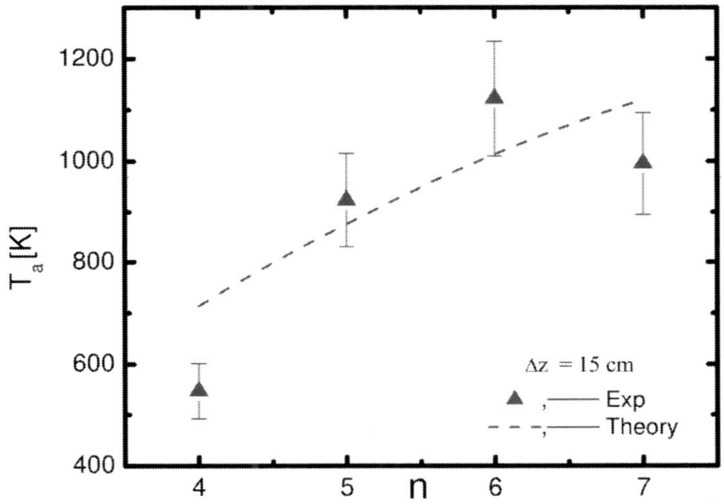

Figure 7. Doppler temperatures of H(n = 4–7) as a function of the upper level quantum number at pressure p = 0.20 mbar and axial distance Δz = 15 cm [T_{gas} = 347 K; T_{vib} = 4840 K; kT_e = 3.48 eV, N_e = 1×10^{10} cm^{-3}; [H]/[H_2] = 7.86 %; E_{eff} = 5.02 V/cm].

The observed result can be tentatively explained if one assumes that three groups of excited H atoms are generated, *viz.*, and "energetic", a "warm" and a "cold" group, originating from different electron induced processes. "Cold" excited hydrogen atoms, with a temperature nearly equal to that of the background gas, are generated by direct electron impact excitation of ground state atoms:

$$\text{H}(1) + \text{e} \to \text{H}(n \geq 2) + \text{e}. \tag{R1}$$

"Warm" excited H atoms can be generated by dissociation processes involving electrons and vibrationally excited H_2 molecules, *viz.*,

$$H_2(X, v=1\text{–}4) + \text{e} \to H_2^*(\varepsilon = 17\text{–}19\text{ eV}) + \text{e} \to \text{H}_{hot}(n=1) + \text{H}_{hot}(n=4\text{–}8) + \text{e}. \tag{R2}$$

Here, H_2^* are hydrogen molecules excited in weakly bound electronic states, with energy ε = 17–19 eV > $E_n + D_{H2} + \Delta\varepsilon_k$, where E_n are the energies of the excited H(n = 4–8) atoms, D_{H2} = 4.48 eV is the dissociation energy of hydrogen molecules and $\Delta\varepsilon_k$ = 0.1–1.8 eV is the range of kinetic energies of the two H atoms resulting from dissociation. The total kinetic energy is therefore $\langle\Delta\varepsilon_{kin}\rangle \approx \Delta\varepsilon_{kin}/2 \approx 0.9$

eV, on the average. These processes can be effective for the high electron temperature (up to 6 eV) and vibrational temperature of $H_2(X)$ molecules, which are typical under the present conditions. Thus, there is a high population in the vibrational levels v = 1–4, for which the transitions (R2) have relatively large Frank-Condon coefficients.

Finally, the group of energetic H (n = 4–7) atoms is likely to be generated from wall electron-ion recombination. Close to the wall, the H^+ ions are accelerated in a space charge sheath and, thus, may acquire high kinetic energy as they fall on the wall. These ions recombine with electrons at the wall and may return back to the gas as fast excited atoms, with kinetic energy $\varepsilon = \varepsilon_{ion} + I - E_n$. (here, I is the ionization energy). At low pressures, the ion mean-free path λ is larger than the radius (for H^+ with energy $\varepsilon_{ion} \geq 0.5$ eV λ is about 3 cm at 0.01 mbar), therefore H^+ ions acquire a kinetic energy equal to the radial potential energy drop, which is of the order of the electron temperature (4–6 eV). As pressure increases this ceases to be the case and energetic atoms disappear as we have observed.

At the higher pressure p = 0.2 mbar, there are therefore two groups of radiating H(n > 2) atoms, a "warm" and a "cold" group, with kinetic temperatures $\boldsymbol{T_{warm}} \approx (2/3)\langle \Delta\varepsilon_{kin} \rangle / 2$ (R2) and $T_{cold} \approx T_g$ (R1), respectively. The population distributions, $\boldsymbol{N_n^{warm}}$ and N_n^{cold}, for both groups have been calculated in the framework of the above mentioned model [13]. The main radiative and collisional processes leading to population and depopulation of the energy levels for "warm" and "cold" atoms have been accounted for using a "single-quantum" approximation [17]. The following rate coefficient $\boldsymbol{R^{warm}}$ for generation of "warm" excited H atoms has been used:

$$R_n^{warm} \approx k_0 \times \exp\left(-\frac{E_{vib}}{kT_{vib}}\right) \times \exp\left\{-\frac{17 - E_{vib} + Ry(1/16 - 1/n^2)}{kT_e}\right\} \times N_e \times [H_2] \tag{2}$$

Here k_0 is a constant; $E_{vib} = 0.54$ eV is the energy of the first vibrational level of $H_2(X)$; T_{vib} is the vibrational temperature of $H_2(X)$ molecules; Ry = 13.6 eV is the Rydberg constant, N_e is the electron density and $[H_2]$ is the number density of hydrogen molecules. We assume that every "warm" atom has the average kinetic energy $\langle \Delta\varepsilon_{kin} \rangle / 2 \approx 0.45$ eV, that is, the kinetic temperature of the ensemble of such atoms is $\boldsymbol{T_{warm}} \approx 3600$ K. As to the rate coefficients for electron impact excitation of the levels $n \geq 2$ from the ground state n = 1, we assume that they scale approximately as n^{-3} [17].

It can now be shown that the interplay of the relative contributions of (R1) and (R2) can produce an increase in the temperature of the ensemble of both populations when n increases. The sum $I_\Sigma(\lambda - \lambda_n)$ of two Gaussian profiles $\boldsymbol{I_n^{warm}}(\lambda - \lambda_n)$ and $I_n^{cold}(\lambda - \lambda_n)$ with not strong (no more than a factor of ~5) difference in temperatures $\boldsymbol{T_{warm}}$ and T_{cold} and with different relative intensities $\boldsymbol{I_n^{warm}} = \boldsymbol{N_n^{warm}} / (\boldsymbol{N_n^{warm}} + \boldsymbol{N_n^{cold}})$ and $\boldsymbol{I_n^{cold}} = \boldsymbol{N_n^{cold}} / (\boldsymbol{N_n^{warm}} + \boldsymbol{N_n^{cold}})$ ($\boldsymbol{I_n^{warm}} + \boldsymbol{I_n^{cold}} = 1$ at $\lambda - \lambda_n = 0$) can be approximated by a single Gaussian profile $I_{FIT}(\lambda - \lambda_n)$ with an ensemble temperature T_n, that is,

$$\boldsymbol{I_\Sigma} \equiv \frac{\boldsymbol{I_n^{warm}}}{\sqrt{\boldsymbol{aT_{warm}}}} \exp\left[-\frac{(\lambda - \lambda_n)^2}{\boldsymbol{aT_{warm}}}\right] + \frac{\boldsymbol{I_n^{cold}}}{\sqrt{\boldsymbol{aT_{cold}}}} \exp\left[-\frac{(\lambda - \lambda_n)^2}{\boldsymbol{aT_{cold}}}\right] \approx \frac{1}{\sqrt{\boldsymbol{aT_n}}} \exp\left[-\frac{(\lambda - \lambda_n)^2}{\boldsymbol{aT_n}}\right] \equiv \boldsymbol{I_{FIT}} \tag{3}$$

Here, $a = 1.85 \times 10^{-13} \lambda_n^2 / \mu$, μ is the mass of the radiating atoms (in atomic units), and the temperatures are expressed in K. The ensemble temperature T_n is given by:

$$\boldsymbol{T_n} = \left[\frac{\left(\boldsymbol{I_n^{warm}}\right)^2}{\boldsymbol{T_{warm}}} + \frac{\left(\boldsymbol{I_n^{cold}}\right)^2}{\boldsymbol{T_{cold}}} + 2\frac{\boldsymbol{I_n^{warm}} \times \boldsymbol{I_n^{cold}}}{\sqrt{\boldsymbol{T_{warm}T_{cold}}}}\right]^{-1} \tag{4}$$

As it can be readily seen, the increase of I_n^{warm} from 0 to 1 leads to the change of T_n from T_{cold} to T_{warm}. It should be stressed that for $T_{warm}/T_{cold} \leq 5$ and $0.1 \leq I_{cold} \leq 0.9$ the mean square deviation between $I_{\Sigma}(\lambda - \lambda_n)$ and $I_{FIT}(\lambda - \lambda_n)$ is much less than 5 %. The calculated variations of kinetic temperature according to equation (4) are shown by the dashed line in figure 7. Good agreement between calculated and measured temperatures is obtained.

Acknowledgment
This work was supported by the Portuguese *Fundação para a Ciência e a Tecnologia* (FCT), under the research contract PTDC/FIS/108411/2008.

References

[1] Tatarova E, Felizardo E, Dias F M, Lino da Silva M, Ferreira C M and Gordiets B 2009 *Appl. Phys. Letters* **95** 181503
[2] Oliveira C, Souza Correa J A, Gomes M P, Sismanoglu B N and Amorim J 2008 *Appl. Phys. Lett.* **93** 041503
[3] Cvetanovic N, Kuraica M M and Konjevic N 2005 *J. Appl. Phys.* **97** 033302
[4] Phelps A V 2009 *Phys. Review E* **79** 066401
[5] Adamov M R G, Obradovich B M, Kuraica M M and Konjevic N 2003 *IEEE Trans. Plasma Science* **31** 444
[6] Cvetanovic N, Kuraica M M and Konjevic N 2005 *J. Appl. Phys.* **97** 033302
[7] Cvetanivic N, Obradovic B M and Kurica M M 2009 *J. Appl. Phys.* **105** 043306
[8] Tomasini L, Rousseau A, Gousset G and Leprince P 1996 *J. Phys. D: Appl. Phys.* **29** 1006
[9] Baravian G, Chouan Y, Ricard A and Sultan G 1987 *J. Appl. Phys.* **61** 5249
[10] Mills R L, Akhtar K 2010 *Int. J. Hydrogen Energy* **35** 2546
[11] Tatarova E, Dias F M, Ferreira C M and Puac N 2007 *J. Appl. Phys.* **101** 063306
[12] Tatarova E, Dias F M and Ferreira C M 2009 *Int. Journal of Hydrogen Energy* **34** 9585
[13] Gordiets B, Pinheiro M, Tatarova E, Ferreira C M and Ricard A 2000 *Plasma Sources Sci. Technol.* **9** 295
[14] Moisan M, Chaker M, Zakrzewski Z and Paszczak 1987 *J. Phys. E: Scientific Instrumentation* **20** 1356
[15] Griem H R 1974 *Spectral Line Broadening by Plasmas* (Academic Press, New York)
[16] Lavrov B P, Kaning M, Ovtchinnikov V L and Ropcke J 1997 Proceedings of *Frontiers in Low Temperature Plasma Diagnostics II* (Bad Honnef, Germany, February 17-21, 1997) 169 – 72
[17] Bibernam L M, Vorob'ev V S and Yakubov I T 1987 *Kinetics of Nonequilibrium Low-Temperature Plasmas* (Consultants Bureau, New York)

Journal of Physics: Conference Series **516** (2014) 012004 doi:10.1088/1742-6596/516/1/012004

Microwave N_2–Ar Plasma Torch

J Henriques, E Tatarova, F M Dias and C M Ferreira
Instituto de Plasmas e Fusão Nuclear, Instituto Superior Técnico, 1049-001 Lisboa, Portugal

E-mail: Julio.henriques@ist.utl.pt

Abstract. This work investigates the spatial structure of a microwave plasma torch driven by an azimuthally symmetric surface wave operating in N_2–Ar mixtures at atmospheric pressure. The main plasma and wave characteristics are obtained in the framework of a self-consistent 2D model that describes the entire spatial structure of this source, including the discharge zone, sustained by the field of the surface TM_{00} mode, and the post-discharge plasma. The set of equations considered to describe the plasma source include: Maxwell's equations; the dispersion equation of the surface mode; the rate balance equations for vibrationally excited states of electronic ground state molecules $N_2(X^1\Sigma_g^+, v)$; the rate balance equations for the triplet $N_2(A^3\Sigma_u^+, B^3\Pi_g, C^3\Pi_u)$ and singlet $N_2(a'^1\Sigma_u^-, a^1\Pi_g, w^1\Delta u, a''^1\Sigma_u^+)$ electronic excited states of nitroge and the Ar(3p^{5}4s) and Ar(3p^{5}4p) excited states of Argon; the rate balance equations for $N^+, N_2^+, N_3^+, N_4^+, Ar^+, Ar_2^+$ ions and electrons; the rate balance equations for the ground state N(^{4}S) and the metastable states N(^{2}P, ^{2}D) of nitrogen atoms; the gas thermal balance equation; and the equation of mass conservation for the fluid. Model calculations of the 2D spatial distributions of species of interest such as charged particles (electrons and positive ions), $N_2(X^1\Sigma_g^+, v)$ vibrationally excited molecules, $N_2(A^3\Sigma_u^+)$ metastables and N(^{4}S) ground state atoms are presented and discussed.

1. Introduction

The development of microwave-based plasma sources applications [1–8] is an issue of growing interest, in particular in what concerns atmospheric plasma torches. One of the main advantages of such discharges is that they make it possible to inject large power densities into the plasma and thus achieve high population densities of active species of interest. Efficient applications of such sources depend on the ability to control the spatial structure of the source and the relevant plasma properties. Fundamental plasma parameters such as gas temperature and active species density profiles need to be determined in order to have a complete plasma characterization.

The development of high-density plasma torches based on microwave propagation is interesting for improving the next-generation plasma-based technologies [7]. In particular, N_2–Ar plasma sources driven by microwaves have advantageous properties for plasma processing technologies, such as the nitriding of surfaces [3]. These sources provide high number densities of active species. Among these, long-lived species like ground-state N(^{4}S) atoms and molecules excited in metastable and vibrational states are of major importance. Moreover, nitrogen atoms play a key role in surface treatment processes [6]. One possible way to control the dissociation rate of nitrogen may be the use of mixtures such as N_2–Ar [4,5].

The spatial structure of the source, i.e., the spatial distribution of active species of interest is crucial for the purposes of treatment processes. However, investigations of different plasma sources have usually focused either on the discharges or their post-discharges only [8]. Since the processes in the post-discharge plasma depend on the source discharge itself, the discharge and the post-discharge must be investigated together [7].

In the present work, a plasma torch excited by a surface wave operating in N_2–Ar mixtures at atmospheric pressure is investigated. The purpose of the work is to describe the physical workings and the spatial structure of this plasma source including its active zone, i.e., the source discharge, which is sustained by the field of the azimuthally symmetric TM_{00} surface mode, as well as the post-discharge plasma resulting from gas flow and species diffusion. A self-consistent, 2D theoretical model is developed to investigate the entire spatial structure of the source and to determine the main plasma and wave characteristics. The 2D spatial profiles of the relevant plasma quantities, such as the densities of charged particles (electrons and positive ions), $N_2(X^1\Sigma_g^+,v)$ vibrationally excited molecules, $N_2(A^3\Sigma_u^+)$ metastable molecules and $N(^4S)$ ground state atoms are obtained and discussed. A large number of elementary processes occurring both in the plasma bulk and at the wall have been taken into account.

The organization of the article is as follows. In Sec. II the theoretical formulation and the system of equations to be solved are presented and discussed. The numerical algorithm used to solve the system of equations is described in Sec. III. The results are discussed in Sec. IV. The conclusions are presented in Sec. V.

2. Theoretical model

The system under analysis is a cylindrical plasma column in a quartz tube ($2a$ = 1.5 cm inner diameter and $2b$ = 1.8 cm outer diameter) in N_2–Ar at atmospheric pressure, sustained by an azimuthally symmetric TM_{00} surface wave mode at 2.45 GHz frequency (see figure 1). For this mode, the finite field components are E_r, E_z and H_φ, in the usual cylindrical coordinates [10]. The wave power flux is mainly concentrated around the plasma–discharge tube interface. The axial power flux P_z results from the radial electric (E_r) and the azimuthal magnetic (H_φ) field components. According to the main mechanisms governing the plasma production and maintenance, the source consists of two different but connected parts, *viz.*, the discharge and the post-discharge regions. Axial gas flow and species diffusion give rise to the latter one.

The system of equations considered to describe the plasma source includes [4, 5, 7, 10, 11]:

i) Maxwell's equations;
ii) The dispersion equation for a pure TM azimuthally symmetric surface mode;
iii) The electron Boltzmann equation in the local approximation (with a correction for non-local terms in the post-discharge region);
iv) The rate balance equations for vibrationally excited states of electronic ground state molecules, $N_2(X^1\Sigma_g^+,v)$;
v) The rate balance equations for $N_2(A^3\Sigma_u^+, B^3\Pi_g, C^3\Pi_u, a'^1\Sigma_u^-, a^1\Pi_g, w'\Delta u)$ excited molecules and for Ar($3p^54s$), Ar($3p^54p$) and N(^2P, ^2D) excited atoms ;
vi) The rate balance equations for $N^+, N_2^+, N_3^+, N_4^+, Ar^+, Ar_2^+$ ions and for electrons;
vii) The rate balance equation for ground state $N(^4S)$ atoms;
viii) The gas thermal balance equation;
ix) The equation of mass conservation for the fluid as a whole.

The above system of equation is solved numerically in a self-consistent manner by iterative procedures (see below). As a result, a self-consistent spatial description of the following quantities is achieved:

electron energy distribution function (EEDF) and its integral quantities; mean power needed for the creation of an electron-ion pair $\theta(r,z)$ in the discharge; electron density $n_e(r,z)$; vibrational distribution function (δ_v) of $N_2(X^1\Sigma_g^+,v)$ molecules (up to v = 45); population densities of the molecular and atomic states $N_2(A^3\Sigma_u^+, B^3\Pi_g, C^3\Pi_u, a'^1\Sigma_u^-, a^1\Pi_g, w'\Delta u)$, Ar(3p^{5}4s), Ar(3p^{5}4p), N(^{2}P, ^{2}D), and N(^{4}S), and of positive $N^+, N_2^+, N_3^+, N_4^+, Ar^+, Ar_2^+$ ions; spatial 2D structure of the microwave electric field E maintaining the discharge and surface wave dispersion characteristics; and gas temperature $T_g(r,z)$.

Figure 1. Plasma torch.

3. Method of solution

A flow chart of the model illustrating the solution algorithm used is depicted in figure 2. The input parameters are the N_2-Ar mixture composition and the gas pressure ($p = NkT_g$, where N is the neutral gas density, T_g is the gas temperature), stimulating frequency $\omega/2\pi$, inner radius of the tube a, dielectric tube thickness $(b - a)$ and its permittivity (ε_d). The calculation starts with the solution of the dispersion equation (DE) assuming an initial Bessel-like profile of the electron density, with a given radially averaged density value, at the position of the launcher (z = 0), a TM_{00} surface mode excitation at 2.45 GHz and a given constant value of the electron-neutral collision frequency ν_{ea}. After this, the propagation constant and the 2D electric field distributions are easily calculated. An arbitrary value for the electric field intensity at the plasma-dielectric interface E_0, is introduced to continue the calculations. The iteration procedure starts with an initial Maxwellian electron energy distribution to solve the integro-differential Boltzmann equation by iterations (loop not shown in figure 2) and with arbitrary 2D Bessel-type profiles of charged particles and vibrationally and electronically excited states. Then, the rate balance equations are solved in 2D-geometry consistently with the calculated EEDF, for the chosen value of E_0 and taking into account the 2D spatial structure of the electric field components. Thus, a 2D matrix with rate coefficients and diffusion coefficients is generated. Then, the non-linear system of rate balance equations for electrons, positive ions and vibrationally and electronically excited species is linearized and solved by iteration. The unknown constants E_0 and ν_{ea} are varied and the iterations proceed until convergence is obtained in this initial plasma slice. Then, the same procedure continues along z taking into account the electron power balance equation at each axial position. The discharge ends at the axial position z_{cr} where the electron density reaches its critical value, i.e., the cut-off density. Beyond this point, i.e., in the post-discharge region, the procedure is similar but much simpler. Since the electric field vanishes here, only the species rate balance equations are solved taking the electron rate coefficients from solutions to the Boltzmann equation with zero applied field and the initial values obtained at z_{cr} for the species populations.

As a result, a self-consistent 2D solution for the EEDF and its integral quantities, the population densities of vibrationally and electronically excited species and positive ions, and the electric field distribution is obtained for the whole volume of the plasma source.

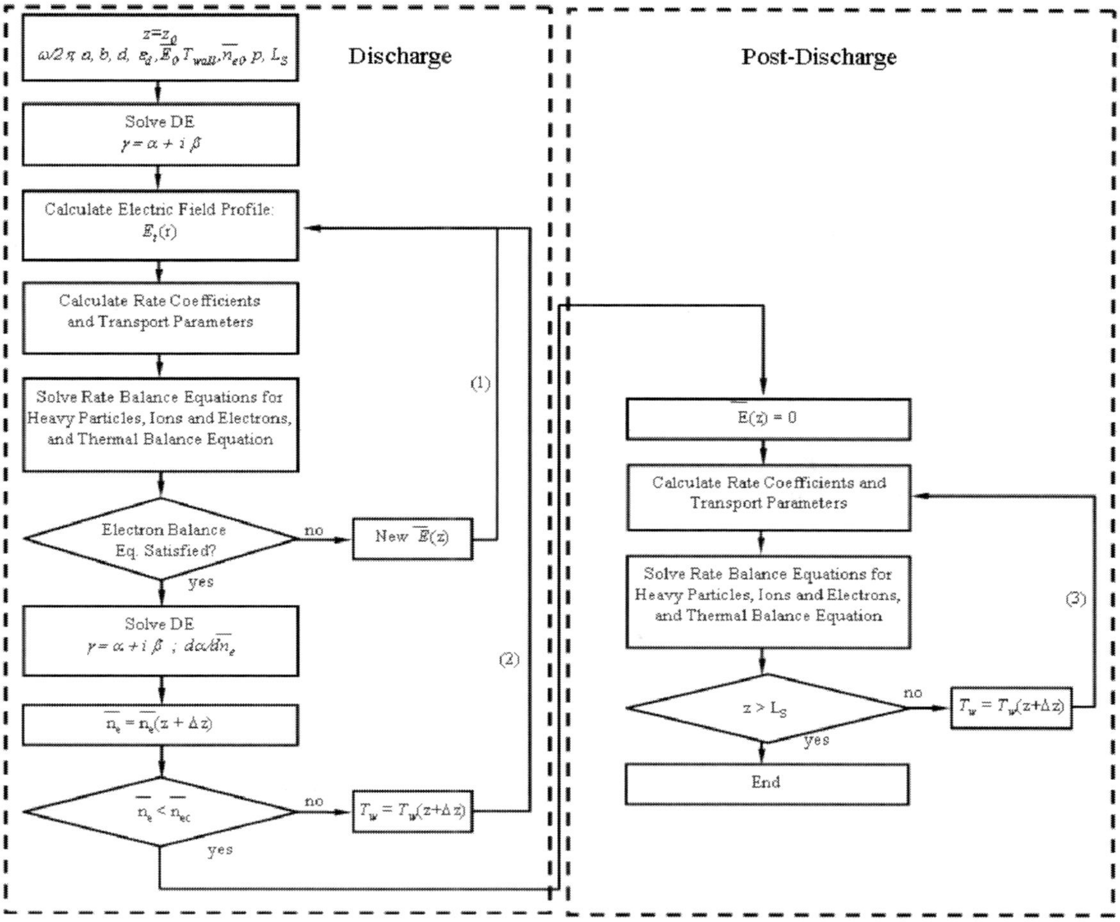

Figure 2. Flow-chart of the model.

4. Results and Discussion

Calculations have been carried out for a 20 % N_2–80 % Ar mixture at atmospheric pressure. The gas flow rate is 500 sccm and the corresponding mean gas velocity is in the range 0.07–0.87 m/s (thus, nearly isobaric conditions prevail).

4.1. Wave-to-Plasma Power Transfer.

As the wave propagates, the wave power flow decreases along the discharge since the wave power is progressively absorbed by the plasma electrons, which dissipate this power in collisions with the gas particles. Thus, the energy transferred from the electrons to the gas particles is subsequently redistributed via numerous elementary processes among the translational (gas temperature) and internal (vibrational, rotational and electronic) degrees of freedom of the particles. Figure 3(a) shows the spatial 2D distribution of the reduced electric field intensity E/N in the discharge zone (up to about 6 cm axial distance from the launcher). The reduced electric field intensity achieves a maximum at the end of the plasma column plasma. The maximum reduced electric field intensity increases from about 6×10^{-17} V cm^2 close to the launcher to about 1×10^{-16} V cm^2 at the end of the discharge zone. The minimum reduced electric field intensity is observed at the wall. This minimum is almost constant, about 1×10^{-17} V cm^2. The 2D map of the mean power needed to sustain an electron–ion pair, θ divided by total neutral density, is shown in figure 3(b). This quantity varies significantly in the radial direction, reaching a maximum at $r/a = \pm 0.5$. Close to the launcher, θ changes nearly by two orders of magnitude. A nearly one order of magnitude change in θ is observed in the axial direction (10^{-12} W – 10^{-11} W) in the central region. The changes of θ/N reflect the strong radial variations of both the electric field intensity and the neutral gas density, as a result of the radial variation in gas temperature, under isobaric conditions (p = const). In fact, the spatial variation of the mean absorbed power per electron essentially reflects that of the reduced electric field E/N sustaining the discharge.

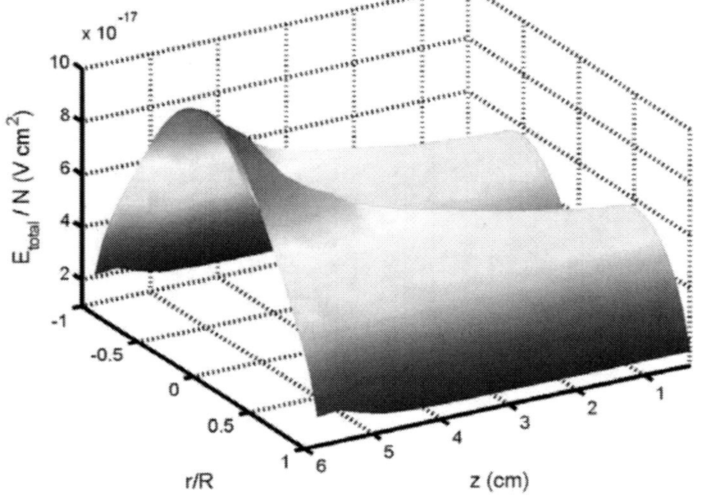

(a)

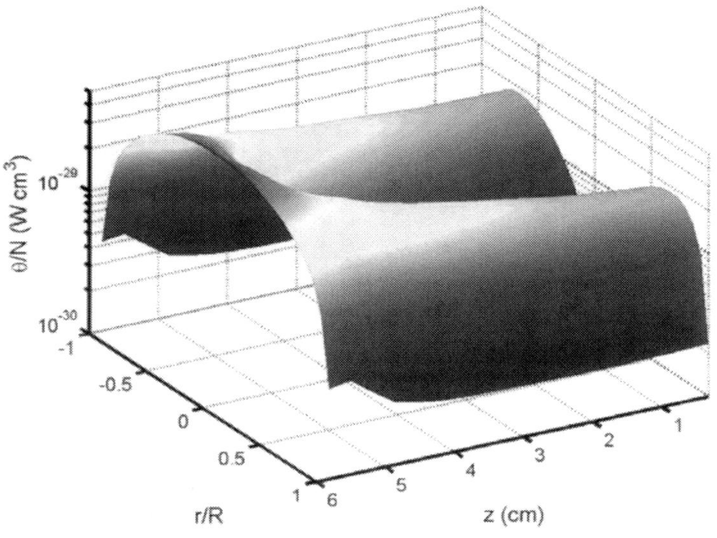

(b)

Figure 3. 2D (r–z) distribution of (a) reduced electric field E_{total}/N and (b) θ/N (p = 1 Atm, P = 600 W, 20 % N_2 – 80 % Ar).

The 2D map of the gas temperature is shown in figure 4. The measured values of the wall temperature have been used as a boundary condition. There are strong variations of gas temperature both in the axial and radial directions. Maximum values of about 5500 K are observed in the discharge zone, close to the launcher. The gas temperature drops radially from about 5000 K at the axis to about 1000 K at the wall. The relatively small variation in gas temperature along the discharge is followed (at about 4 cm axial distance from the launcher) by a sharp drop in the post-discharge, where it decreases from 2000 K to nearly room temperature at the end.

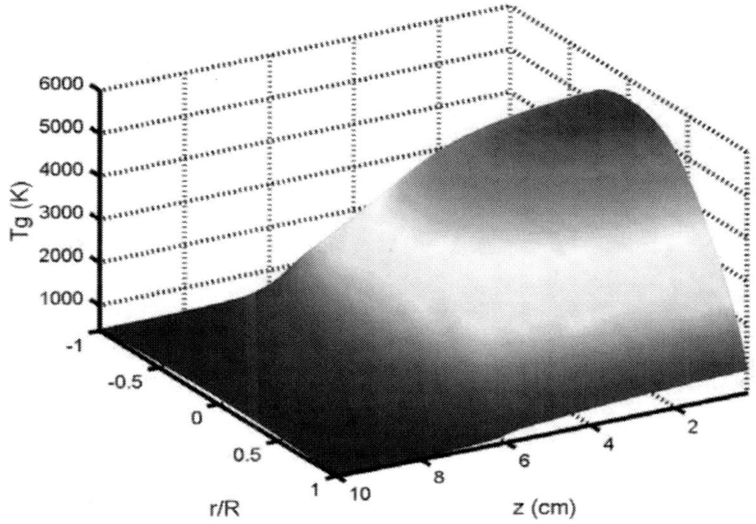

Figure 4. 2D (r–z) distribution of the gas temperature (p = 1 Atm, P = 600 W, 20 % N_2 – 80% Ar).

4.2. Charged Particles

The spatial density distribution of the electronic density is shown in figure 5. There are small axial changes in the discharge zone (up to about 6 cm axial distance) but, as expected, a rapid decrease of about two orders of magnitude occurs in the post-discharge region. Electron densities higher than 10^{13} cm^{-3} are observed close to the launcher.

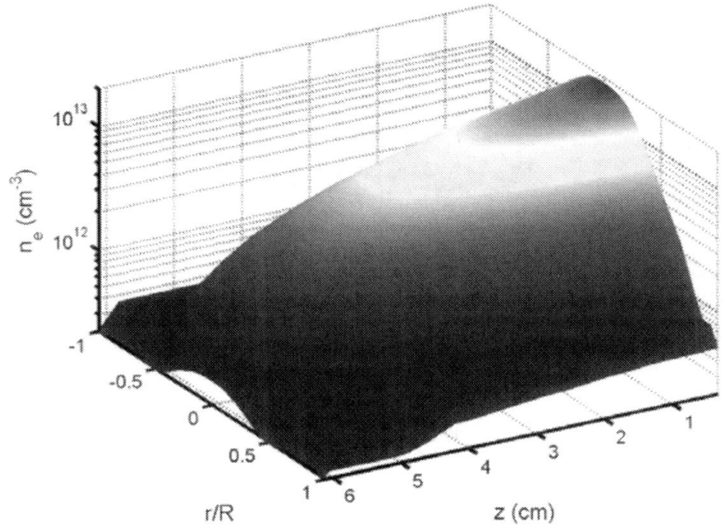

Figure 5. 2D (r–z) distribution of the electron density (p = 1 Atm, P = 600 W, 20 % N_2 – 80 % Ar).

Spatial distributions of the relative densities of the dominant positive ions Ar^+, N_2^+ and N_3^+ are shown in figures 6(a) to 6(c), respectively. As seen, the distributions of Ar^+ and N_2^+ have similar behaviours, showing almost no changes in the central zone of the discharge, but dropping very fast close to the tube wall and in the post-discharge region. According to the calculations, Ar^+ ions are the dominant positive ions (about 60 % of all ions) in the central zone of the discharge discharge, up to 6 cm axial distance. The main sources of Ar^+ ions are step-wise electron ionization processes from Ar metastables and pooling reactions between excited Ar atoms (see ref. 9). Positive N_2^+ ions constitute only about 35 % of the total ion density population in the discharge. These ions are principally created by associative ionization processes involving $N_2(A^3\Sigma_u^+)$ and $N_2(a'^1\Sigma_u^-)$ metastables (see ref 9).

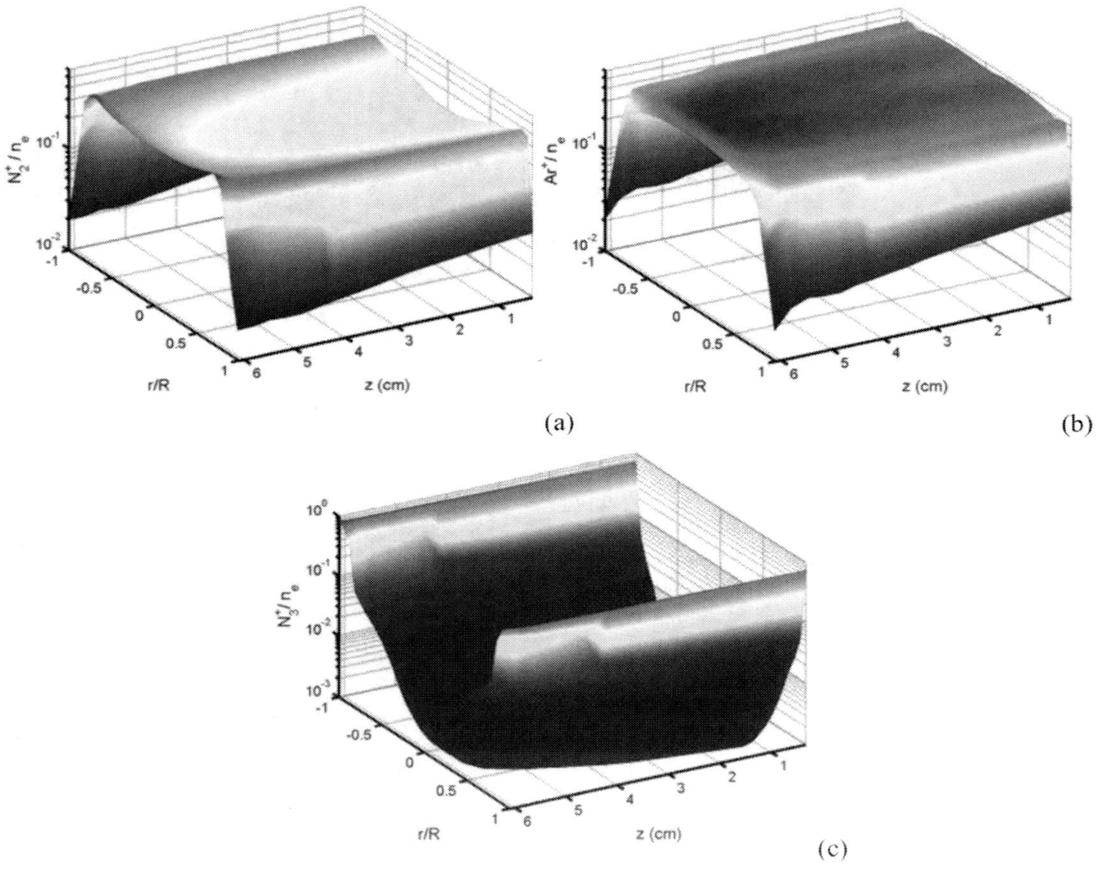

Figure 6. 2D (r–z) distributions of (a) Ar^+, (b) N_2^+ and (c) N_3^+ relative populations. (p = 1 Atm, P = 600 W, 20 % N_2 – 80 % Ar)

Close to the tube wall N_3^+ ions are the most populated. In the post-discharge, however, N_4^+ ions constitute over 70 % of the total ion population. This is a result of a strong increase in the relative contribution of associative ionization processes involving $N_2(A^3\Sigma_u^+)$ and $N_2(a'^1\Sigma_u^-)$ metastables in this region. The contribution of this channel reaches values up to 70 % of the total positive ion yield.

Three-body recombination processes between Ar atoms and Ar^+ ions provide also significant amounts (up to 30 %) of Ar_2^+ ions in the post-discharge.

4.3. Metastable and Vibrationally Excited Molecules

Microwave discharges containing nitrogen generate numerous particles in metastable states, which store energy and make it available for different processes such as ionization, chemical reactions and gas heating. Of particular importance is the triplet state $N_2(A^3\Sigma_u^+)$, which has a long radiative lifetime (about 2 s) and is usually highly populated. The spatial variation of the relative (to the total concentration of N_2) population of $N_2(A^3\Sigma_u^+)$ state is shown in figure 7(a). Close to the launcher, the $N_2(A^3\Sigma_u^+)$ relative population reaches levels higher than 10^{-4}. The decrease in *E/N* and electron density both axially and radially causes a decrease in the $N_2(A^3\Sigma_u^+)$ relative population. The $N_2(A^3\Sigma_u^+)$ population can be enhanced by the effective excitation transfer from $Ar(^3P_0, ^3P_2)$ to N_2 yielding $N_2(C^3\Pi_u)$ via the fast radiative decay of $N_2(C^3\Pi_u)$ to $N_2(B^3\Pi_g)$ and the subsequent radiative and collisional transitions from this state to $N_2(A^3\Sigma_u^+)$. The decrease in $N_2(A^3\Sigma_u^+)$ population close to the wall and to the end of the plasma source is determined by the increase of neutral density in these regions. Under nearly isobaric condition, the decrease in temperature towards the wall and the column end results in a neutral density increase and, consequently, in an increase of the $N_2(A^3\Sigma_u^+)$ quenching rate by neutral particles. The spatial variation of the relative population of the $N(^2P)$ state is shown in figure 7(b). This atomic metastable state follows closely the behaviour of the molecular state $N_2(A^3\Sigma_u^+)$, and its relative population reaches values higher than 10^{-4} close to the launcher. These two metastable states are closely correlated: quenching of $N_2(A^3\Sigma_u^+)$ by nitrogen atoms is an important mechanism for the creation of $N(^2P)$ and $N_2(A^3\Sigma_u^+)$ is produced in collisions between $N(^2P)$ and vibrationally excited $N_2(X^1\Sigma_g^+, v)$ states, with $v = 9$ and 10.

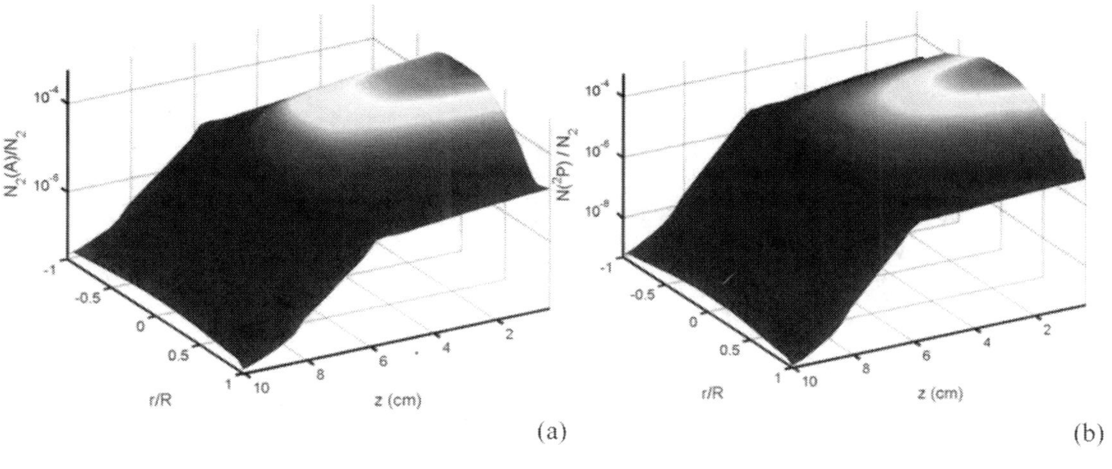

Figure 7. Spatial distribution of the relative metastable (a) $N_2(A^3\Sigma_u^+)$ and (b) $N(^2P)$ population densities (p = 1 Atm, P = 600 W, 20 % N_2 – 80 % Ar).

Journal of Physics: Conference Series **516** (2014) 012004 doi:10.1088/1742-6596/516/1/012004

At the high degrees of ionisation (10^{-3}–10^{-5}) and reduced electric fields typical of microwave discharges, the calculated vibrational distribution functions (VDFs) are considerably "excited" and exhibit significant changes both radically and along the discharge length. Figures 8(a) and 8(b) show the axial and radial evolutions of the vibrational distribution function at the axis and at the gap, respectively. High vibrational excitation of the lower and intermediate vibrational levels is observed. There are small changes along the discharge region followed by a sharp decrease in the population of the higher and intermediate levels, with vibrational quantum numbers $v > 5$, in the post-discharge plasma. This sharp decrease in the population density of the vibrational levels ($v > 5$) is a consequence of the simultaneous effects of the decrease in degree of ionization and the higher V–T depopulation rates by N_2–N collisions due to the gas temperature decrease [13]. The observed increase in the population density, for levels up to 30, close to the wall is a consequence of the simultaneous effects of the increase in electric field and the decrease in gas temperature.

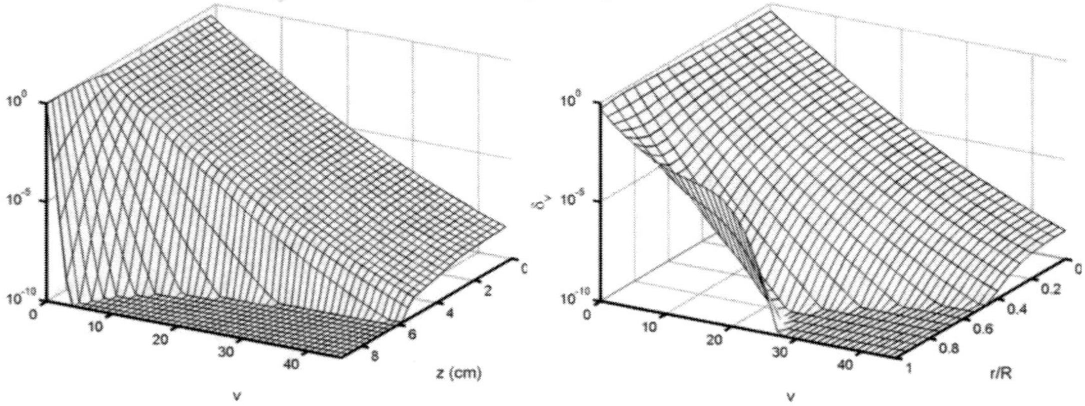

(a) (b)

Figure 8. VDF: (a) axial variation at the centre of the tube and (b) radial variation at the gap. (p = 1 Atm, P = 600 W, 20 % N_2 – 80 % Ar, r/a = 0)

4.4. Nitrogen [N(^{4}S)] Atoms

The spatial distribution of the degree of dissociation, *i.e.*, of the N(^{4}S) atom concentration relative to that of N_2, is shown in figure 9. Direct electron impact dissociation is the main source channel of ground state atoms in the discharge zone up to about 5–6 cm axial distance. The decrease in electron density and reduced electric field causes a decrease of the dissociation degree towards the plasma column end as well as across the discharge. A smooth axial variation is observed in the discharge zone, which is followed by a sharp drop of nearly 3 orders of magnitude in the post-discharge. A high dissociation degree (higher than 10^{-2}) is observed close to the launcher.

5. Conclusions

A 2D theoretical model for a microwave plasma torch driven by an azimuthally symmetric surface wave in N_2-Ar mixtures at atmospheric pressure has been developed. The model incorporates the description of both the discharge plasma, sustained by a pure TM_{00} surface mode, and the flowing afterglow. Maxwell's equations, the electron Boltzmann equation and the rate balance equations for the most important species – $N_2(X^1\Sigma_g^+, v)$ and $N_2(A^3\Sigma_u^+, B^3\Pi_g, C^3\Pi_u, a'^1\Sigma_u^-, a^1\Pi_g, w^1\Delta_u)$ molecular states, Ar($3p^54s$) and Ar($3p^54p$) atomic states, $N^+, N_2^+, N_4^+, Ar^+, Ar_2^+$ ions, and N(^{4}S,^{2}P,^{2}D) atoms – were consistently solved to yield the species spatial distributions both in the radial and the axial directions. Strong spatial correlations are shown to exist between the density distributions of plasma electrons and electronically excited states of molecules and atoms. The model predicts high populations in vibrational and metastable states of nitrogen molecules as well as high degree of nitrogen dissociation, of up to 5 %, in the discharge zone.

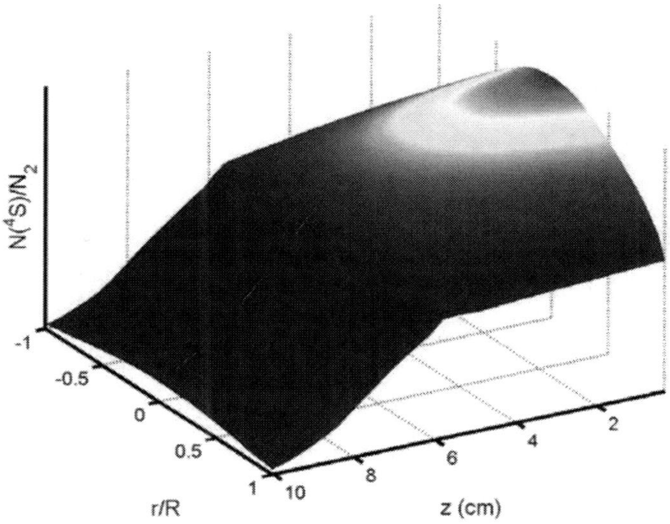

Figure 9. Spatial distribution of the degree of dissociation (p = 1 Atm, P = 600 W, 20 % N_2 – 80 % Ar).

Acknowledgement
This work was supported by the Portuguese *Fundação para a Ciência e a Tecnologia* (FCT), under the research contract PTDC/FIS/108411/2008.

References

[1] Chen Zh, Liu M, Tang L, Hu P and Hu X 2009 *J. Appl. Phys.* **106** 013314
[2] Foster J, Tomas M and Neuber A A 2009 *J. Appl. Phys.* **106** 063310
[3] Ricard A, Oseguera-Pena J E, Frank L, Michel H and Gantois M 1990 *IEEE Trans. Plasma Sci.* **18** 940
[4] Henriques J, Tatarova E, Guerra V and Ferreira C M 2002 *J. Appl. Phys.* **91** 5622
[5] Henriques J, Tatarova E, Dias F M and Ferreira C M 2002 *J. Appl. Phys.* **91** 5632
[6] C M Ferreira, E Tatarova, V Guerra, B Gordiets, J Henriques, F M Dias and M Pinheiro 2003 *IEEE Trans. Plasma Sci.* **31** 645
[7] Henriques J, Tatarova E, Dias F M and Ferreira C M 2008 *J. Appl. Phys.* **103** 103304
[8] Guerra V, Dias F M, Loureiro J, Sá P A, Supiot P, Dupret C and Popov T *2003 IEEE Trans. Plasma Sci.* **41** pp 542 – 552
[9] Henriques J, Tatarova E and Ferreira C M 2011 *J. Appl. Phys.* **109**, 023301
[10] Ferreira C M, Tatarova E, Henriques J and Dias F M 2009 *J. Phys. D: Appl. Phys.* **42** 194016
[11] Ferreira C M and Moisan M 1988 *Physica Scripta* **38** 382
[12] Golant B E, Jilinski A P and Sakharov S A 1977 The Essentials of Plasma Physics (Moscow: Atomizdat)
[13] Guerra V, Tatarova E, Dias F M and Ferreira C M 2002 *J. Appl. Phys.* **91** 2648
[14] Vereschchagin K A, Smirnov V V and Shakhatov V A 1997 *Tech. Phys.* **67** 487
[15] Kabouzi Y, Graves D B, Castaños-Martínez E and Moisan M 2007 *Phys. Rev. E* **75** 016402
[16] Tatarova E, Dias F M, Ferreira C M, Guerra V, Loureiro J, Stoykova E, Ghanashev I and Zhelyazkov I 1997 *J. Phys. D: Appl. Phys.* **30** 2663

Journal of Physics: Conference Series **516** (2014) 012005 doi:10.1088/1742-6596/516/1/012005

Excitation of mercury atoms in nitrogen post-discharge

F Krčma[1], I Bocková[1], V Mazánková[1], I Soural[1], A Hrdlička[2], V Kanický[2]

[1] Faculty of Chemistry, Brno University of Technology, Purkyňova 118, 612 00 Brno, Czech Republic
[2] Faculty of Science, Masaryk University, Kotlářská 2, 611 37 Brno, Czech Republic

E-mail: krcma@fch.vutbr.cz

Abstract. The work presents results obtained during spectroscopic observations of nitrogen DC flowing post-discharges at the total gas pressure of 1000 Pa and at the discharge current of 100 mA. Mercury traces were introduced into the system using auxiliary pure nitrogen flow enriched by mercury vapor. A very low mercury concentration of 3.7 ppb was introduced into the system before the active discharge. The strong quenching of nitrogen pink afterglow was observed but no mercury lines were recorded. Moreover, the vibrational distributions of nitrogen excited states were nearly unchanged. Based on these results, the new experimental set up was created. The introduction point of mercury vapor with higher concentration of 600 ppm was movable during the post discharge up to decay time of 40 ms. Besides three nitrogen spectral systems (first and second positive and first negative), NO^{β} and NO^{γ} bands, the mercury line at 254 nm was recorded at these conditions. Its intensity was dependent on the mercury vapor introduction position as well as on the mercury concentration. No other mercury lines were observed. The creation of mercury $^3P_1$ state that is the upper state of the observed mercury spectral line is possible by the resonance excitation energy transfer form vibrationally excited nitrogen ground state $N_2(X\ ^1\Sigma_g^+,\ v = 19)$. The observed results should form a background for the development of a new titration technique used for the highly vibrationally excited nitrogen ground state molecules determination.

1. Introduction

Nitrogen post-discharges in various configurations have been subjects of many studies during last fifty years [1-6]. Besides laboratory and technological plasmas, the nitrogen post-discharge is studied also in the kinetics of upper Earth atmosphere (corona borealis [7-9]) and these processes are also taken into account in some extraterrestrial systems, for example in Titan atmosphere [10, 11]. The neutral nitrogen molecule can form many electronic states. Due to its symmetry, all vibrational levels of the ground state $N_2(X\ ^1\Sigma_g^+)$ and also the first eight levels of the first electronically excited state $N_2(A\ ^3\Sigma_u^+)$ are metastables. Besides them, there are some other strongly metastable highly excited states (a' $^1\Sigma_u^-$ – 67 739 cm^{-1} and a $^1\Pi_g$ – 68 951 cm^{-1} [12]). All these states conserve the excitation energy for very long time. The excitation energy transfer during the collisions among these species as well as atomic recombination processes lead to formation of some radiative states and the visible light emission can be observed up to one second after switching off an active discharge depending on the experimental conditions, mainly on pressure and nitrogen purity.

The first period (up to about 3 ms) of the post-discharge in the pure nitrogen is characterized by a strong decrease of the light emission. After that, the strong light emission at the decay time of about 5–14 ms after the end of an active discharge, known as a pink afterglow, can be observed [13, 14] (for photography see in [15]). The pink afterglow is manifested by a strong increase of the pink light emission at the decay times of about 6–8 ms in pure nitrogen, while the yellow-orange color is characteristic for the other parts of the nitrogen afterglow. The nitrogen pink afterglow can be characterized as a secondary discharge because the electron concentration strongly increases due to various collisionally induced ionization processes [16, 17], and thus many discharge properties are similar to those in the active discharge. The electron density measurements during the afterglow show the strong increase of the free electron concentration during this post-discharge period [18]. The effect of the nitrogen pink afterglow can be studied only in nitrogen of very high purity; various traces (especially carbon traces and oxygen traces) quench it [19, 20]. The present work extends our experiments of the nitrogen post-discharge kinetics changes caused by mercury impurity [15]. A simplified kinetic model of the post-discharge processes running in the studied system is given, too.

2. Experimental setup

The DC flowing post-discharge was used for the experimental study. A simplified schematic drawing of the experimental setup is given in figure 1.

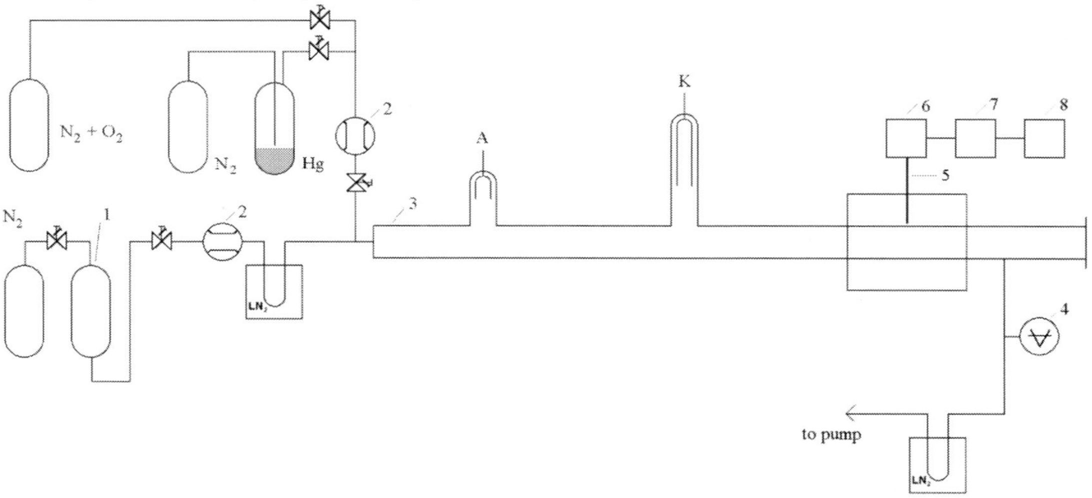

Figure 1: A simplified scheme of the experimental setup: 1 – catalyzer Oxiclear; 2 – mass flow controller; 3 – Pyrex/Quartz discharge tube (900 mm long, inner diameter of 13 mm); 4 – capacitance gauge; 5 – quartz optical fiber; 6 – monochromator Jobin Yvon Triax 550; 7 – CCD; 8 – PC.

The active discharge was created in a Pyrex or Quartz discharge tubes with a 120 mm electrode distance at current of 100 mA and pressure of 1000 Pa. Hollow molybdenum electrodes were placed in the side arms of the main discharge tube to minimize their sputtering and also to suppress the light emitted in the electrode regions. Nitrogen was of 99.999%–99.9999% purity and it was further cleaned by Oxiclear and LN_2 traps. No oxygen or carbon traces were detectable in the pure nitrogen discharge (concentrations under 1 ppm and 0.01 ppm, respectively [20, 21]). The reactor system was pumped continuously by a rotary oil pump separated from discharge tube by another LN_2 trap to suppress oil back flow and keep high nitrogen purity needed for the experiments.
A sealed PE test tube with liquid Hg of 99.995% purity was purged with the pure N_2 flux of 7.5 ml min^{-1} ensuring a saturated Hg vapor (0.2 μg min^{-1} Hg at laboratory temperature of 30 °C in a steady state). The mercury vapor condensation was observed in the MFC valve which controlled its flow into the discharge. Due to this fact, a special procedure was developed to verify the exact quantity of mercury in the discharge [15]. The mercury vapor concentration in the discharge was determined as

(3.70 ± 0.05) ppb. As some oxygen traces were detected in the mercury flow, the synthetic air was used to verify the oxygen influence. The appropriate oxygen concentration of 528 ppm was calculated using the spectra of NO^{β} system (see correlation presented in figure 5).

The spectra of post-discharge were measured by Jobin Yvon monochromator TRIAX 550 with the 1200 grooves per mm grating coupled with CCD multichannel detector. The emitted light was led to the entrance slit of the monochromator by the multimode quartz optical fiber movable along the discharge tube without any cut off optical filters. The optical fiber holder (length of 6 cm, optical fiber is mounted at its center) had to be filled by liquid nitrogen. Thus the reactor wall temperature was possible to cool down of 77 K around (± 3 cm) the observation point. The temperature of decaying plasma was calculated as 400 K at the ambient wall temperature and 100 K at liquid nitrogen wall temperature using the simulated nitrogen 1st positive ($N_2(B\,^3\Pi_g) \rightarrow (A\,^3\Sigma_u^+)$) 2-0 band spectrum [22].

Nitrogen 1st ($N_2(B\,^3\Pi_g) \rightarrow (A\,^3\Sigma_u^+)$) and 2nd ($N_2(C\,^3\Pi_u) \rightarrow (B\,^3\Pi_g)$) positive and nitrogen 1st negative ($N_2^+(B\,^2\Sigma_u^+) \rightarrow (X\,^2\Sigma_g^+)$) systems were recorded in all spectra. The bands of NO^{β} system ($NO\,(B\,^2\Pi) \rightarrow NO\,(X\,^2\Pi)$) and NO^{γ} system ($NO\,(A\,^2\Sigma^+) \rightarrow NO\,(X\,^2\Sigma)$) systems dominantly originating at vibrational level 0 were observed, too. No other molecular emissions were observed. Further, no mercury lines were recorded during the discharge or post-discharge in the measured spectral region of 300–800 nm even in quartz tube. The relative vibrational populations at the selected nitrogen levels were calculated using all measurable emission band intensities. The transition probabilities and wavelengths of the transitions were taken from *Gilmore's* tables [23].

3. Results

Populations at the selected vibrational levels are presented in figures 2–4 as a function of decay time and wall temperature. The pink afterglow effect is well visible in all cases in pure nitrogen. At the decreased wall temperature, the pink afterglow maximum is displaced for longer time, in comparison to the afterglow at the ambient temperature and its maximal emission takes of about 10 ms duration. The molecular ion $N_2^+(B\,^2\Sigma_u^+)$ population (see figure 4) decrease during the post-discharge after pink afterglow is about one order higher than for the other states.

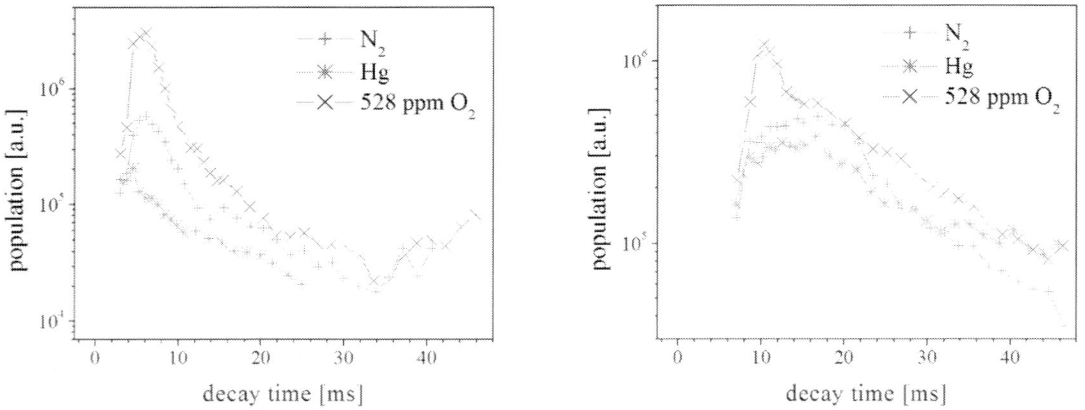

Figure 2: Population profiles at the $N_2\,(C\,^3\Pi_u,\ v = 0)$ vibrational level during the post-discharge in pure nitrogen and pure nitrogen containing traces of oxygen and mercury at the ambient (left) and liquid nitrogen (right) wall temperatures.

The most visible effect of oxygen presence is strong enhancement of the pink afterglow emission. The maximum is about five times higher than in pure nitrogen and it is strongly located in short time interval between 10 and 13 ms. After that, the populations of neutral nitrogen states show nearly the same dependence as in the pure nitrogen, only the populations are slightly higher. The population of molecular ionic state is lower than in pure nitrogen in the time interval of 13–30 ms, after that the population is nearly the same as in pure nitrogen.

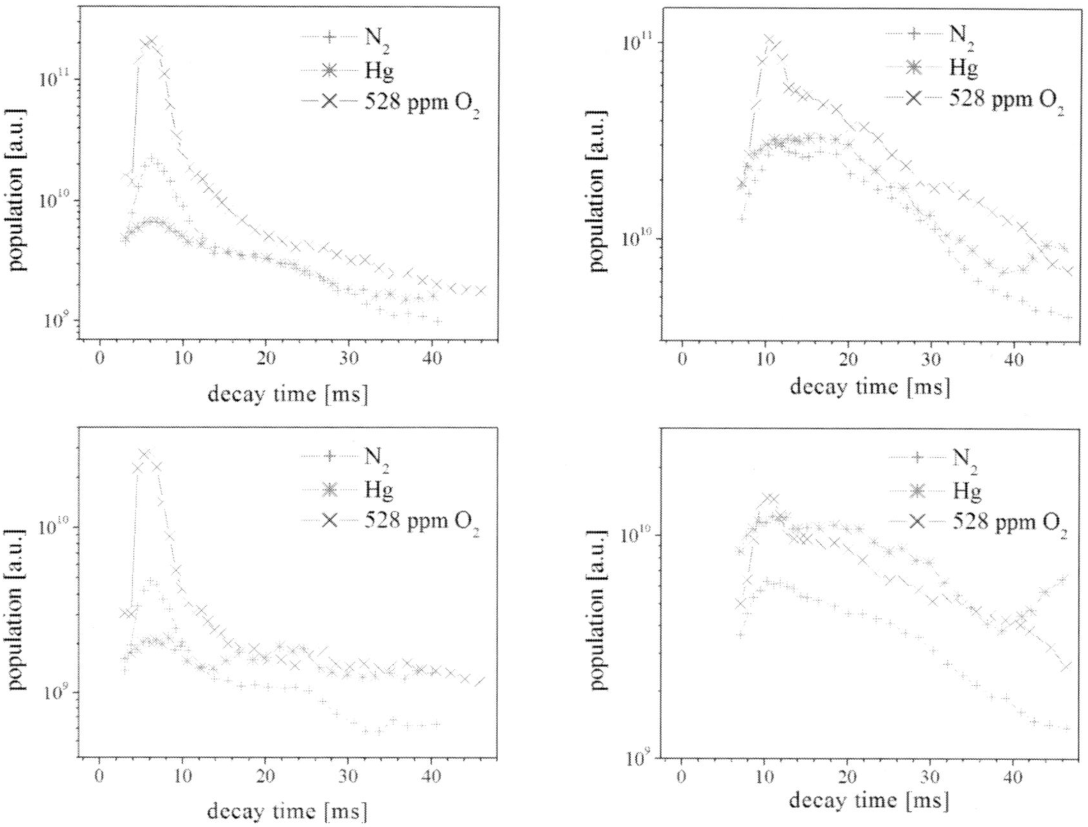

Figure 3: Population profiles at the vibrational levels 2 (top) and 11 (bottom) of N_2 (B $^3\Pi_g$) state during the post-discharge in pure nitrogen and pure nitrogen containing traces of oxygen and mercury at the ambient (left) and liquid nitrogen (right) wall temperatures. The vibrational levels are selected by the main mechanisms of their population – see the section 4.

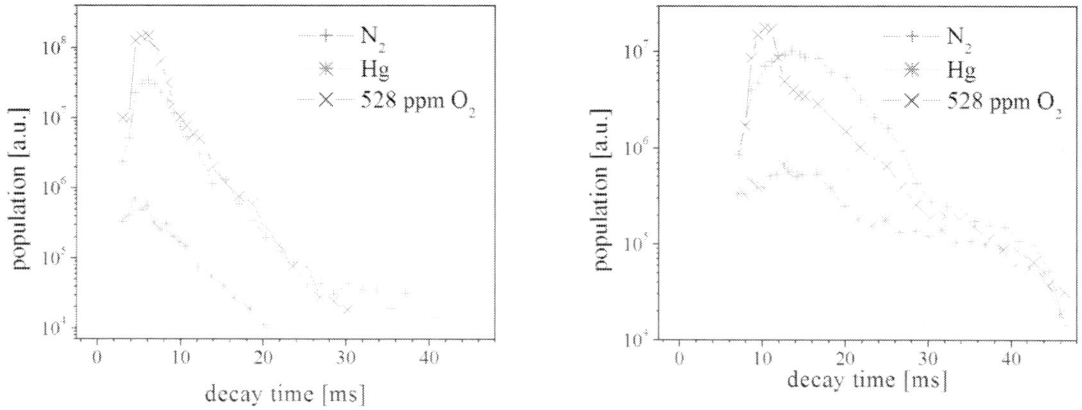

Figure 4: Population profiles at the molecular ionic N_2^+ (B $^2\Sigma_u^+$, v = 0) level during the post-discharge in pure nitrogen and pure nitrogen containing traces of oxygen and mercury at the ambient (left) and liquid nitrogen (right) wall temperatures.

The situation is completely changed when mercury traces are added. The pink afterglow effect is nearly the same or slightly more intensive as in pure nitrogen at lower N_2 (B $^3\Pi_g$) state levels, at levels over the predissociation limit (v > 12) as well as in the N_2 (C $^3\Pi_u$) state the pink afterglow is slightly quenched. The populations at levels just under the predissociation limit, i.e. levels N_2 (B $^3\Pi_g$, v = 11 and 12) show the significant increase of populations in the time interval of 10–12 ms but not such strong as in nitrogen-oxygen mixture. After that the populations are even the highest among the studied gas mixtures. The population increase at the later decay times in the case of all vibrational levels of N_2 (B $^3\Pi_g$) state is surprising. The profile of molecular ion population shows strong pink afterglow quenching, the populations at pink afterglow position is more than one order lower than in pure nitrogen. But at the latest decay times over 30 ms, the populations are nearly the same as in the other gas mixtures.
The intensities of NO^β bands are presented in figure 5. No data are available for the pure nitrogen due to its really high purity. Although the dependencies in the decay time interval of 10–35 ms are different, the figure 5 shows that the oxygen presence in mercury flow was well estimated. The difference of both dependencies should be caused by the increase of nitrogen dissociation by mercury atoms or by some other process (probably mainly wall processes) that increases the atomic nitrogen particles concentrations.

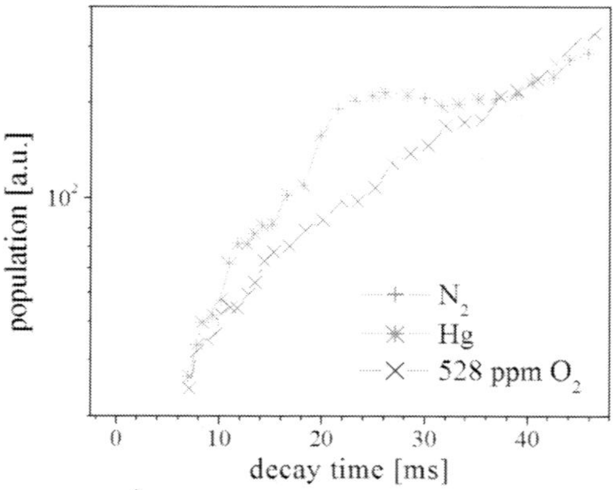

Figure 5: Intensity profile of the NO^β 0-8 band head during the post-discharge in pure nitrogen and pure nitrogen containing traces of oxygen and mercury at liquid nitrogen wall temperature.

The vibrational distributions of N_2 (B $^3\Pi_g$) state are presented in figure 6 at three selected decay times. The distributions show the same results as presented in figure 3. The vibrational distributions are nearly the same in all three gas mixtures at the earliest afterglow. At the pink afterglow maximum (middle figures), the distributions in pure nitrogen and in N_2-Hg mixture are generally the same, only at levels 11 and 12 the population is significantly higher when mercury is added; the populations in nitrogen-oxygen mixture are significantly higher at all levels. In later afterglow, the populations at higher levels (v > 12) are increased by oxygen but quenched by mercury. At lower levels (v <13) the populations are increased by both oxygen and mercury. The population enhancement by mercury increases with the increase of decay time and at later times it is higher than the increase caused by oxygen.

4. Kinetic model
The nitrogen afterglow kinetics is a really complicated problem. The mechanisms that populate the radiative states of a neutral molecule and a molecular ion are different and they must be discussed separately. Further, the kinetics in both nitrogen mixtures is discussed. The main processes populating the observed states are schematically given in figure 7. The kinetic model was proposed recently [15] but some additions to this model are included as a result of the experimental data given in the last section.

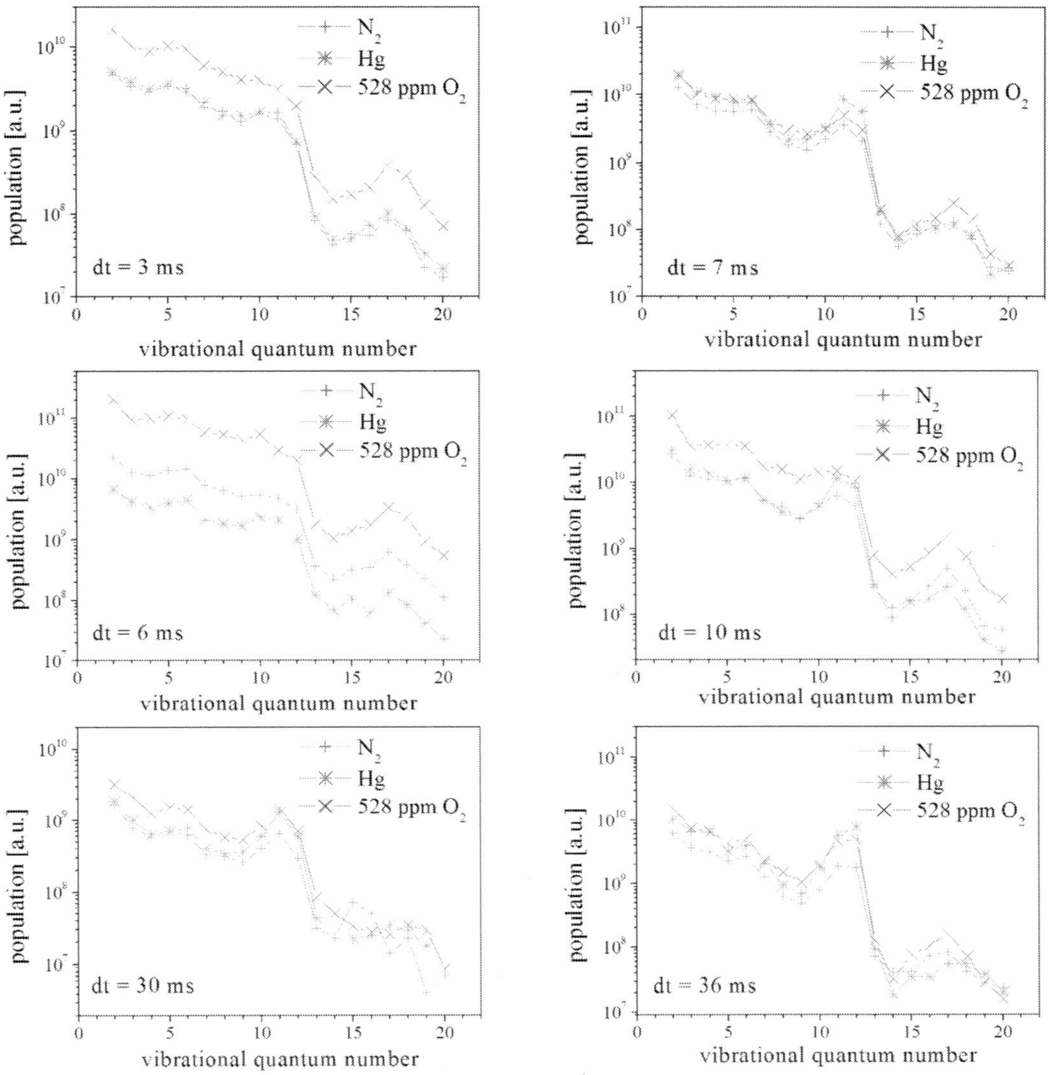

Figure 6: The vibrational distributions of the $N_2(B\ ^3\Pi_g)$ state during the post-discharge. The upper figures correspond to the position between the active discharge and pink afterglow maximum; middle figures describe distribution at the pink afterglow maximum; bottom figures correspond to the late post-discharge. Distributions at ambient (left) and liquid nitrogen (right) wall temperatures are shown. (dt marked in figures means the decay time)

4.1 Neutral nitrogen

The $N_2(B\ ^3\Pi_g)$ and $N_2(C\ ^3\Pi_u)$ states are dominantly created by pooling reactions of lower metastable states, especially by the vibrational excited ground state and by the lowest 8 levels of $N_2(A\ ^3\Sigma_u^+)$ state [2-4, 20]. The higher ground state vibrational levels populated by v-v process are also precursors for $N_2(A\ ^3\Sigma_u^+)$ state creation because its concentration at the end of the active discharge is not high enough [13, 24]. The contribution of other metastable states, especially metastable singlet states, to the pink afterglow creation is not fully understood. Besides pooling reactions, the atomic nitrogen ground states three body recombination significantly contributes to vibrational populations, mainly at levels $N_2(B\ ^3\Pi_g$, v = 10–12). More complex description of the mechanisms was given recently [13, 25].

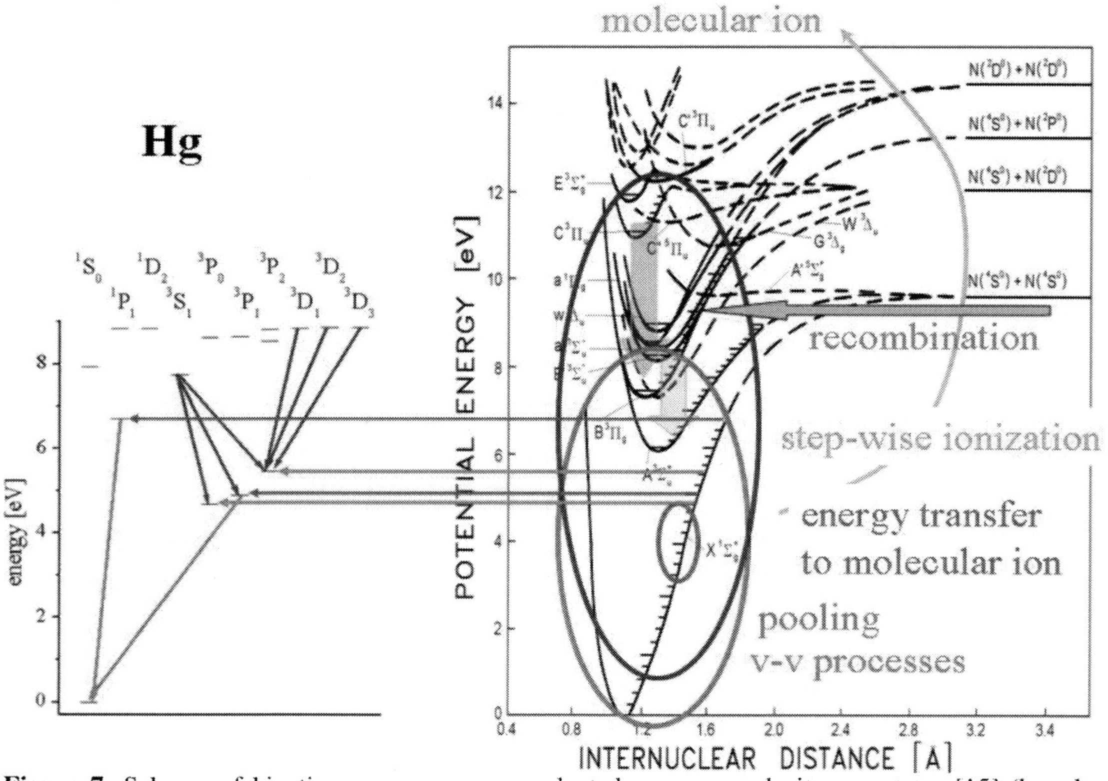

Figure 7: Scheme of kinetic processes among selected mercury and nitrogen states [15] (based on energies given by [12, 27]).

4.2 Nitrogen molecular ion

The kinetics of the molecular ion radiative state is more complicated and it can be explained in a two-step scheme. The charged particles concentration is very low before the pink afterglow [18], but during the pink afterglow it significantly increases. So the first step is the molecular ion creation. This process is known as a step-wise ionization [16, 17]. In its principle, the highly excited neutral metastable molecules (excited both electronically and vibrationally) can have energy sufficient for the ionization during their mutual collisions. After molecular ion creation, the excitation to the radiative state must be completed. Recent studies have demonstrated that the main process responsible for the population of the radiative $N_2^+(B\ ^2\Sigma_u^+)$ state is the collisionally induced energy transfer from the vibrationally excited neutral ground state [21].

4.3 Nitrogen with oxygen traces

Kinetics of pure nitrogen with a small addition of oxygen can be generally described by the following simplified scheme. The first process is atomic recombination that forms the NO $(B\ ^2\Pi)$ state, preferentially at vibrational level v = 0. This state is an origin of the measured NO^β spectral system. The NO can be effectively back dissociated during collisions with nitrogen ground state molecules excited at least to level v = 13. This process significantly influences the v-v pumping in nitrogen and thus the formation of electronically excited nitrogen states is quenched. The influence on formation of the lowest $N_2(B\ ^3\Pi_g)$ levels is smaller because they can be formed during collisions of lower vibrationally excited ground states. The formation of nitrogen molecular ion is strongly decreased, too, and thus the nitrogen 1st negative system is effectively quenched. High vibrationally excited ground state nitrogen molecules (over v = 22) can also excite the NO species into the NO $(A\ ^2\Sigma^+)$ state that is an origin of NO^{γ} spectral system. Such process was confirmed by our observations but the results are not presented here. The NO dissociation leads to the increase of atomic nitrogen concentration during the post-discharge. This process is confirmed by the enhancement of populations at $N_2(B\ ^3\Pi_g$, v = 10–12) levels in the nitrogen-oxygen gas mixture.

The reactions with molecular oxygen can be neglected because we can assume high molecular oxygen dissociation in an active discharge (the residence time of about 8 ms) and the recombination forming molecular oxygen during the afterglow can be neglected due to its low concentration.

4.4 Nitrogen with mercury traces

The influence of mercury atoms on the post-discharge kinetics is more simple because the mercury atomic energy levels lay mostly at higher energies (see figure 7) where nitrogen molecular states are not metastable, and thus their populations are very low. Due to this fact no mercury lines from visible and near UV part of the spectrum (marked by blue arrows in figure 7) were observed. Thus only the following reactions (marked by dark blue arrows in figure 7) between nitrogen metastables and mercury atoms can be taken into the model.

$N_2(X\ ^1\Sigma_g^+, v = 19) + Hg(^1S_0) \rightarrow N_2(X\ ^1\Sigma_g^+, v = 0) + Hg(^3P_1)$ R1

$N_2(X\ ^1\Sigma_g^+, v = 28) + Hg(^1S_0) \rightarrow N_2(X\ ^1\Sigma_g^+, v = 0) + Hg^*(^1S_0)$ R2

$N_2(A\ ^3\Sigma_u^+, v = 4) + Hg(^1S_0) \rightarrow N_2(X\ ^1\Sigma_g^+, v = 0) + Hg^*(^1S_0)$ R3

Reaction R1 leads to the formation of the excited $Hg(^3P_1)$ mercury state that is the origin of the resonance line at the wavelength of 254 nm. R2 and R3 reactions produce the excited mercury $^1S_0$ state that is the origin of the resonance line at 185 nm. Both these allowed transitions are marked as red arrows in figure 7. The experimental data presented in previous section do not confirm these reactions directly. R1 and R2 reactions partially stop the v-v processes in the nitrogen ground state, R3 reaction leads to the depopulation of $N_2(A\ ^3\Sigma_u^+)$ state. Thus the pooling reactions or step-wise ionization processes could not efficiently create highly excited electronic states of nitrogen and pink afterglow is quenched. The production of the lowest $N_2(B\ ^3\Pi_g)$ vibrational levels is possible by the reactions of lower excited ground state metastables that are not influenced by mercury presence, and thus it is possible to observe some remains of the pink afterglow at these levels (see figure 3).

The other channels originating from adjacent metastable nitrogen vibrational levels could play some role but they are not creating any radiative states. These reactions are marked by magenta arrows in figure 6.

5. Introduction of mercury directly into the post-discharge

The new experimental device was built up based on the precedent results. This new set up allows introduction of mercury vapor at higher concentration (contemporary up to 600 ppm) directly into the selected post-discharge positions (i.e. at different decay times) using a thin capillary in the post-discharge tube oriented from the pumping side. Thus there is no influence on the discharge and post-discharge up to the mercury introduction position. To introduced mercury vapor the auxiliary pure nitrogen flow was used.

The emission of mercury line at 254 nm was observed as it is demonstrated by figure 8 where mercury line is recorded in the spectrum of the second order. No mercury emission was observed at the earlier decay times (before the introduction point) as it is demonstrated by figure 9. It can be also seen that there is about 2 ms delay between the mercury introduction and mercury line presence. After very fast increase of mercury line emission, relatively fast decrease of its intensity can be observed; the peak takes about 5 ms. The mercury line intensity remains more or less independent on the decay time behind this peak. The same mercury line emission profile was observed at all post-discharge positions but it was slightly different at the introduction positions before the pink afterglow. This is probably directly connected to the kinetics of $N_2(X\ ^1\Sigma_g^+$, v = 19) precursor state. The other proposed mercury line (at 185 nm) was never recorded during the experiments up to now.

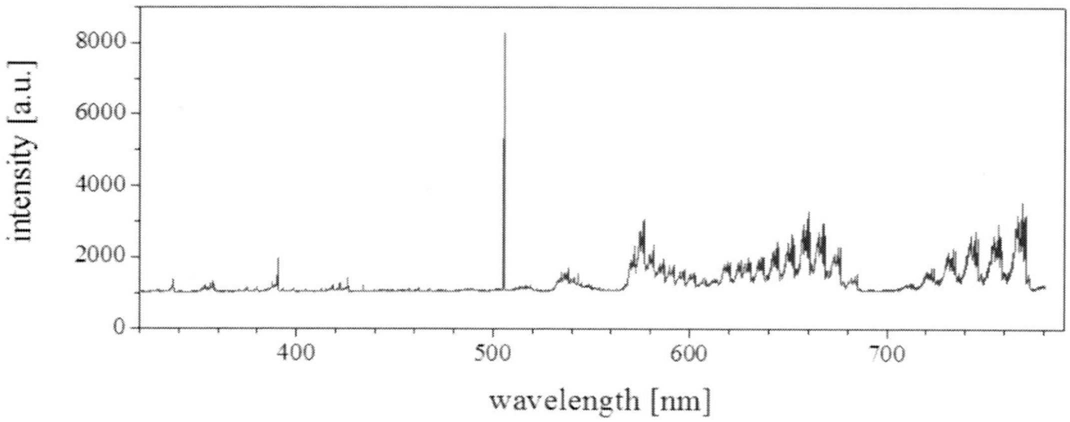

Figure 8: Example of nitrogen post-discharge spectrum at the decay time of 31 ms with mercury line at 254 nm observed in the second order spectrum. Mercury was introduced at decay time of 28.5 ms.

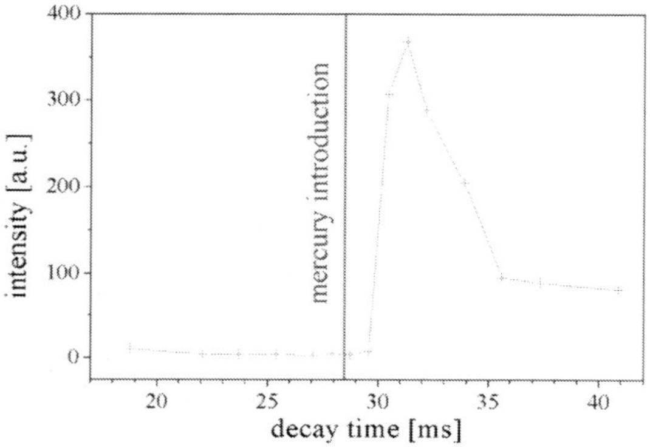

Figure 9: Mercury 254 nm line intensity during the post-discharge at selected position of mercury introduction.

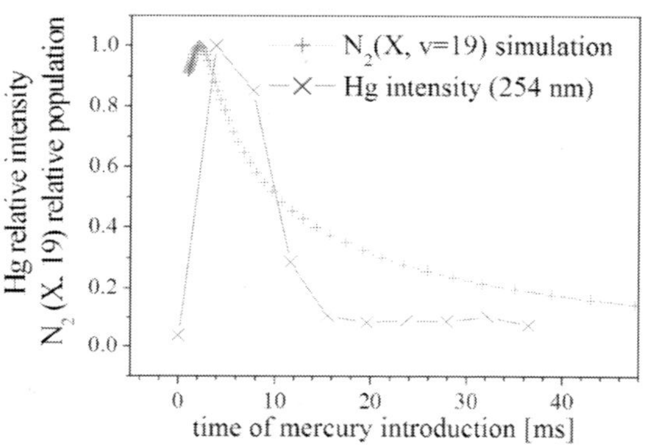

Figure 10: Comparison of mercury line maximal intensity at different positions of introduction and calculated population of precursor nitrogen state.

As we proposed in the kinetic model, the mercury line at 254 nm should directly reflect the concentration of its precursor, i.e. concentration of $N_2(X\ ^1\Sigma_g^+,\ v = 19)$ molecules. To confirm this hypothesis, we measured maximal mercury line intensity from response peaks ~~like~~ as it is presented in figure 9 as a function of the mercury introduction position into the post-discharge. The result is presented by a blue curve in figure 10. The obtained mercury intensity profile is compared to the calculated concentration of the precursor nitrogen state. The qualitative agreement is well visible in figure 10. The correspondence is not ideal due to two main facts. First of all, the time resolution of mercury line response peak (see figure 9) is limited and thus the maximal intensity presented in figure 10 could not fully reflect the real maximum of mercury line intensity peak. The second problem is that mercury amount was fixed and thus it is not sure if it was possible to reflect the precursor concentration correctly. The last experiments showed that the mercury line intensity was directly proportional to its concentration at low mercury concentrations but saturation effect was observed. The saturation should be reached to obtain a good correlation between mercury line intensity and nitrogen metastable precursor concentration. The proposal that mercury titration into the post-discharge should be used for the monitoring of highly vibrationally excited nitrogen ground state concentration should be a conclusion of these most recent experiments.

Journal of Physics: Conference Series **516** (2014) 012005 doi:10.1088/1742-6596/516/1/012005

6. Conclusions

The influence of mercury traces on the pure nitrogen post-discharge kinetics was experimentally studied in the DC flowing afterglow in Pyrex and Quartz reactors at ambient and decreased wall temperatures. It was confirmed that mercury de-excites the $N_2(X\ ^1\Sigma_g^+,\ v = 19)$ vibrational level, the second proposed process de-excitating higher metastables was not confirmed, yet. Thus nitrogen v-v processes and various pooling reactions are not effective enough to create the nitrogen pink afterglow effect. The mercury spectral line at 254 nm was observed at the mercury titration directly into the post-discharge using higher mercury concentration. The correlation between mercury line maximal intensity and population at nitrogen precursor level was confirmed. Based on these results the possible way for monitoring of nitrogen metastables was proposed.

Acknowledgement

Authors thank to Dr. Rostislav Červenka for the AAS analyses needed for determination of mercury traces concentration. This work was supported by the Czech Science Foundation, contract No. 202/05/0111, 205/08/1106, and 104/09/H080.

References

[1] Beale G E and Broida H P 1959 *J. Chem. Phys.* **31** 1030-1034
[2] Piper L G 1988 *J. Chem. Phys.* **88** 231-239
[3] Piper L G 1988 *J. Chem. Phys.* **88** 6911-6921
[4] Piper L G 1989 *J. Chem. Phys.* **91** 864-873
[5] Guerra V, Sa P A and Loureiro J 2001 *J. Phys. D, Appl. Phys.* **34** 1745-1755
[6] Krčma F 2005 *Acta Phys. Slovaca* **55** 453-460
[7] Lie-Svendsen Ø, Rees M H, Stamnes K and Whipple E C 1991 *Planet. Space Sci.* **39** 929-943
[8] Kirillov A S and Aladjev G A 1995 *Adv. Space Res.* **16** 105-108
[9] Morrill J S and Benesch W M 1996 *J. Geophys. Res. Space Phys.* **101** 261-274
[10] Keller C N, Anicich V G and Cravens T E 1998 *Planet. Space Sci.* **46** 1157
[11] Dimitrov V and Bar-Nun A 2004 *Prog. Reac. Kin. Mechanisms* **29** 1-41
[12] Lofthus A and Krupenie PH 1977 *J. Phys. Chem. Ref. Data* **6** 113-307
[13] Sa P A, Guerra V, Loureiro J and Sadeghi N 2004 *J. Phys. D, Appl. Phys.* **37** 221-231
[14] Loureiro J, Sa P A and Guerra V 2006 *J. Phys. D, Appl. Phys.* **39** 122-125
[15] Kanický V, Otruba V, Hrdlička A, Krásenský P and Krčma F *Journal of Atomic Analytical Spectrometry* **22**, 754-760
[16] Polak L S, Slovetskii D I and Sokolov A S 1972 *Opt. Spectrosc.* **32** 247-251
[17] Paniccia F, Gorse C, Cacciatore M and Capitelli M 1987 *J. Appl. Phys.* **61** 3123-3126
[18] Janča J, Tálský A and El Kattan N 1978 *Folia Physica* **27** 23-36
[19] Krčma F and Babák L 1999 *Czech. J. Phys.* **49** 271-288
[20] Krčma F, Mazánková V and Soural I 2007 *Publications of the Astronomical Observatory of Belgrade* **82**, 133-147
[21] Krčma F and Protasevich E T 2003 *Post-discharges in Pure Nitrogen and in Nitrogen Containing Halogenated Hydrocarbon Traces* (Tomsk: Tomsk Polytechnic University Publishing)
[22] Krčma F, Mazánková V, Soural I and Šimek M 2007 *Proc. Int. Conf. on Phenomena in Ionized Gases* (Prague 2007) pp 1722-1725
[23] Gilmore F R, Laher R R and Espy P J 1992 *J. Phys. Chem. Ref. Data* **21** 1005-1107
[24] Pintassilgo C D, Loureiro J and Guerra V 2005 *J. Phys. D, Appl. Phys.* **38** 417-430
[25] Sadeghi N, Foissac C and Supiot P 2001 *J. Phys. D, Appl. Phys.* **34** 1779-1788
[26] NIST Atomic Spectra Database: http://physics.nist.gov/PhysRefData/ASD/index.html

Structure and electrochemical properties of track membranes with a polymer layer obtained by plasma polymerization of acetylene

L I Kravets[1], S N Dmitriev[1], V Satulu[2], B Mitu[2], G Dinescu[2]

[1] Joint Institute for Nuclear Research, Flerov Laboratory of Nuclear Reactions, Joliot-Curie Str. 6, 141980 Dubna, Russia
[2] National Institute for Laser, Plasma and Radiation Physics, Atomistilor Str. 111, 77125 Magurele, Bucharest, Romania

E-mail: kravets@lnr.jinr.ru

Abstract. The structure and electrochemical properties of poly(ethylene terephthalate) track membrane modified by acetylene plasma were studied. It was found that polymer deposition on the surface of a track membrane by plasma polymerization of acetylene results in the creation of composite membranes that, in the case of formation of a thin semipermeable layer, possess an asymmetry of conductivity in electrolyte solutions – a rectification effect. It is caused by the reduction of the pores diameter due to the plasma polymer that results in changing the pore geometry, and as well due to the existence of an interface between the initial membrane and the polymer layer which have various concentrations of carboxyl groups.

1. Introduction

Membrane processes have found wide application in many areas of science and technology, such as gas separation, desalination of water, pervaporation, separation and isolation of individual solutes, and purification and concentration of biologically active substances [1]. Prominent among the variety of the membranes used in these processes are polymer membranes. However, the properties of existing membranes are frequently inconsistent with the requirements of industrial processing technologies, since the range of polymers suitable for the manufacture of membranes is limited. To extend the application area of commercial membranes, research works on the modification of their properties are performed. The most popular technique used for this purpose is the treatment of membranes in low-temperature plasma [2]; an important advantage offered by this process is the possibility of modifying a thin surface layer, which alters membrane properties, namely, the adsorption, transport, and selectivity properties. This possibility substantially extends the application area of membranes. The bulk of the membrane matrix remains intact in this case, which is undoubtedly very important from the viewpoint of retention of its mechanical and physicochemical properties.

The low-temperature plasma treatment entails a number of physicochemical processes, depending of the discharge type (frequency) and the nature of the plasma gas, which makes it possible to control in a desired mode the structure and the chemical composition of the surface of polymer membranes [3]. Chemical processes that occur during modification in the surface membrane layers are largely determined by the gas-phase composition of plasma. Two kinds of plasma are used, the plasmas of simple gases that do not result in the formation of polymers (non-polymerizing gases) and the plasma produced by discharge in vapors of any organic or organoelement compound that leads to polymer formation [4]. When various organic vapors are used for the plasma treatment of membranes, a thin polymer film is deposited on their surface as a result of polymerization. In this case, composite membranes consisting of the porous substrate and a plasma deposited polymer layer are produced. Depending on the plasma treatment time and the pore diameter of the parent membrane, composite ultra- and nanofiltration membranes and reverse osmosis membranes can be obtained [4]. In the latter case, a thin semipermeable layer, which completely covers the pores, is deposited on the surface. The possibility for controlling the thickness of the plasma deposited layer, which determines the selective properties of membranes, and a wide choice of organic compounds suitable for this process make this method especially promising. The surface properties of composite membranes obtained via plasma modification depend on the chemical compound used.

In this paper the structure and electrochemical properties of the membranes composed of a porous substrate and a polymer layer deposited by plasma polymerization of acetylene were studied, where the substrate was a poly(ethylene terephthalate) track membrane (PET TM).

2. Experimental details

In the present experiments, PET TM samples with a thickness of 9.5 μm and an effective pore diameter of 65 nm (pore density 3×10^{9} cm^{-2}) were used. To produce the membrane, poly(ethylene terephthalate) film with a thickness of 10.0 μm (Lavsan, Russia) was irradiated by krypton positive ions, accelerated to ~ 3 MeV/nucleon at the cyclotron. Then the ion-irradiated film was additionally sensitized with ultraviolet irradiation with maximum emission intensity at 310 nm. Chemical etching was performed in an alkaline (NaOH) aqueous solution (1.0 mol/L) at 75 °C for up to 4 min.

Acetylene was used as a monomer to obtain a polymer film by the plasma polymerization method. The deposition of the polymer thin film on the membrane surface from the acetylene was done in a plasma-chemical reactor using a RF-discharge in parallel plate configuration at the frequency of 13.56 MHz, gas pressure of 12 Pa, discharge power of 20 W, and treatment duration of 300 and 600 s. Before treatment, the chamber was vacuumed down to a residual pressure of 6.5 Pa. The acetylene vapors were carried into the reaction chamber by an argon gas flow. Disk-shaped membrane samples, with an area of 56 cm^2, were positioned on the grounded electrode at a 4-cm distance from the powered electrode. The treatment procedure and the scheme of this installation were described previously in [5]. Only one side of the membrane was subjected to the plasma treatment.

The characteristics of the initial membrane and of membranes with the deposited layer of plasma polymer were determined by different procedures. The amount of the polymer obtained by plasma on the membrane surface was defined by the gravimetric method. The thickness of the plasma polymer on the membrane surface was measured with an electronic counter of thickness Tesa Unit (Austria). The gas flow rate through the membranes was determined at an adjusted pressure drop. Gas consumption was measured by a float-type flow meter. From the obtained data, the effective pore diameter was calculated using the Knudsen equation. The study of the samples microstructure and the determination of the pore diameter at the membrane surface were conducted by SEM using a JSM-840 (JEOL). The changing of the wettability of the membrane was evaluated by measurement of the water contact angle. For this purpose the sessile drop method with a horizontal microscope equipped with a goniometer was applied. Measurements of the current-voltage characteristics of the membranes were carried out in direct current regime, in the voltage range of −1 to +1 V, using a PC-controlled potentiostat Elins P-8S with a scan rate of 100 mV/s. A two-chambered cell with Ag/AgCl electrodes, containing a water solution of potassium chloride (KCl) of identical concentration on both sides of the membrane was used for this purpose. The procedure of the measurements is detailed elsewhere [5].

The chemical composition of the polymer obtained by plasma polymerization of acetylene was studied with ESCA and FTIR-spectroscopy. The ESCA spectra were recorded with a Riber SIA-100 spectrometer using the MAC-2 analyzer (MgKα, 100 W). The peaks' positions (the binding energy values) were calibrated to the C_{1s} standard peak (284.6 eV). FTIR reflection spectra were measured on a Bruker Equinox 50S Fourier-transform IR spectrometer with a MIRacle™ single reflection horizontal ATR attachment, using a ZnSe crystal, over the range of 400–4000 cm^{-1}; the data of 500 scans were collected with a scanning step size of 2 cm^{-1}. In order to record the ESCA and FTIR-spectra of the plasma polyacetylene, the polymer was deposited on a 10×15-mm silicon plate.

3. Results and discussion

The parameters of the membrane subjected to the acetylene plasma are presented in the table 1. The results show an increase of sample mass during plasma treatment on the PET TM surface. This increase is explained by the deposition of the layer of plasma polymerized acetylene (PPAc). The thickness of the membrane increases, and the effective pore diameter in this process decreases. The increasing thickness and decreasing effective pore diameter indicate that the deposition of polymer takes place both on the surface of the membrane and on the walls of its pores. Electron-microscopic research of the modified membranes (figure 1) shows that deposition of the polymer at the chosen gas pressure and discharge power occurs mainly on the sample surface. For a membrane treated for 300 s, only an insignificant reduction of the pore diameter on the membrane surface is observed (figure 1b). The membrane pores are open in this case, and the thickness of the deposited polymer layer on the surface is 70 nm (table 1). An increase in the plasma treatment time up to 600 s leads to an increase in the thickness of the deposited polymer layer to 150 nm. The membrane pores on the plasma treated side are covered by a PPAc layer in this case (figure 1c). SEM investigation of the cross-section of this membrane (figure 2c) shows that a thin PPAc layer is formed on its surface that blocks the pores. From the electron microscopy data, it also follows that for the modified membranes the pore diameter on the untreated side remains unchanged. Moreover, results of the parameters' testing of modified membranes show that an effective pore diameter of the membrane formed at the plasma treatment time of 300 s decreases insignificant. On the contrary, the membrane formed at the plasma treatment time of 600 s has an effective pore diameter of 26 nm. It's indicates that the pore diameter of a polymer layer deposited by plasma is much smaller than the ones of the initial membrane and pores get the asymmetric form.

Table 1. Change of the membrane characteristics during treatment by acetylene plasma.

Parameters	Initial membrane	Plasma treatment time (s)	
		300	600
Relative increase in the mass (%)	–	3.7	8.2
Thickness of deposited layer (nm)	–	70	150
Air flow rate at ΔP=3×10^4 Pa (ml/min cm^2)	165.0	95.0	10.0
Effective pore diameter (nm)	65.0	54.5	26.0
Water contact angle (deg)	65	67	65

From the SEM images of the membrane surface and cross-sections it was difficult to determine how the plasma polymer is deposited inside the PET TM pores. It was also difficult to present the pore geometry in the polymer layer. We can only suppose how the plasma polymer fills in the membrane pores. We suppose the plasma polymer coating tapers off quite rapidly in thickness as one descends into the pores from the pore entrances, but it was not possible to see this in the SEM images. Besides, we assume that short treatment time in plasma (300 s) leads to formation of a thin polymer film on the membrane surface and inside the pore entrances that does not cause significant decreasing the pore diameter. Increasing the deposition time to 600 s causes additional polymer growth on the surface that leads to significant reduction of the pore diameter in the plasma polymer layer. Probably, extended growth of the polymer film using longer deposition time may cover totally the pores on the membrane surface. Also, the research on the surface properties shows that for the modified TM no improvement of the wettability of the surface is observed. Indeed, if for the initial membrane the water contact angle is 65°, then for the modified membrane the value of the water contact angle is almost the same at 66°.

The analysis of the ESCA spectra of the PPAc film deposited directly on the silicon plate in the plasma reactor shows the presence of peaks related to atoms of carbon and insignificant contents of oxygen. The latter is possible due to the presence of residual oxygen in the vacuum reaction chamber and subsequent oxidation of PPAc on air that is characteristic for polymers synthesized by plasma [4]. A detailed analysis of C_{1s} of the spectrum of PPAc samples shows that it has a complex character due to the presence in the polymer of C–C/C–H (284.4 eV) and $-CH_2-$ (283.5 eV) bonds as well as some bonds of C–O (286.1 eV) and O–C=O (288.9 eV). The presence of the last two groups means that part of the carbon atoms could be bound to oxygen. The deconvolution of the O_{1s} peak indicates that the predominant groups are O=C (531.4 eV), C–O–C (532.8 eV) and –O–C– (534.6 eV). There is a small amount of ions O^{2-} (529.9 eV) in the spectrum. It also means that part of carbon atoms is connected to oxygen. Apparently, these groups of atoms exist as functional end groups.

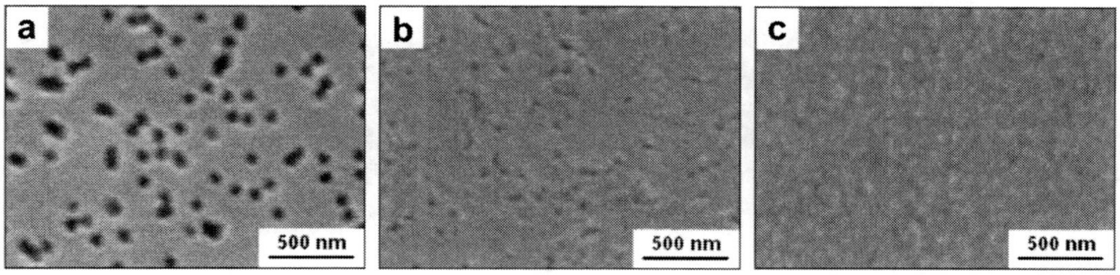

Figure 1. SEM photographs of the initial PET TM surface (a) and membranes treated by acetylene plasma for 300 (b) and 600 s (c).

Evaluation of the chemical composition of the plasma polymer by FTIR-spectroscopy shows absorptions compatible with the structure of ethylene. So, there are a band at 1600 cm^{-1} in the spectrum which corresponds to stretching vibrations of the bonds C=C in olefins, a band at 1300 cm^{-1}, corresponding to in-plane bending of C–H in $-CH=CH_2$, a band at 900 cm^{-1}, corresponding to in-plane bending of the bond C–H in $RCH=CH_2$. Besides, in the spectrum there are absorption bands which are characteristic for acetylene. So, the spectrum contains a band at 3300 cm^{-1} which corresponds to stretching vibrations of the bond C–H in H–C≡; bands at 2198 cm^{-1} and 2102 cm^{-1} corresponding to stretching vibrations of triple bond C≡C in R'–C≡C–R and H–C≡C–R as well as a band at 609 cm^{-1} which can be related to in-plane bending of groups H–C≡C–. Apparently, these groups exist as end ones on the polymer macromolecules. Data of the FTIR-spectroscopy thus confirm that fact that olefin groups are the basic structural element of the polymer synthesized from acetylene plasma. In addition to that, in the spectrum one can observe an absorption band corresponding to stretching vibrations of bond C–H in satiated hydrocarbons: at 2953 cm^{-1} – asymmetric vibrations in CH_3-groups, at 2928 cm^{-1} – asymmetric vibrations in CH_2-groups and 2870 cm^{-1} – symmetric vibrations in CH_3-groups. The spectrum also has the absorption bands corresponding to in-plane bending of CH_3 and CH_2-groups: 1452 cm^{-1} and 1376 cm^{-1}, and a band at 750 cm^{-1} which answers rocking vibrations of CH_2-groups. Apparently, these groups are formed as a result of cross-linking and recombination processes in the discharge. In the spectrum one can observe also an absorption band at 1020 cm^{-1} due to the stretching vibrations of groups C–O, a band at 1150 cm^{-1}, corresponding to stretching vibrations of C–O–C and absorption at 1720 cm^{-1}, corresponding to stretching vibrations bond C=O, most likely, of carboxyl groups. The presence of the carboxyl groups in the surface PPAc layer is proved also by the results of research by the ESCA method.

The deposition of polymer on the PET TM surface by plasma polymerization of acetylene thus leads to formation of composite membranes consisting of two layers. One of them, which corresponds to the initial membrane polymer matrix, has cation-exchange carboxyl groups on the surface [6]. In KCl solutions (pH of which is 6.0), dissociation of surface COOH-groups leads to formation of negatively-charged segments on the polymer macromolecules. The presence of a negative charge on the macromolecular segment causes the surface layer to swell and a polyelectrolyte gel to form. The second layer synthesized by the plasma polymerization of acetylene also contains some cation-exchange carboxyl groups, as our results have shown. In the electrolyte solutions, dissociation of surface COOH-groups leads to formation of negatively-charged segments on the polymer chains too and formation of a polyelectrolyte gel. Certainly, the volume of the gel in this case is much less than the volume of the gel on the PET surface because the concentration of COOH-groups is lower on the plasma polymer surface.

Figure 2. SEM photographs of the cross-sections taken at an angle 40° of the initial PET TM (a) and the membranes treated by acetylene plasma for 300 (b) and 600 s (c).

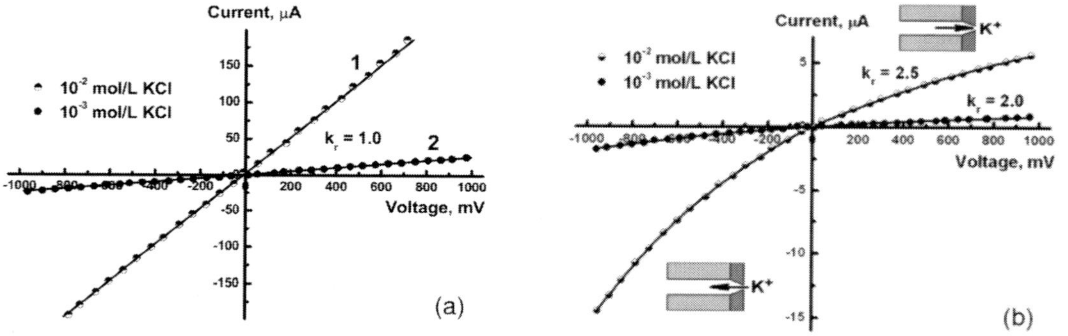

Figure 3. Current-voltage characteristics of the initial PET TM and membrane treated by plasma for 300 s (a) and membrane treated by plasma for 600 s (b) in KCl solution with concentrations of 10^{-2} and 10^{-3} mol/L. On the inserts a membrane orientation is specified in the electric field.

The results of measuring the current-voltage characteristics of the membranes (figure 3) show that the conductivity of the initial track membrane does not depend on the current direction. The pores have symmetric (cylindrical) form in this case (figure 2a). The conductivity of the membrane treated by plasma for 300 s also does not depend on the current direction because the membrane structure does not change too much in this case – the pores also have practically a symmetric form. On the contrary, the analysis of current-voltage characteristics of the membrane treated by plasma for 600 s shows that its conductivity depends on the current direction. The size of rectifying effect can be characterized by a rectification coefficient (k_r) calculated as the ratio of values of the current in the opposite directions at a potential of 1 V. The research shows that the rectification coefficient for the membrane treated by plasma for 600 s in KCl solution with a concentration of 10^{-2} mol/L equals 2.5, and in the solution with a concentration 10^{-3} mol/L it is 2.0 (figure 3b). The obtained results can be explained as following. Firstly the pores of this membrane have asymmetric form. Besides, the pore diameter on the plasma treated side of membrane is much lower than the pore diameter of the initial membrane. Secondly in the composite membrane there is an interface between two layers – the initial membrane and the plasma polymer, which have various concentrations of carboxyl groups on their surface layer. Obviously, presence of this interface between layers may also be the reason of the appearing of the conductivity asymmetry for track membranes with the layer of plasma polyacetylene on the surface. Therefore, appearing the asymmetry of conductivity for the composite membranes with a thin semipermeable PPAc layer on the surface can be interpreted as follows. Under a direct current (DC) reverse bias condition, when PPAc layer is inverted to the anode (figure 3b, lower insert), a significant decrease in the resistance of the system is observed. It is caused by the change of the transference number of counter ions in each layer of the membrane and appearance of the effect of concentration polarization. So, K^+ ions in this case easily pass the gel phase of the PPAc layer as the transfer number for the cations in gel with anion segments is higher than in a solution. It leads to their accumulation in a wide area of pores of PPAc layer. Ions of Cl^- moving in an opposite direction also are accumulated in a wide part of the pores of this layer because the transfer number for the anions through the gel phase with anion segments is lower than in a solution of electrolyte. Increase of the local concentration of KCl in the left (wide) part of the pore channels of PPAc layer on the boundary of this layer and PET TM promotes the transfer of current.

Under the DC forward bias condition, when the PPAc layer is inverted to cathode (figure 3b, upper insert), one can observe increase of resistance of the system. It is caused by the following. At such orientation of the composite membrane there is a removal of K^+ ions from the PPAc layer through gel phase that leads to decrease in the local concentration of these ions on the interface of this layer and PET TM. Their replenishment from the volume of electrolyte solution which is in the PET TM pores is complicated, since the transfer number of K^+ ions in the pores of the initial membrane is a little bit lower than the transfer number of these ions in the gel phase of the PPAc layer. A similar phenomenon is observed for Cl^- ions – they are removed from pores, and their replenishment from the electrolyte solution through the gel phase of the PPAc layer is complicated. This also causes decrease of the local concentration of Cl^- ions in the wide part of the pore channels of the PPAc layer – on the boundary of this layer and the PET TM that leads to decreasing the current in the system.

4. Conclusion

Summarizing the obtained results, one can conclude that the conductivity of the initial PET track membrane in the electrolyte solutions does not depend on the current direction. The deposition of the polymer on its surface by the plasma polymerization of acetylene, resulting in formation of a thin semipermeable layer that blocks the pores, leads to the creation of a composite membrane possessing asymmetry of conductivity – a rectification effect. This effect is caused by an important reduction of the pore diameters in the plasma polymer that results in changing the pore geometry, as well as an existence of an interface between the initial membrane and the polymer layer, which separates regions with different concentrations of carboxyl groups. Membranes with such properties can be used to create chemical and biochemical sensors.

Acknowledgement

This work was supported by the grant (No 08-08-12207) from Russian Foundation for Basic Research and the project Partnership CNMP 72226 financed by the Romanian Authority for Scientific Research.

References

[1] Brock T D *Membrane Filtration: A User's Guide and Reference Manual*, Madison: Science Tech., 1983
[2] Garbassi F, Morra M and Ochiello E *Polymer Surface from Physics to Technology*, New York: John Wiley & Sons, Inc., 1994
[3] Kravets L I, Dmitriev S N and Gilman A B 2009 *High Energy Chemi*stry **43** 181
[4] Yasuda H *Plasma Polymerization*, Orlando: Academic, 1985
[5] Kravets L, Dmitriev S, Lizunov N, Satulu V, Mitu B and Dinescu G 2010 *Nucl. Instr. and Meth. in Phys. Res. B* **268** 485
[6] Dmitriev S N, Kravets L I, Sleptsov V V and Elinson V M 2002 *Desalination* **146** 279

DC Underwater Capillary Discharge With Symmetrical Hole: Diagnostics and Pumping Effect

Ivana Halamová[1,2], Anton Nikiforov[2], František Krčma[1], Christophe Leys[2]

[1]Institute of Physical and Applied Chemistry, Faculty of Chemistry, Brno University of Technology, Purkynova 118, 612 00 Brno, Czech Republic
[2]Department of Applied Physics, Ghent University, Jozef Plateaustraat 22, 9000 Ghent, Belgium

E-mail: xchalamova@fch.vutbr.cz

Abstract. Capillary discharge investigated by this work was created in the reactor using positive half-cycle of AC high voltage up to 2 kV. Electric field created between two electrodes which were separated by the dielectric ceramic barrier with a symmetrical cylindrical hole (diameter of 1 mm, length of 5 mm) in it. When voltage reached the value sufficient for liquid breakdown, the discharge appeared initially in bubbles of evaporated solution in the hole vicinity and spread further into the liquid volume. After the rise of discharge, two kinds of plasma channels (streamers) propagated towards electrodes from the pin-hole: longer positive streamers on the side with the cathode (analogically as in the positive corona discharge) and shorter negative streamers (like negative corona discharge). These streamer kinds differed especially in the energy dissipation originating from different electron velocities in plasma channels due to electron collisions with positive particles accelerating or decelerating electron avalanches and it gave rise of a significant pump effect.

1. Introduction

Electrical discharges in electrically conductive liquids can be generated in several ways and various electrode configurations have been studied. Among the various configurations to generate plasma in a conductive water solution only the diaphragm and the capillary discharge scheme allow generation of plasma which is not in contact with electrodes. In these discharge configurations plasma is generated in one or more holes in a dielectric separating two water vessels. In capillary discharges, the length-to-diameter ratio of the holes is >> 1 [1]. If this ratio is about 1, the discharge is called diaphragm discharge [2].

Recent research showed that periodic plasma generation in an asymmetrical capillary induces a flow of liquid between the two vessels, which might be utilized to design a micro pump [3]. It is assumed that these underwater electrodeless discharges are initiated by the breakdown of a water vapour bubble inside the hole, the bubble being nucleated due to Ohmic heating of the conductive liquid filling the dielectric hole [4].

This work presents our research focused on the pump effect using the symmetrical geometry of the capillary. Parameters such as initial solution conductivity or input power on this effect as well as diagnostics by optical emission spectroscopy and electrics characteristics have been investigated.

Journal of Physics: Conference Series **516** (2014) 012007 doi:10.1088/1742-6596/516/1/012007

2. Experimental

The experimental apparatus consisted of an AC high-voltage power supply and a batch reactor (Figure 1, left). The high-voltage power supply gave the AC output voltage (up to 3.5 kV) directed by diode for using only positive or negative half-cycle of voltage. Two circle electrodes made of stainless steel were separated by the dielectric barrier thickness of 5 mm and made from shapal ceramic. Both electrode spaces were connected by only a small hole (diameter of 1 mm) in this ceramic diaphragm (Figure 1. right). Discharge breakdown appeared just in this orifice and plasma channels propagated from this spot towards electrodes. Due to the application of voltage, two different kinds of plasma streamers appeared on the opposite sides of the dielectric barrier [4]. On the side with the positive electrode, shorter plasma channels fill a spherical volume around the orifice. In the other reactor part containing the cathode, longer plasma channels propagated from the pin-hole towards the negative electrode. [3][5]

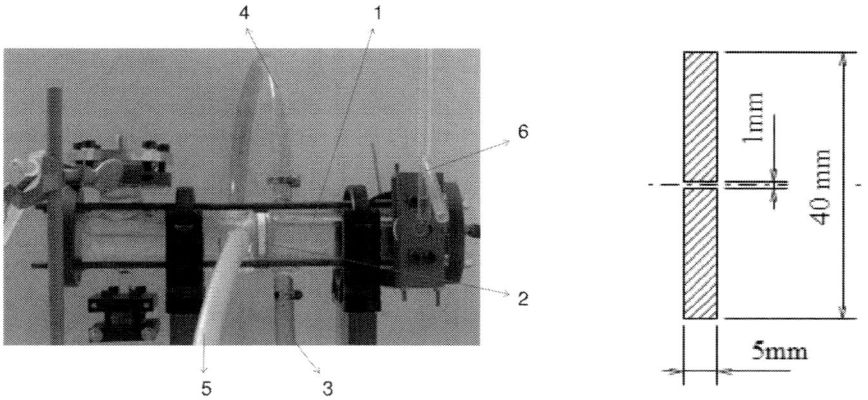

Figure 1. Photo of discharge reactor (1 – reactor vessel, 2 – ceramic dielectric barrier with symmetrical capillary hole, 3 – cooling inlet, 4 – cooling, 5 – cooling outlet, 6 – aqueous solution outlet) and scheme of ceramic symmetrical capillary.

Each part of the reactor contained 100 ml of treated solution. Water solutions contained distilled water and a definite amount of supported electrolyte NaCl providing particular solution conductivity (50 – 1000 $\mu S \cdot cm^{-1}$). There was provided the weighing of flowing drops of aqueous solution during the pump effect. Parameters such as initial solution conductivity or input power on this effect as well as diagnostics by optical emission spectroscopy and electrical characteristics have been investigated.

3. Results and discussion

Emission spectroscopy was used to identify the various species that are formed. Spectroscopic measurements of the plasma induced emission performed without time resolution indicate that the discharge might be an efficient source of OH, H and O radicals, which are among the strongest oxidants available. Figure 2 shows emission radiation produced by the discharge in spectral range of 200–900 nm for three different conductivities of water solution.

The average electrical characteristics of discharge, featuring values of power and voltage, are presented in figure 3 for different initial conductivities of aqueous solution. There was observed similar results for positive half-cycle and negative half-cycle of applied voltage.

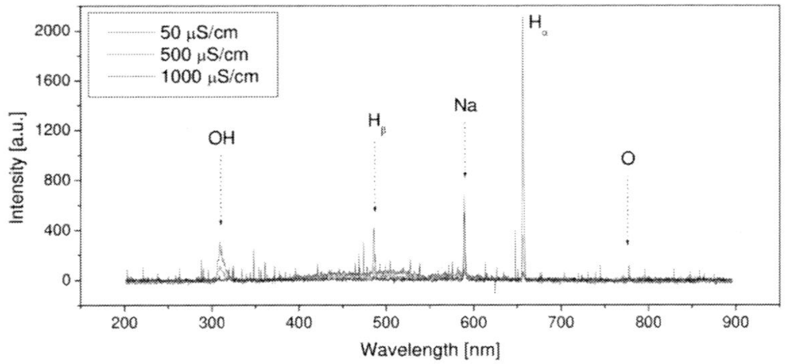

Figure 2. Optical emission spectra of the underwater discharge emission for different solution conductivities (50 µS/cm – blue line, 500 µS/cm – red line, 1000 µS/cm – black line) containing NaCl electrolyte and obtained for electric output 33 W using positive half-cycle of passing voltage.

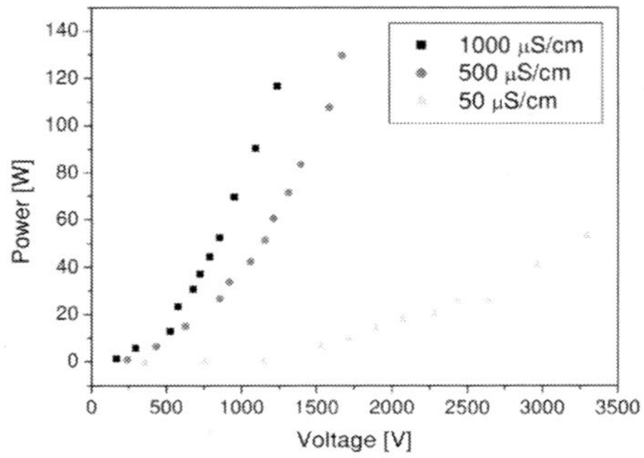

Figure 3. Electrical characteristics for three different water solution conductivities during the positive half-cycle of passing voltage.

Typical examples of the time evolution of the amount of discharge treated aqueous solution coming out at highest outlet during 20 minutes after switching on the discharge for various solution conductivities are displayed in figures 4, 5 and 6.

There was obtained only one pumping effect for output power 10 W for solution conductivity of 50 µS/cm and for glowing discharge (with plasma formation). There was no pumping effect for lower values of applied voltage without plasma (Figure 4).

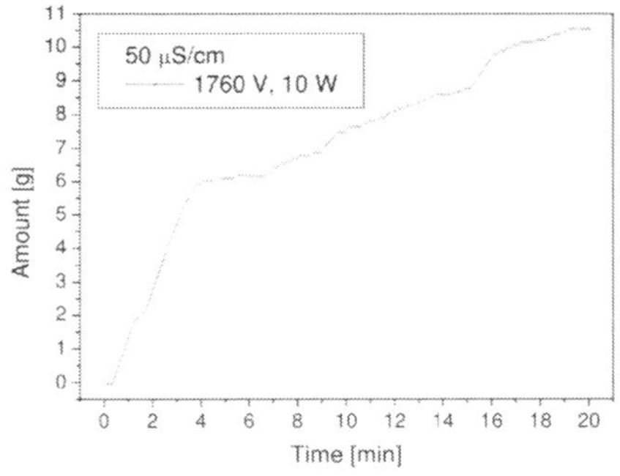

Figure 4. Time evolution of the amount of discharge-treated aqueous solution coming out at the outlet for positive half-cycle of voltage. Used solution conductivity of 50 µS/cm (electrolyte NaCl).

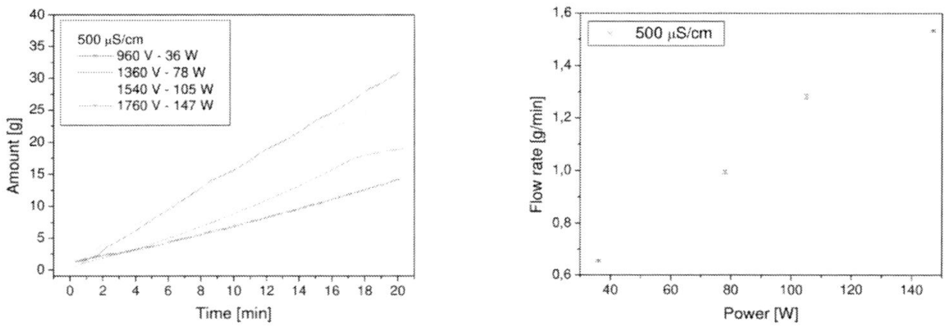

Figure 5. Time evolution of discharge-treated aqueous solution amount coming out at the outlet for positive half-cycle of voltage on the left and dependence of aqueous flow rate on applied power on the right side. Used solution conductivity of 500 μS/cm (electrolyte NaCl).

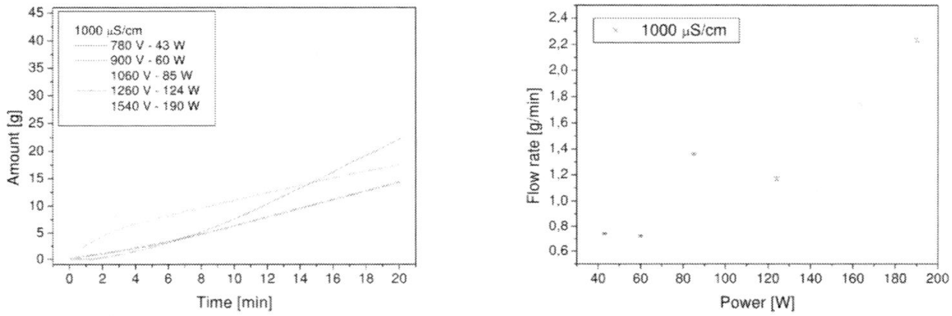

Figure 6. Time evolution of discharge-treated aqueous solution amount coming out at the outlet for positive half-cycle of voltage on the left and dependence of aqueous flow rate on applied power on the right side. Used solution conductivity of 1000 μS/cm (electrolyte NaCl).

The dependences of outlet amounts of aqueous solution on applied power for solution conductivity of 500 μS/cm (Figure 5, left) and 1000 μS/cm (Figure 6, left) were measured. Presented plots were realized for different values of applied voltage with formation of plasma and it indicates that higher used voltage gives larger outflow and it records a linear dependences. There is no major modification between results for positive and negative half-cycle of voltage.

Figures 5 and 6 (right) show the dependence on applied power of the aqueous flow rate resulting from the DC plasma generation in symmetrical capillary. Among these two different conductivities (500 and 1000 μS/cm), the highest and most stable plasma-induced aqueous flow rate is obtained for a solution conductivity of 1000 μS/cm and used power of 190 W in figure 6 right.

Acknowledgements

This work has been supported by the Czech Science Foundation, project No. 104/09/H080.

References

[1] F. De Baerdemaeker, M. Šimek, J. Schmidt and Ch. Leys, *Plasma Sources Sci. Technol.* **16**, (2007) 341-354

[2] Z. Stara, F. Krcma, *Czech. J. Phys.* **54**, C1–6

[3] F. De Baerdemaeker, M. Šimek, Ch. Leys, W. Vestraete, *Plasma Chem Plasma Process.* **27** (2007) 473-485

[4] I.P. Kuzhekin, *Proc. 9th Int. Symp. On High Voltage Engineering*. (1995) 8073-1-8073-3

[5] A. Nikiforov, Ch. Leys, *Plasma Sources Sci. Technol.* **16**, (2007) 273-280

Journal of Physics: Conference Series **516** (2014) 012008 doi:10.1088/1742-6596/516/1/012008

Destruction of Direct Blue 106 Dye in Underwater Discharge

L Němcová[1], F Krčma[1], A Nikiforov[2], C Leys[2]
[1] Faculty of Chemistry, Brno University of Technology, Purkynova 118, Brno 612 00, Czech Republic
[2] Department of Applied Physics, Ghent University, Jozef Plateaustraat 22, Ghent 9000, Belgium

E-mail: xcnemcova@fch.vutbr.cz

Abstract. An application of underwater discharge is one possible way how make the destruction of organic dyes. This contribution presents results of Direct Blue 106 destruction in discharge generated in bubbles. The initial conductivity value of 30 $\mu S.cm^{-1}$ was obtained using electrolyte; the starting dye concentration was 20 $mg.l^{-1}$. The DC voltage from 1.5 kV to 3.0 kV was applied to generate the discharge at the mean current of 10–30 mA. The system was bubbled through the high voltage capillary electrode by He, Ar and N_2 at the constant gas follow of 200 sccm. The dye destruction rate was directly proportional to the applied discharge current, so the highest efficiency was reached at the current of 30 mA. The destruction rate was strongly dependent on the filling gas. While using He and Ar only 4% destruction was obtained during the 20 minutes treatment at 10 mA of discharge current but the decomposition of 52 % was reached if nitrogen was introduced into the high voltage electrode. The destruction efficiency of about 40 % (He, Ar) and over 60 % in nitrogen was reached at discharge current of 30 mA. This enormous difference was probably connected not only to the production of hydrogen peroxide that seems to be usually the main oxidative specie in under water discharges but also atomic and excited nitrogen particles, both atomic and molecular, can have very positive effect in the dye destruction. The detailed study of the kinetic mechanisms leading to the Direct Blue 106 dye destruction will be a subject of the further studies.

1. Introduction

Direct Blue 106 dye belongs under the group of Organic Synthetic Dyes. By definition, dyes are colored, ionizing and aromatic organic compounds which show an affinity towards the substrate to which it is being applied. One characteristic of dye is that the dyes must get completely or at least partially soluble in which it is being put to. The rule that we apply to other chemicals is similarly applicable to dyes also. For example certain kind of dyes can be toxic, carcinogenic or mutagenic and can pose as a hazard to health. Direct Dyes are one of the most popular dye groups. They have a wide range of applications because they are often cheap and easy to apply. Their chemical composition consists of salts of complex sulfuric acids. They are usually soluble in water and they show a good affinity for vast fibre types and thus they are applied to numerous substrates for example to textiles, leather, plastic, paper, etc. in liquid form. Due to their wide range of applications, the waste water containing these dyes can be a great environmental problem. It is easier to decompose these dyes directly in the liquid phase. The dye concentration in waste water is usually not very high and thus the dye decomposition should be completed in liquid phase directly. Unfortunately, many of organic dyes are not possible to decompose using microorganisms due to the presence of multiple bounds mainly in

Journal of Physics: Conference Series **516** (2014) 012008 doi:10.1088/1742-6596/516/1/012008

the benzene ring (see Fig. 1). Thus, and due to the increasing demand for purification of dye containing industrial wastewaters by advanced oxidation processes (AOP), a large number of papers are published on the subjects. In AOP, generally reactive, strongly oxidizing ·OH radicals play the main role in destruction of the dye molecules. The decoloration of the dye is generally followed by spectrophotometry at the absorbance maximum between 350 and 700 nm [1-2].

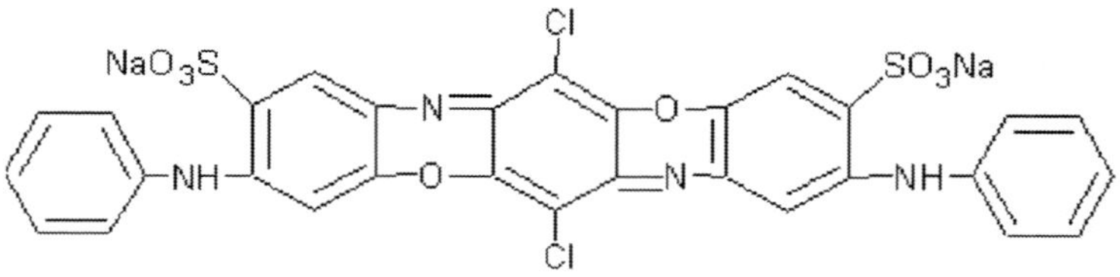

Figure 1. Direct blue 106 chemical structure

Underwater discharge is a promising environmental process to water treatment. Nowadays there are more different configurations suitable for the undewater discharge ignition. Besides well described pulsed corona like systems in point to plane [3] and coaxial configurations [4] and so called electrode less configurations (diaphragm [5] and capillary [6]), the discharge created in the bubbles introduced into the systems can be applied. All these discharges generate the non thermal plasma very similar to gas corona discharge but it presents several physical differences. There is a lower ionic mobility in the bulk liquid; on the other hand, the electron density and electron collision frequency are much higher than in the gas phase. This underwater bubble discharge configuration is relatively new one and fully combines both gas and liquid phase discharges. The gas bubbles are introduced into the system by thin stainless steel capillary that play simultaneously a role of HV pin electrode. Thus the HV pin electrode is covered by a thin gas layer and the discharge is generated in the gas phase in pin to plane configuration. The discharge streamers (plasma channels) generated in the gas phase are long enough (up to 1 cm) and thus they introduce into the liquid phase and further propagate in it [7, 8].

2. Experimental technique

The principal scheme of the set-up for the generation of the underwater discharge in gas bubbles is presented in fig. 2.

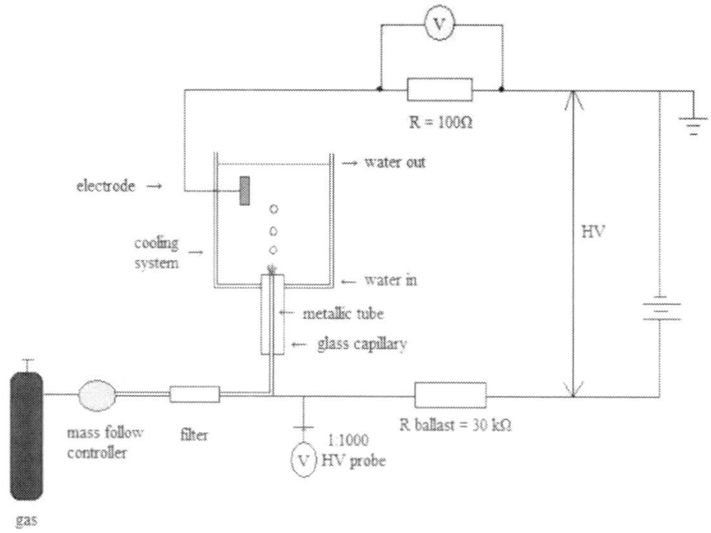

Figure 2. Scheme of the experimental set-up

The discharge reactor consisted of perspex discharge chamber (volume of 1 L), on the bottom of which a glass capillary (the inner diameter of 1.2 mm, length of 50 mm) is placed. Liquid was connected through a shunting resistance of 100 Ω to the ground potential by stainless steel electrode placed at the upper part of the chamber. The reactor was water cooled in order to avoid thermal destruction of hydrogen peroxide at temperatures above 30 °C [9]. HV electrode was a stainless steel tube (the diameter of 0.5 mm) which was placed inside glass capillary. The gas (Ar, He or N_2) has been applied through this tube in order to produce bubbles. Gas flow in the system was supplied by mass flow control system (MKS 4000). Flow rate of used gases was fixed at 200 sccm in all experiments. The electric discharge was located directly on the surface of the metallic tube inside of bubbles. The DC power supply for the discharge sustaining has been connected to the reactor through a ballast resistor 30 kΩ. The applied voltage was varied from 1.5 to 3.0 kV at mean current from 10 mA to 30 mA.

3. Results and discussion

The kinetic plots for discharge currents of 15 and 30 mA are presented in Figure 3. Discharge generated at 10 mA was not stable and thus the destruction of direct blue dye was not sufficiently effective as it is demonstrated in Table I. One can see that in the case of He and Ar plasmas the destruction efficiency of the direct blue dye molecules was much lower than in the case of N_2. Maximal destruction of about 64 % was achieved in N_2 plasma at the highest applied current. We suppose that this effect is connected to the formation of HNO_3 in case of N_2 due to oxidation of N_2 that leads to the decrease of solution pH down to 2. Used dye has lower stability in acidic solutions with pH<6 and the oxidative ability of plasma is stronger in acidic conditions (e.g. the oxidative potential of the hydroxyl radical is 2.70 V at pH 3.0 and 2.34 V at pH 9.0). A new set of experiments will be arranged to confirm this hypothesis.

As it was pointed in Introduction, the underwater discharge is a source of different reactive particles (aqueous electrons, OH, O, H radicals, O_3, H_2O^*, HO_2, and many other reactive species) as well as UV radiation. Besides them, also the temperature in the bubble surrounding should be taken into account. Gas temperature inside the bubble estimated by fitting the emission spectra of OH bands can reach 1500-2000 K [10] and thus it can influence the destruction of organic compounds due to thermal dissociation. In order to estimate an importance of different mechanism in chemical destruction of the dye additional experiments with scavenger of OH radicals [11] have been carried out. The full understanding of the dye destruction mechanism and its kinetic is thus very complicated task that is impossible to solve without further studies including the exact analyzes of the discharge products by various analytical techniques.

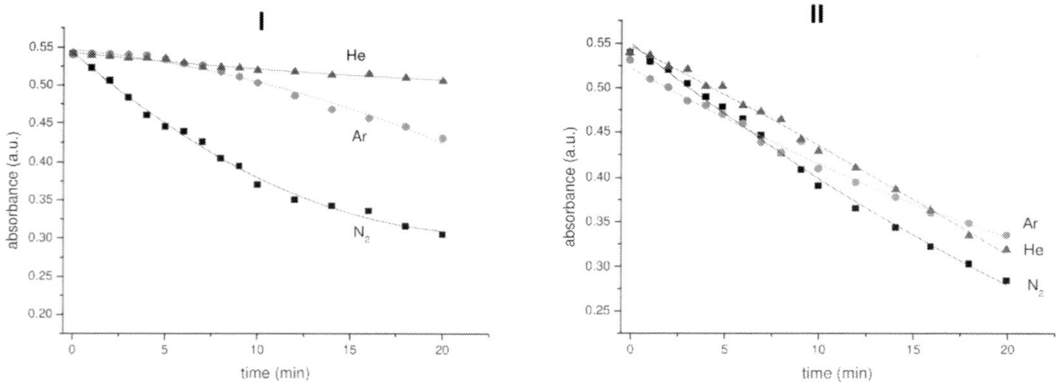

Figure 3. Kinetic plots of Direct Blue 106 dye destruction at the current of 15 mA (I) and 30 mA (II)

Table I. Efficiency of the Direct Blue 106 destruction after 20 minutes plasma treatment.

Current [mA]	**Decomposition of the dye [%]**		
	He	Ar	N_2
10	4	4	52
15	6	15	53
20	21	24	55
25	32	30	59
30	36	43	64

4. Conclusion

This work has been focused on chemical efficiency of the electric discharge in water solution generated in gas bubbles. The influence of the used gas (He, Ar, N_2) was studied at the constant gas flow. The destruction of Direct Blue 106 dye was observed as a function of applied discharge current. A great difference between used gases was found. By using lower current of 10 mA and 15 mA (He, Ar) the plasma was not fully stable and efficiency of decomposition was very low of a few percents. It increased up to 40 % at the current of 30 mA in both cases. On the other hand a 52 % destruction of dye during 20 minutes was observed in N_2 bubbles even at very low current of 10 mA and it increased up to about 65 % at 30 mA. The possible explanation of this great difference is in the creation of HNO_3 acid molecules when nitrogen is introduced into the system. These molecules decrease the pH of solution that accelerates the dye decoloration.

Acknowledgement

This work was partially supported by the Interuniversity Attraction Poles Program of the Belgian Science Policy, project No. PSI-P6/08 and by project of Specific University Research at Brno University of Technology, project No. FCH-S-10-8.

5. References

[1] Vojnárovits L, Pálfi T and Takács E 2007 *Radiation Phys. Chem.* **76** 1497-1501
[2] Ward D B, Tizaoui C and Slater M J 2006 *Chem. Eng. Process.* **45** 124-139
[3] Joshi A A, Locke B R, Arce P and Finney W C (1995) *J. Hazard. Mater.* **41** 3-30
[4] Šunka P, Babický V, Člupek M,Lukeš P, Šimek M, Schmidt J and Černák M (1999) *Plasma Sources. Sci. Technol.* **8** 258-265
[5] Stará Z, Krčma F, Nejezchleb M and Skalný J D (2009) *Desalination* **239**, 283-294
[6] DeBaedemaeker F, Šimek M, Schmidt J and Leys C (2007) *Plasma Sources. Sci. Technol.* **16** 341-354
[7] Nikiforov A Y 2009 *IEEE Trans. Plasma Sci.* **36** 872-876
[8] Malik M A, Ghaffar A and Malik S A 2001 *Plasma Sources. Sci. Technol.* **10** 82-91
[9] Schumb W C, Satterfield Ch N and Wentworth R L 1955 *Hydrogen peroxide*, Am. Chem. Soc. Monograph Ser., New York
[10] Bruggeman P, Schram D, Gonzales M A, Rego R, Kong M G and Leys C (2009) *Plasma Sources. Sci. Technol.* **18** 025017
[11] Nakui H, Okitsu K, Maeda Y and Nishimura R (2007) *Ultrasonics Sonochem.* **14** 627-632

Journal of Physics: Conference Series **516** (2014) 012009 doi:10.1088/1742-6596/516/1/012009

Modified by air plasma polymer tack membranes as drainage material for antiglaucomatous operations

T V Ryazantseva[1], L I Kravets[2], V M Elinson[3]

[1] Saratov State Medical University, Bolshaya Kazachia Str. 112, 410012 Saratov, Russia
[2] Joint Institute for Nuclear Research, Flerov Laboratory of Nuclear Reactions, Joliot-Curie Str. 6, 141980 Dubna, Russia
[3] Tsiolkovsky Moscow State Aviation Technological University, Orshanskaya Str. 3, 121552 Moscow, Russia

E-mail: tvroko@gmail.com

Abstract. The morphological and clinical studies of poly(ethylene terephthalate) track membranes modified by air plasma as drainage materials for antiglaucomatous operations were performed. It was demonstrated their compatibility with eye tissues. Moreover, it was shown that a new drainage has a good lasting hypotensive effect and can be used as operation for refractory glaucoma surgery.

1. Introduction

Interest in the treatment of refractory glaucoma by implantation of drainages on the basis of various materials is getting enormous due to a high degree incidence of this pathology and its serious complications. The refractory glaucoma accounts about 40 % of all glaucoma pathology. Contraindication for antiglaucomatous operation with the use of an explantodrainage was the absence of intraocular pressure normalization and tonografic indices in refractory glaucoma patients. However, unsuccessful antiglaucomatous operations result, as a rule, from proliferation of connective tissue and obstruction of surgically formed outflow tracts [1–3]. The history of usage of auto- and allodrainages has revealed a number of shortcomings such as a shallow anterior chamber, long-standing postoperative hypotonia, macular edema in an early follow-up, growth of a course connective-tissue capsular, rejection of a transplant, and necrosis of a scleral flap and the conjunctiva, to name but a few [4–6]. Therefore, search for new drainage materials for antiglaucomatous operations is getting quite an urgent problem. Most ophthalmologists of late have given preference to polymer drainages which are easily sterilized, can be kept for a long time and have no immunoallergic effects. Ophthalmological clinic of Saratov State medical University was the first to implant a new drainage [7] in one's capacity a poly(ethylene terephthalate) track membrane was (PET TM) chosen due to a big number of their advantages: low resistance to the flow of the filtered medium, high filtration selectivity, low adsorption of resolved substances, transparency, lack of toxicity, small weight, high strength and elasticity. As nontoxic, this drainage has a relevant certificate and a permit for usage granted by the Department supervising the introduction of new medications and medical equipment.

In this paper we report on experience of using the polymer track membranes with a modified by air plasma surface as a drainage material for antiglaucomatous operations. For experiments poly(ethylene terephthalate) track membranes were used. By preliminary tests it has been found that the membranes of this type are stable to biodestruction and can remain in the intrascleral space for several years. For modification the membrane surface, a treatment by air plasma was applied.

2. Experimental details

In the experiments, PET TM samples with a thickness of 10.0 μm and an effective pore diameter of 0.4 μm (pore density 5×10^7 cm^{-2}) were used. To produce the membrane, poly(ethylene terephthalate) film (Lavsan, Russia) with a thickness of 10.0 μm was irradiated by krypton positive ions, accelerated to ~ 3 MeV/nucleon at the cyclotron. Then the ion-irradiated film was sensitized with ultraviolet irradiation with a maximum emission intensity at 310 nm. Chemical etching was performed in an alkaline (NaOH) aqueous solution with concentration of 3.0 mol/L at 80 °C for during 4 min. The plasma treatment was performed at a plasma chemical set-up providing a 13.56 MHz RF-discharge. Nitrogen, air and oxygen as well nitrogen and oxygen with various rations were used as plasma-forming gases. The gas pressure, discharge power and treatment time were varied. The treatment procedure and the scheme of the plasma set-up described in [8]. The both sides of the membranes were subjected to the plasma treatment. The characteristics of the initial and modified membranes were determined with the help of mutually complementary methods described in detail in [8].

Experimental research of the drainage material (for this purpose the membrane samples previously were sterilized by boiling in distilled water or in an autoclave at temperature up to 150 °C was conducted by its implanting in the anterior chamber and sclera layers of the rabbits (33 eyes) for one, two and six months. The clinical analyses 24 patients with refractory glaucoma (24 eyes) after surgical treatment have been performed during six months after the operation. Observation involved patients with refractory glaucoma with different etiology who had gone through two to five antiglaucomatous operations. In this group, deep sclerectomy was followed by a separation of a conjunctive and scleral flap and then by a separation of a conjunctive and scleral flap and then by an incision of deep-lying scleral layers. Next was basal iridectomy. The outer end of the drainage was fixed to the inner scleral layers, and the inner one was implanted the anterior chamber. The operation was completed by a suture placed on the scleral flap and by a continuous suture on the conjunctival flap.

3. Results and discussion

The investigations of the air plasma effect on the PET TM allowed to establish the following. The first, effect of the air plasma on the PET TM leads to etching both the external membrane surface and the surface pore layer that accompanied by decreasing a membrane thickness and increasing an effective pore diameter table 1). The results of pore's etching of track membranes are illustrated by figure 1. A surface of an initial membrane and modified in plasma membrane are submitted on the microphotographs. It is possible to see, the pore diameter on the surface of modified membranes is increased. As received results shown, the etch rate depends on a discharge parameters, i.e. gas pressure and discharge power – increase in the gas pressure and discharge power lead to rise of the etch rate. The etch rate also depends on the composition of the plasma-forming gas. So, introduction of oxygen into the plasma-forming gas and increasing its concentration lead to increase of the etch rate. The use of pure oxygen as a plasma-forming gas allows one to increase essentially the etch rate. This permits to make the process of gas discharge etching more intensive. The increase of the effective pore diameter causes the increase of porosity, as a results, the modified membranes have better air and water flow rate characteristics. The filtrate volume is higher for the modified membranes (table 1). The second, depending on the choice of the discharge parameters, etching can be executed either in a part of the channel or along the whole length of the pore channels; anyway, the asymmetric track membranes possessing high porosity and flow rate are formed. It is possible use the gas discharge etch method to change the TM structure. The optimum choice of the treatment conditions in plasma allow to produce the membranes with a high flow rate without changing the selective properties of the membranes.

Another observation is that the plasma effect leads to the formation of a thin layer with increased content of carboxyl groups. This makes the membranes surface hydrophilic. So, the analysis of the FTIR-spectra of the membrane samples processed by air plasma shows an increase in the intensity of the absorption band at 1720 cm^{-1} corresponds to stretching vibrations of the carbonyl in carboxyl groups. The formation of carboxyl groups at the treatment in air plasma at the membrane surface layer can be explained the following way. At the effect of active particles of plasma, a destruction of polymer chains on the surface occurs. A break of the bonds C–O (binding energy is equal to 376 kJ/mol) and C–C (binding energy is equal to 335 kJ/mol) energetically looks more probable. As a result of secondary reactions, end carboxyl groups are formed. The radicals generated at breaking the bonds C–C, apparently, make a recombination, probably with formation of intermolecular bonds. Besides, the plasma prompts morphological changes of the surface relief. Numerous craters appear on the initially smooth surface. It becomes rough. This phenomenon is probably caused by different etch rates for crystalline and amorphous regions. If varying the treatment conditions and duration, the sizes of the craters change. The sizes of the craters change, when varying the conditions of the treatment and duration of the discharge effect.

Table 1. Change of the characteristics for membrane with pore diameter 0.4 μm in the process of treatment by air plasma during 10 min.

Gas pressure (Pa)	Discharge power (W)	Thickness (μm)	Effective pore diameter (μm)	Initial water flow rate at ΔP=4×10^4 Pa (ml/min cm^2)	Water contact angle (deg)	Increasing of COOH-groups content (%)
–	–	10.0	0.395	5.4	65	–
0.15	70	9.9	0.405	6.7	20	4
0.25	70	9.8	0.425	7.6	25	7
4.0	70	9.4	0.445	8.4	30	19
10.5	70	9.0	0.460	9.1	35	28

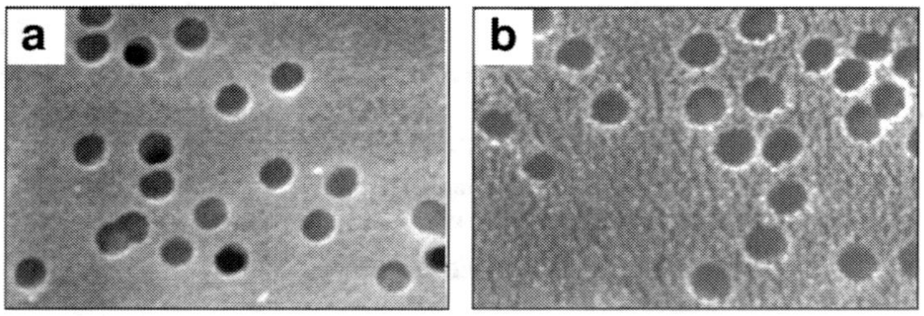

Figure 1. Microphotographs of the surface of the initial PET TM (a) and the membrane after air plasma treatment during 10 min at the gas pressure of 0.26 Pa and discharge power of 100 W (b).

Development of erosion of the membrane surface and their hydrophilization stipulate increase of wettability – the value of the water contact angle for modified membranes essentially decreases (table 1). As our experiments shown, increase of the carboxyl group's content in the membrane surface layer leads to increase of a negative charge on the pore surface. Really, the measurement of the charge of the pore surface shows that at the air plasma treatment of membrane there is an increase of the density of the negative charge on the walls of its pores. So, if the value of charge of the pore surface for the initial membrane is equal to 1.17×10^{-2} Coul/m^2, then for modified membrane (at the gas pressure of 13.5 Pa and discharge power of 300 W) this value is equal to 1.62×10^{-2} Coul/m^2. It is stimulate (and also development of a roughness of a surface) reduction of proteins adsorption of and others multipliers of intraocular liquids, blood cells, and also allows to avoid formation around of a drainage rough a connective tissue capsule.

The technique of an experimental research was used for 35 patients. In 20 (57.1 %) of these the disease lasted from one to five years, in 15 (42.9 %) over five years. Before this time 17 (48.6 %) patients had underwent antiglaucomatous operations without the use of drainage, 12 (34.3 %) had other surgical interventions, and 6 (17.1 %) persons were operated for the first time. Eight (22.9 %) of 35 patients had traumatic, 4 (11.4 %) – photogenic, 14 (40.0 %) – uveal, and 9 (25.7 %) – vascular glaucoma. In 20 (57.1 %) the anterior chamber angle was opened, and in 15 (42.9 %) it was closed. Ten (28.6 %) subjects had initial, 15 (42.9 %) – advanced, 7 (20.0 %) – far advanced and 3 (8.6 %) – terminal glaucoma. Mean intraocular pressure (IOP) was 38.3 ± 5.2 mm Hg. Tomographic indicators: Po = 46.1 ± 4.2 mm Hg; C = 0.09 ± 0.01 mm3/min/mm Hg; F = 2.51 ± 0.1 mm3/min; Bl = 512.2 ± 71.4. The operation was preceded by a thorough gonioscopy and choice of a surgical access.

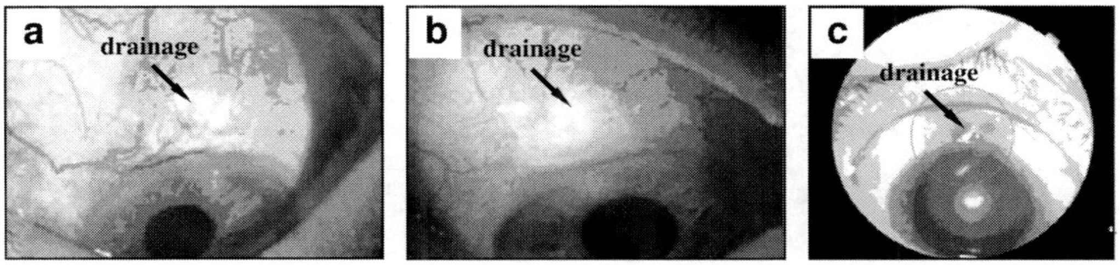

Figure 2. Eyes of the patients after surgical treatment (a, b) and optic coherent tomography of eye after operation (c).

Experimental research has shown that no signs of either inflammation, cell infiltration, neovascularization of cornea and iris or anterior and posterior synechia have been found in eyes during six months after operation (figure 2). Histological research data demonstrate that the reaction of the inflammatory cells to drainage is minimal. The implant does not resolve and keeps its porous structure during up to six months. Fibrosis capsule around the drainage is not exposed. There was not a single case of reaction to drainage. After operations with modified track membranes in all cases filtering blebs were split and plane. In all cases stable hypotensive result was achieved. The follow-up involved 35 patients. A day after the operation, the IOP decreased to 15.5 + 3.5 mmHg. During the follow-up (from 1 month to 6 months postoperatively) the IOP was within the physiological norm in 35 (100 %) persons. Operations stabilized visual field boundaries in 35 (100 %) persons. Visual acuity improved on the average by 0.1-0.4.

4. Conclusion

Thus, the morphological and clinical studies of modified by air plasma track membranes as drainage clearly demonstrates its compatibility with eye tissues. Moreover, a new drainage has a good lasting hypotensive effect and can be used as operation for refractory glaucoma surgery.

References

[1] Netland P A and Walton D S 1993 *Ophthalmic. Surg.* **24** 723
[2] Lee W B, Brandt J D, Mannis M J, Huang C O and Robin G J. 2003 *Cornea* **22** 178
[3] Sidoti P A, Dunphy T R, Baerveldt G, LaBree L, Minckler D S and Lee P P 1995 *Ophthalmology* **102** 1107
[4] Englert J A, Freedman S F and Cox T A 1999 *Am. J. Ophthalmol.* **127** 34
[5] Hill R, Ohanesian R, Voskanyan L and Malayan A 2003 *Br. J. Ophthalmol.* **87** 673
[6] Aung T 1998 *Ophthalmology* **105** 2117
[7] Saprykin P I and Ryazantseva T V 1997 Patent RU2089149
[8] Dmitriev S N, Kravets L I, Sleptsov V V 1998 *Nucl. Instr. and Meth. In Phys. Res. B* **142** 43

Journal of Physics: Conference Series **516** (2014) 012010 doi:10.1088/1742-6596/516/1/012010

Investigation of the NBI heated plasma on the Globus-M tokamak with the use of Thomson scattering diagnostics.

G.S.Kurskiev[1], S.Yu. Tolstyakov[1], V.K.Gusev[1], M.M.Kochergin[1], N.A.Khromov[1], A.D. Melnik[1], V.B.Minaev[1], E.E.Mukhin[1], M.I.Patrov[1], Yu.V.Petrov[1], N.V. Sakharov[1], V.V.Semenov[1], I.Yu. Senichenkov[2]

[1]Ioffe Physico-Technical Institute, RAS, St. Petersburg, Russia
[2]St. Petersburg State Polytechnical University, St. Petersburg, Russia

Gleb.Kurskiev@gmail.com

Abstract. Electron temperature and density spatial distribution dynamics on the Globus-M (R =0.36 m, a=0.24 m, B =0.4 T, I= 0.2 MA) spherical tokamak was investigated during the NBI heating. The key tool of this research was the upgraded Thomson scattering diagnostics with variable intervals between probing pulses (20 pulses in total). The scattering points were placed along the major plasma radius from the inner to the outer plasma border (10 points in total). Simulation of the electron transport in the L and H-mode, based on experimentally measured $n_e(R)$ $T_e(R)$ profiles was carried out using the code ASTRA for the OH and NBI dischargers.

1. Thomson scattering diagnostics at the Globus-M tokamak

The Thomson scattering diagnostics with a variable delay between probing pulses [1] became the key instrument to perform the dynamic measurements of the electron temperature and density profiles. The Nd-glass laser (the basic wavelength 1.054 μm) can generate up to twenty 2-4 J pulses with intervals of 0.3 -1000 ms. During 2008-2009 the Thomson scattering (TS) diagnostics upgrade was made. The layout of the experiment is shown in figure.1. A new observation point and high quality collection lens allowed to pick up the scattered radiation from inner to outer plasma border with minimal light losses. The TS system was equipped with the five novel 4-channel filter polychromators. It allowed us to decrease the measured electron density down to $10^{18}m^{-3}$ owing to the optical scheme optimization and the reduced noise amplifiers for avalanche photodiodes (APD). A new magnetic drive can move the beam-marker (7 in figure 1) along the laser axis to align collection system and promptly check the position of the ten TS measurement points without vacuum break. The filter polychromators and the

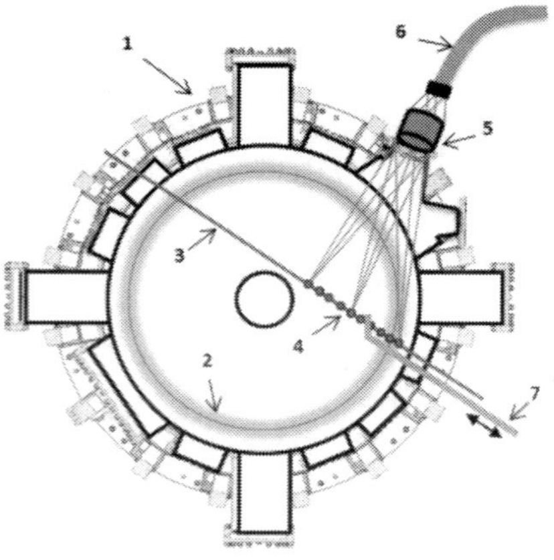

Figure.1. TS light collection system. 1- Globus-M vacuum vessel, 2 – outer plasma border, 3- laser beam, 4 – TS measurement points, 5- collection lens, 6- fiber line, 7- magneict drive

registration system were placed to the special thermo stabilized room. The optical connection was performed though the fiber lines with low losses.

2. Experiment and modelling.

1.5D code ASTRA [2] was used for experimental data processing. The following experimental data was used in modeling: plasma current, the shape of the plasma boundary (taken from the equilibrium reconstruction EFIT [3]), spatial distribution of the electron temperature and density, measured with

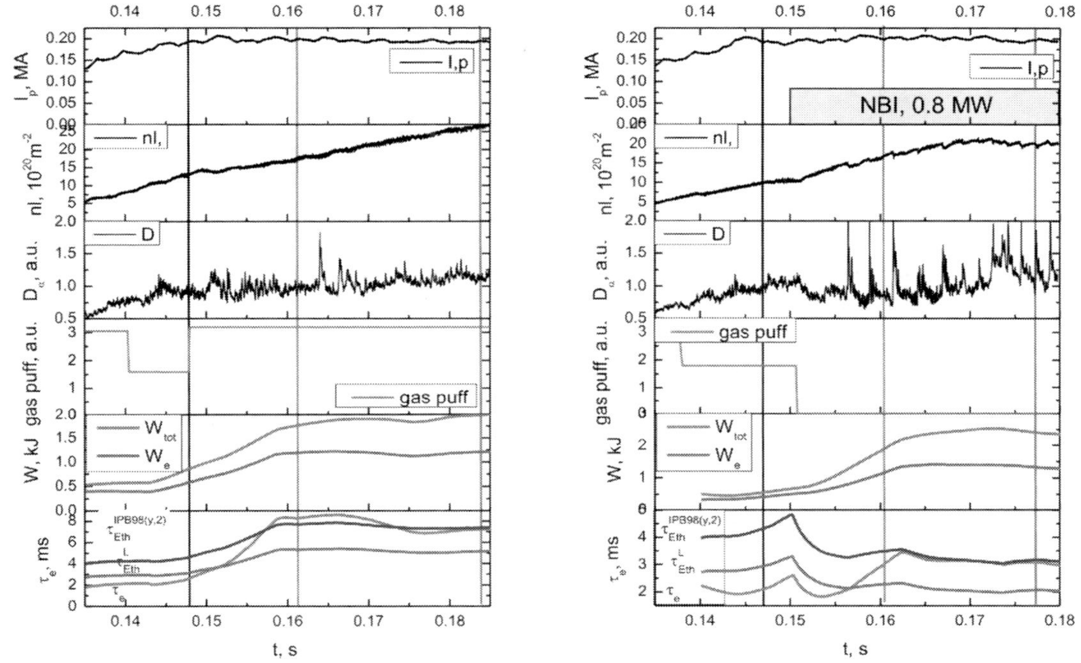

a) #26505 b) #26238

Figure.2. Main plasma characteristics. a) OH discharge. b) NBI discharge. Plasma current, line average density measured along the vertical chord at the position R=0.42 m , D_α emission, gas puff rate, plasma total stored energy and electron energy content, confinement time and scalings for H and L modes.

temporal resolution of 5 ms between pulses. The transport model consists of a system of the diffusion equations for electron and ion temperatures, the equation for the poloidal flux and the Grad-Shafranov equilibrium equation. The plasma conductivity and ion heat diffusion coefficients were derived by the neoclassical code NCLASS [4]. The equation for plasma density is not considered in this presentation. The experimental density spatial distribution was taken for modeling. The effective plasma charge was taken in such way that calculated loop voltage should be equal to the loop voltage measured in the experiment. Two discharges are considered in this work, ohmically heated and with NBI injection of hydrogen atoms with energy 27 keV [5]. The temporal distributions of the main discharge parameters are shown in figure 2. The L-H transition is observed in both discharges at the time 150 ms. In the case of NBI heated discharge the L-H is clearly observed: one can see the density rise and drop of the D_α emission. In the case of the OH discharge the moment of the transition is not clearly seen due to the strong gas puff. But the moment of the transition can be observed in the electron temperature and density spatial distribution changes. It also should be noticed that the effective plasma charge changes from 1.3 to 2.3 during the NBI, that may be the result of plasma-wall interaction growth.

3. Discussion and conclusions.

Electron temperature and density profiles and the calculated electron heat diffusivity are shown in figure 3 for 3 time slices: for L-mode, after L-H transition and in the end o the discharge. The shown profiles corresponds to the vertical markers in figure 2. The density rise and the formation of the "flat top" density profile are observed during the L-H transition in both cases, that is usual for H-mode. The density rise is the result of the drop of the diffusivity at the plasma edge and formation of the transport barrier for particle diffusion [6]. Electron heat diffusivity decreases by 3-4 times in OH discharge and by 2-3 times in the discharge with NBI. During the density rise, electron temperature rises too, and reaches maximum value after 10 ms after the L-H transition moment. At this moment one can see the high gradient region in the temperature profiles in the point r/a~0.65 , that is typical for both discharges. The electron heat diffusivity drops at this point by 6 times in OH discharge, and by 10 times in NBI discharge. This allows to suggest an internal transport barrier (ITB) formation there. As the density rises further, the temperature decrease is observed, so the energy content doesn't change. The confinement time coincides with IPB98(y,2) scaling. Usually L-H transition occurs when the heat power in the ion channel exceeds the value of the threshold power P_{th} [7]. In the NBI discharge the

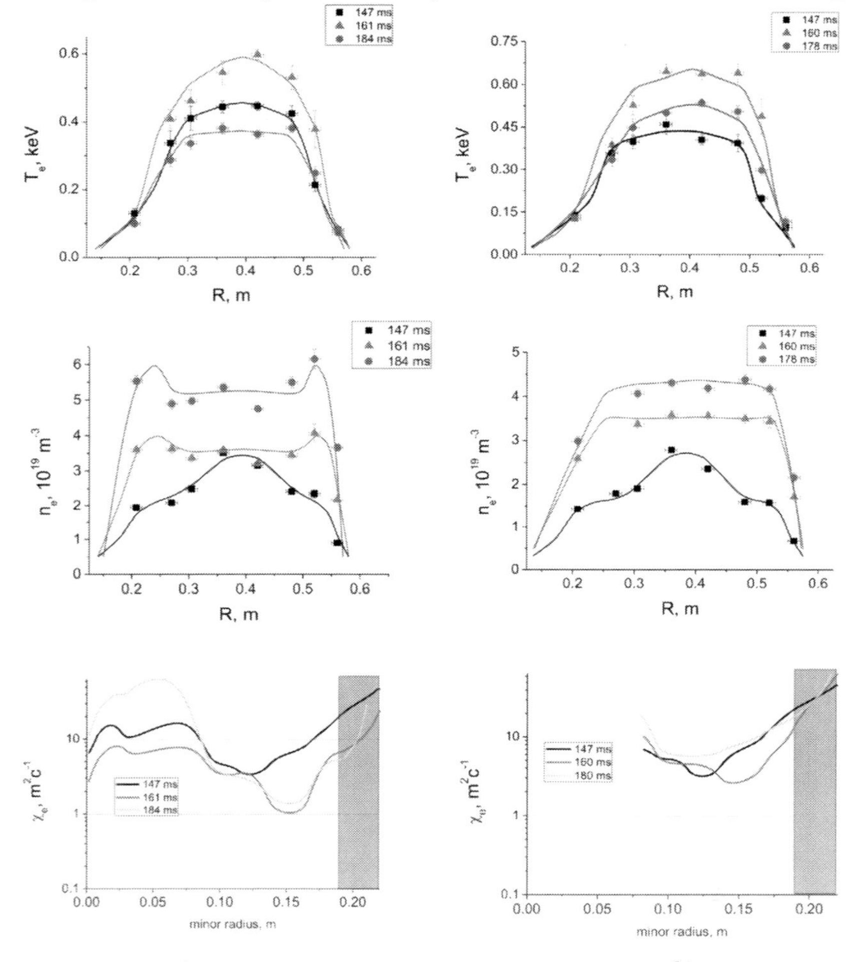

a) b)

Figure.3..Measured electron temperature and density spatial distribution and calculated electron heat diffusivity – a) OH, b) NBI.

value of the P_{th} is equal to 72 kW during the L-H transition. The part of the NBI heating power is 28 kW at the moment of transition. The moment of the L-H transition in the OH discharge is suggested between 147 ms and 152 ms, according to the TS data. During this period the heat power in the ion channel rises from 63 kW to 80 kW, and reaches the value of 70 kW at the 150 ms. The value of P_{th} derived in this discharges is higher than prediction of the scaling obtained on MAST and NSTX [6], P_{th} =20 kW.

The measurements performed with upgraded TS diagnostics at the Globus-M tokamak allowed to obtain electron heat diffusivity spatial distribution. The transport analyses was performed for the OH and NBI heated discharges, and in both cases occurrence of the ITB formation in the region r/a~0.65 was found during L-H transition. Also it was discovered that the electron heat diffusivity rises toward the edge of the plasma. In this case further investigation is required. Influence of the spatial distribution of Z_{eff} and radiation losses should be more carefully analyzed. Values of electron temperature, density and neutrals density at the plasma boundary are also required.

References

[1] Tolstyakov S Yu et al. 2006 *Tech. Phys.* **51** 846-852
[2] Pereverzev G V and Yushmanov P N 2002 *Max-Plank IPP report* 5-98
[3] Lao L L et al. 1985 *Nucl. Fusion* **25** 1611
[4] Houlberg W A et al. 1997 *Phys. Plasmas* **4** 3230
[5] Tolstyakov S Yu et al. 2010 *Proc. of 37th EPS Plasma Phys.* P-5.137
[6] Senichenkov I Yu et al 2008 *Proc. of 35th EPS Plasma Phys.* P-2.097
[7] ITER Physics Basis. 2007 *Nucl. Fusion* **47** S18

Advanced probe diagnostics for measurement of electromagnetic properties of turbulent structures at the plasma edge of the TJ-II stellarator

K Kovařík[1,2], I Ďuran[1], M Hron[1], C Hidalgo[3], D Carralero[3], M A Pedrosa[3] and TJ-II team[3]

[1]Institute of Plasma Physics AS CR, v. v. i., Prague, Czech Republic, Association EURATOM/IPP.CR
[2]Faculty of Mathematics and Physics, Charles University in Prague, Czech Republic
[3]Laboratorio National de Fusión, Associación EURATOM-Ciemat, Madrid, Spain

E-mail: kovarik@ipp.cas.cz

Abstract. Improved understanding of properties of the current filamentary structures in the edge plasma region is believed to allow better insight into development and possible control of the Edge localised modes (ELMs) and consequently, mitigation of their impact on the plasma performance and the first wall structures. We have focused on measurements of electric and magnetic properties of the filaments and electromagnetic features of the edge turbulence on TJ-II. We used a probe consisting of two 3D coil sets, two Hall sensors and 4 Langmuir tips. We present the first results of measurements with this probe on TJ-II stellarator.

1. Probe description

The new probe diagnostics combining a set of magnetic sensors and Langmuir tips in single probe head was recently commissioned on TJ-II. The sensing elements within the magnetic part of the probe are the three sets of magnetic sensors (3D inner coils set, 2D Hall probe, 3D outer coils set) and one platinum thermo-resistor PT-100 for temperature monitoring inside the probe head, see figure 1. The probes housing is made out of the non-magnetic stainless steel tube with outer diameter of 10 mm and the wall thickness about 0.5 mm. This housing provides electrostatic shielding of especially Hall sensors from the plasma and creates also a support structure for all the probe's components. The tube is slotted along its main axis so that it can be opened by removing the lid. The reasons for this configuration are to enhance penetration of the oscillating magnetic field inside the probe head through its stainless steel housing and to have a good access to all the components.

Each 3D coil set is labeled as inner or outer in the sense of its deepness of insertion in the vacuum chamber, i. e. the inner coils are in the front of the probe head with respect to plasma and outer are at the rear part of the probe head (side of wire output). The second part of labeling denotes orientation of the measuring element. Radial coils measure the magnetic field component along the main axis of the probe, parallel and perpendicular sensors have the sensing plane parallel and perpendicular to the plane of the slit respectively. Hall probe HP-P and HP-L are perpendicular and parallel respectively.

Signals of Hall probes are preprocessed by special electronics which provides adjustable amplification and stabilized source of supply current of 50 mA.

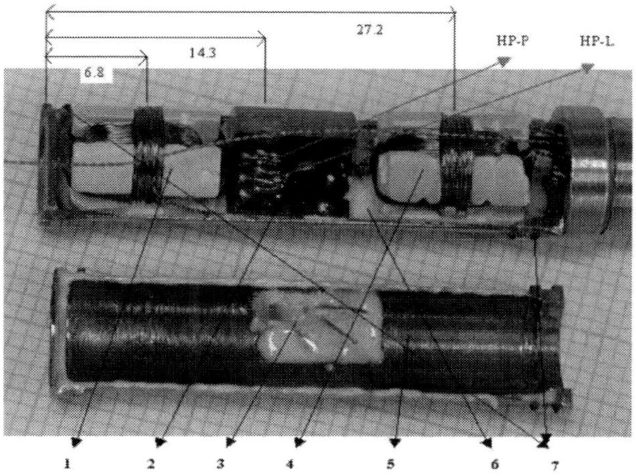

Figure 1. Detailed view of the magnetic part of the probe head (1: Inner 3D coil set, 2: 2D Hall probe, 3: PT-100 thermo-resistor, 4: Outer 3D coil set, 5. Removable non-magnetic stainless steel lid, 6: Special high vacuum compatible glue, 7: Rift used to accommodate several turns of wire for fixing the removable lid (5) to the rest of the probe's body).

Electrostatic part of the probe consisted of 4 Langmuir tips made from Tungsten. They are arranged at the corners of the square and pointing radially to the plasma core. Two of them are operated in ion saturation and two in floating potential regime.

2. Calibration of the probe

2.1. Coils

We performed calibration of all the coils within frequency range 1 – 300 kHz using the test bench which consists of RLC oscillation circuit, Rogowski coil for current measurement and long straight wire. Capacitor biased to 100V is discharged through resistor and inductance into the long straight wire creating a well defined calibration magnetic field. Output current with an amplitude of a few hundreds of Amps and waveform of damped sinusoid is generated. Frequency of output current is selected by proper values of capacity *C* and inductance *L*.

The effective area of all coils have significant flat top up to about 100 kHz. Basic parameters of the coils measured during the calibration are summarized in table 1.

Table 1: The basic properties of individual coils

Coil	R [Ω]	L [μH] at 1 kHz	No of turns	Total area [cm^2]
C_{par}^{inner}	10.2	211.8	154	95
C_{par}^{outer}	7.4	111	110	63
C_{perp}^{inner}	7.8	88.9	102	46
C_{perp}^{outer}	7.2	112	113	61
C_{rad}^{inner}	4.8	39.6	75	38
C_{rad}^{outer}	4.2	32.6	60	37

2.2. Hall Probes

We performed absolute calibration of the Hall sensors in DC magnetic field provided by permanent magnets and measured by absolutely calibrated Lake Shore Hall probe. The permanent magnets provide calibration magnetic field up to 350 mT. The measured sensitivity of the Hall sensor HP-P is 383 mV/T and HP-L is 343 mV/T.

3. Installation on TJ-II

Probe was installed at stellarator TJ-II at the upper manipulator "Sonda D" on 21st May 2010. Slit of the magnetic part of the probe was oriented paralel to the toroidal magnetic field. So all the parallel sensors measure the poloidal component of the magnetic field and perpendicular sensors measure the toridal component of the magnetic field. Signals of both 3D sets of coils were lead as three wires with shieldings connected together and used as common return signal. This type of connection was used for the whole signal track from the probe itself up to data acquisition system placed at the north wall of the stellarator hall. This wiring system provides significant reduction of the noise.

4. First results

4.1. Magnetic field profile measured by Hall sensors

We have measured toroidal and poloidal component of the magnetic field vector using the Hall sensors on shot-to-shot basis. Average values of the the magnetic field for each position are given in figure 2. The full red line is the theoretical model of the magnetic field on TJ-II.

The theoretical model is regrettably available only for inner part of the plasma. But, we have measured mostly in the plasma edge or out of the plasma. It is due to the fact that the Hall probes are about 17 mm from the leading edge of the probe-head and the probe cannot be moved too deep in plasma column without disturbing it. It is clear that the extrapolation of the theoretical model fits the measured values very well. Slight disproportion between theory and measurement for toroidal field is probably caused by improper orientation of the sensor. We assume the plane of the Hall sensor is exactly perpendicular to toroidal magnetic fied, but it have inclination of up to a few degrees. This error denotes inclination of about 5 degrees which is at the edge of calibration precision.

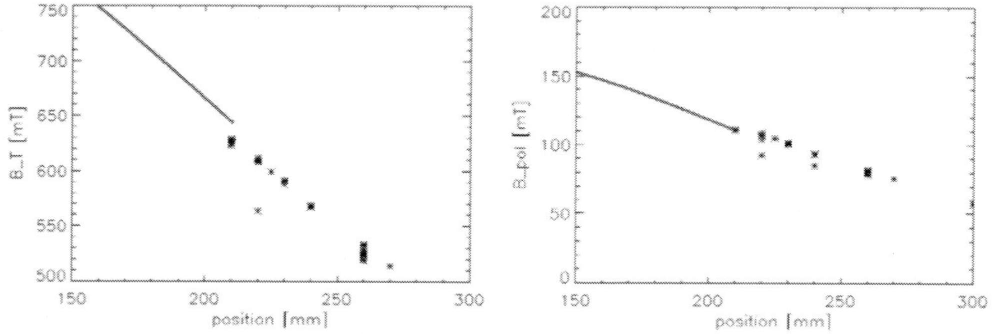

Figure 2. Comparison of measured profile of magnetic field (black stars) with theoretical model (red lines) in toroidal (left) and poloidal (right) direction.

4.2. Turbulence observation during discharge #25130

The phase sequence of the discharge #25130 is as follows. Magnetic field flat-top phase begins at 1000 ms and ends at 1300 ms. Plasma begins at 1020 ms with ECRH heating. At time 1090 ms, the ECRH is switched off and the phase of NBI starts. The NBI heating is finished at time 1160 ms and fast decay of the plasma follows.

Left pannels of Fig. 3 contain raw signals of inner radially oriented coil (upper) and Langmuir probes in ion saturation (middle) and floating potential (bottom) regime. Right pannels show the frequency spectra of the signals in phase of ECRH (green) and NBI (red). Noise spectrum is plotted in black. The Hall sensor frequency range is limited by 30 kHz by construction of the sensors themselves and therefore it is not used for studies of magnetic field fluctuations.

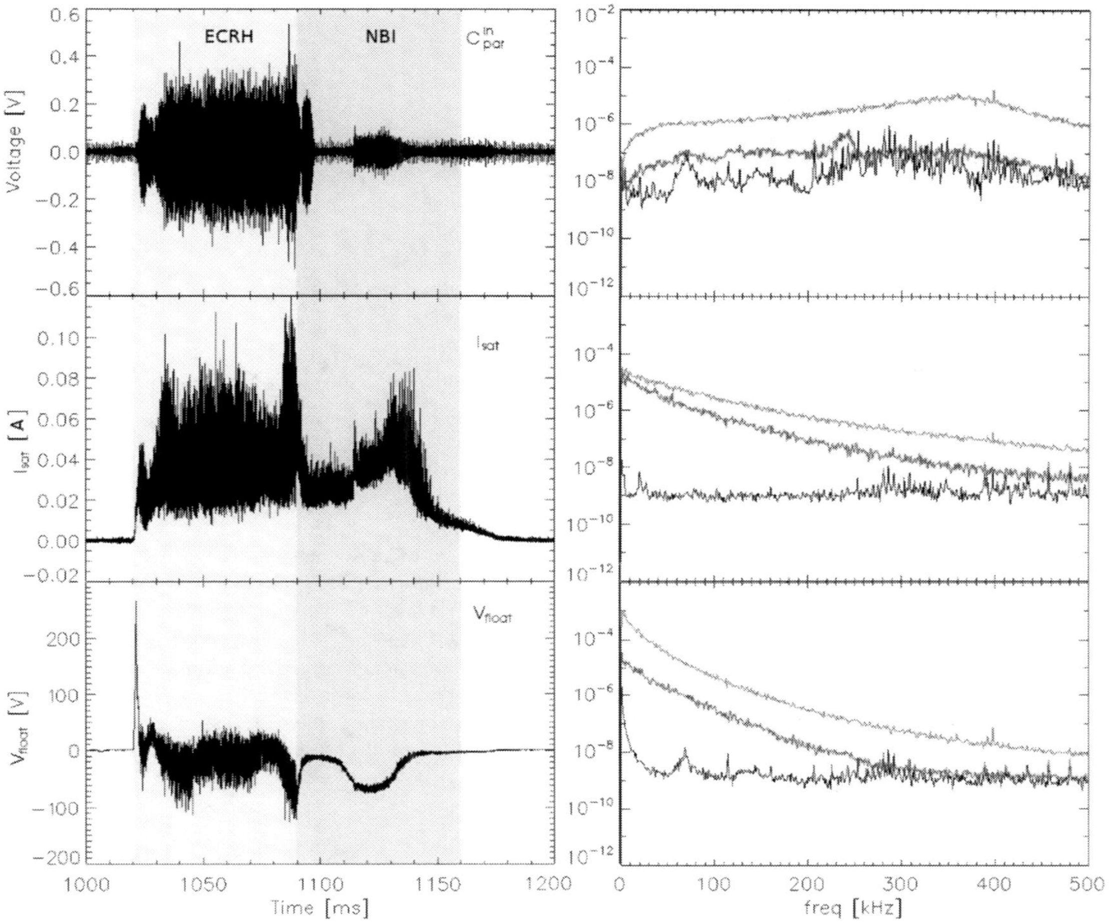

Figure 3. Raw signals (left) and frequency spectra (right) of the signals of inner radial coil (top), Ion saturation current (middle), and floating potential (bottom) during phase of ECRH (green) and NBI (red). Noise spectrum is plotted in black.

Level of both electrostatic and magnetic turbulence is significantly higher in ECRH phase compared to NBI phase of the discharge.

No particular features in frequency spectra indicating presence of significant instabilities or turbulent modes are observed neither in magnetic nor in electrostatic field.

5. Summary

Advanced probe diagnostics consisting of two 3D sets of coils, two Hall sensors and 4 Langmuir probes was engineered, manufactured, absolutely calibrated and successfully installed on TJ-II stellarator. Measurement of magnetic field profile agrees with the theoretical model. We have used this probe also to measure electromagnetic properties of turbulence in the edge plasma of TJ-II.

Acknowledgement

The work was supported by project of Ministry of Industry and Trade of Czech Republic 2A-TP1/101, the AS CR #AV0Z20430508, MSMT #7G09042 and EURATOM. The views and opinions expressed herein do not necessarily reflect those of the European Commission.

Adhesive properties of PTFE modified by DC discharge

A Gilman, M Piskarev, M Yablokov, A Kechek'yan and A Kuznetsov
Enikolopov Institute of Synthetic Polymer Materials, Russian Academy of Sciences, ul. Profsoyuznaya, 70, 117393, Moscow, Russia

E-mail: plasma@ispm.ru

Abstract. It is shown that low-pressure DC discharge treatment of PTFE films at both the anode and cathode leads to substantial enhancement of the contact properties of the polymer surface and alters considerably the adhesion properties of PTFE film surface. A procedure is developed for determination the adhesion characteristics of the modified surface of plasma treated thin polymer films using Scotch® 810 adhesive tape. The procedure comprises of a study of the surface coating created by means of physical vapor deposition of an aluminum layer with thickness of the order of 100 nm, an adhesive joint of the film with Scotch® 810 tape, and T-peel testing of the specimen. Using this procedure, the peel resistance of the initial and the modified PTFE film on both the anode and the cathode in DC discharge is measured.

1. Introduction

Polytetrafluoroethylene (PTFE) has unique chemical and physical properties: it neither swells nor dissolves in common solvents; it is resistant toward oxidants, acids, and alkalis and exhibits good dielectric properties over wide temperature and frequency ranges [1]. However, some applications concerning the use of PTFE as an engineering material demand that contact properties of the polymer surface should be improved, a task of vast practical importance. It is well known that the treatment by low-temperature plasma is a very effective method for enhancement of the polymer films contact properties such as wettability, the adhesive strength at the polymer/metal interface, the adhesive ability, the removability of applied printed matter etc. It should be noted that some changes take place on the polymer surface or within a thin surface layer with a thickness of 10 to 1000 nm as estimated by different authors. But the bulk of polymer remains intact, retaining the inherent mechanical and physicochemical properties [2, 3]. Therefore, it is of particular interest to determine experimentally the adhesion characteristics of the modified thin PTFE films.

There are various processes for surface modification of PTFE usig high energy chemistry techniques: electron-beam irradiation [4], bombardment with argon ions [5, 6], treatment by means of radiofrequency (13.56 MHz) discharges in O_2, Ar, N_2, He, H_2, etc.[6, 7], microwave (2.45 GHz) discharge [8], or the afterglow of RF discharge [9]. A combination of plasma and ozone treatment followed by grafting of chemically different substances onto the polymer surface can also be used [10]. These techniques make it possible to improve the contact properties of the PTFE surface, however they exhibit high values of the water contact angle (θ) (usually not less than $\theta = 55°$).

The most effective means of surface modification of polymers is found to be the treatment with a direct-current (DC) discharge. Such treatment allows separating the effect of the active plasma components on the polymer surface [3, 11]. Hence follows our interest to study the surface

Journal of Physics: Conference Series **516** (2014) 012012 doi:10.1088/1742-6596/516/1/012012

modification of PTFE in a low-pressure direct-current discharge and to examine the adhesion characteristics of these films.

There are several investigations in literature devoted to the adhesion properties of polymers modified by low-temperature plasmas [12–15].

Wade et al. [12] studied the adhesion properties of commercial polymer films of 8–12 μm thickness (polycarbonate, polysulfone, poly(ethylene terephthalate), polystyrene, polypropylene, poly(vinylidene fluoride)) modified in 13.56 MHz RF discharge in various gases such as oxygen, helium, and mixture of 96 % CF_4 and 4 % O_2. The adhesion strength at the metal/modified surface boundary was determined by ASTM D3359-02. Scotch® 810 pressure-sensitive adhesive tape is applied on and subsequently removed from the Al layer. As a measure of the adhesion strength, the ratio of the Scotch® 810 tape area covered with Al to the total area was used. The adhesion was evaluated qualitatively as very good, good, or poor depending on the amount of removed aluminum.

The adhesion characteristics of 50 μm FEP films modified in H_2, O_2, N_2 and Ar RF discharges were studied in [13]. In order to do this, a copper layer of ~30 μm thickness was applied onto the film surface in two steps. First a chemical vapor deposition was applied to create a layer of ~0.2 μm and then the thickness was increased to 30 μm by electrochemical plating. The adhesion of copper to the polymer surface was measured by T-peel test with Shimadzu Instron AGS 100-A tester. The peel strength was 27–199 mN/5 mm depending of the gas nature. The disadvantage of this method is the complicated two-step wet procedure of copper deposition. In it acids (sulfuric and hydrochloric), tin-palladium catalyst, copper sulfate, and a special polishing agent are used. After that the obtained sample is subjected to washing and drying at 80 °C for 12 h.

Sandwiched specimens were prepared from plates of PET, polypropylene, poly(vinylidene fluoride), etc. treated by means of air RF discharge. The thickness of the plates was 4 mm. A two-component polyurethane resin (Betamate 7000/7050, Dow Chemicals) with a layer thickness of 0.1 mm was applied. The tests were realized on a Zwick Z030 machine [14]. The shear stress was measured with an Instron 4301 machine [15].

It should be noted however that all of the aforementioned methods allow comparing only relative values of the adhesion strength. Thus, development and application of procedure for measuring the absolute values of adhesion strength is of high interest.

2. Experimental

Samples of commercial PTFE film of 40 μm thickness are used. The procedure for modification of the film surface in a DC discharge is given in details in [16]. The film samples is mounted on the anode or cathode and is treated by air discharge for 15–60 s at a gas pressure of 10–20 Pa and discharge currents of 20–100 mA.

To characterize the surface properties the value of the contact angle (θ) is used. It is determined with a goniometer (error ±1°). Two testing liquids were used – bidistilled water and glycerol. By the obtained result the work of adhesion (W_a), the total surface energy (γ), and its polar (γ^p) and dispersion (γ^d) terms are calculated [17].

The composition and the chemical structure of the surface of the initial and the plasma-treated PTFE films are studied by means of IR spectroscopy and ESCA. The IR reflection spectra are measured with Bruker Equinox 50S Fourier-transform IR spectrometer. It has MIRacle™ single reflection horizontal ATR attachment and uses ZnSe crystal working over the range 400–4000 cm^{-1}. The data of 500 scans with a scanning step of 2 cm^{-1} is collected. Band assignment is chosen according to [18]. ESCA spectra are measured on a Shimadzu ESCA 3400 spectrometer with a monochromatic MgK_α source (5 kV, 20 mA). The peak positions (binding energies) were calibrated with respect to the standard C_{1S} peak (284.6 eV) [19].

A special procedure is developed to measure the adhesion characteristics of the films. An aluminum layer of ~100 nm thickness is applied by physical vapour deposition. The layer is deposited through a mask onto the central part of the samples of the initial and the modified (at the anode or cathode) PTFE film. The layer is with 35 mm length and covering the entire width of the sample. A specimen of this type is sketched in figure 1.

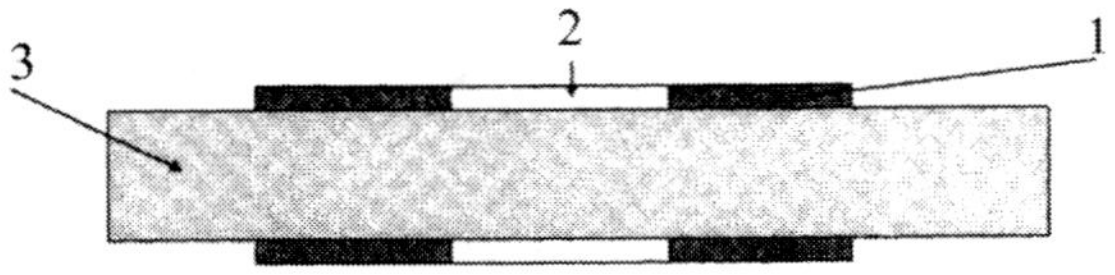

Figure 1. Schematic top view of the specimen for the measurement of adhesion; (1) – PTFE film, (2) – aluminum layer, and (3) – Scotch® 810.

Aluminum is simultaneously deposited onto the plasma-treated and the unchanged specimens. Then, Scotch ® 810 pressure-sensitive adhesive tape (19 mm width), used to measure the adhesion of paints according to ASTM D3359-02, is rolled with rubber roller of d = 30 mm onto the specimen over its entire length. The force on the roller axis during rolling is 29.4 N.

T-peel test (figure 2) is performed on an Autograph AGS-10 KNG universal testing machine (Shimadzu) at a crosshead speed of 100 mm/min. As a result a curve presenting the change in the peel resistance (**A**) along the sample length is obtained. The experimental data are averaged over 20 runs. The representativeness of this procedure is established on the basis of the peel strength of two Scotch® 810 tape specimens joint by their adhesive layers (autohesion strength – A_a). When adhesive layers (two adhesive tape layers in the given case) are brought into contact with one another, the layer thickness doubles. Since the peel resistance is proportional to the adhesive-layer thickness, the value of **A** for the Scotch® 810/ Scotch® 810 joint is $A_a/2$, and is equal to 198±5 N/m.

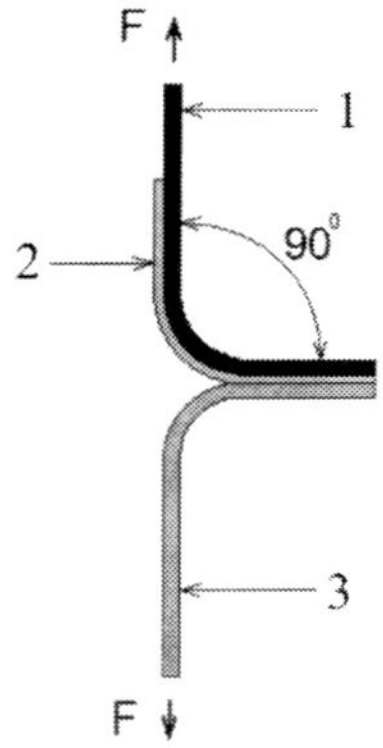

Figure 2. Schematic of the T-peel test for the modified film: (1) PTFE film, (2) Al layer, (3) Scotch® 810 tape. F denotes the applied peeling force.

3. Results and discussion

The experimental values of the contact angles (θ) of water and glycerol for PTFE films treated at the anode and cathode of a DC discharge are presented in table 1. The discharge current is 50 mA and the treating time is 60 s. The work of adhesion, the total surface energy, and its polar and dispersion terms calculated are also given. The data show that the plasma treatment leads to substantial reduction in θ, increase in both W_a and γ, and manifold increase in γ^p. The results indicate that treatment of PTFE films in DC discharge leads to significant enhancement of the contact properties of the polymer surface. The values of θ are substantially lower compare to the results obtained after treatment with RF discharge. It is also seen that PTFE films modified at the anode have lower values of θ and higher values of W_a, γ and γ^p than those treated on the cathode.

Table 1. Contact angles, work of adhesion, and surface energy of PTFE films

Polymer	DC discharge treatment	θ (deg)		W_a (mJ/m^2)		γ (mJ/m^2)		
		By water	By glycerol	By water	By glycerol	γ	γ^p	γ^d
	–	120	106	36.4	45.9	13.18	0.03	13.15
PTFE	On the anode	33	26	133.9	120.4	61.5	42.0	19.5
	On the cathode	49	40	120.6	112.0	50.7	29.3	21.4

An IR spectroscopic investigation of PTFE films shows that after DC discharge treatment at the anode and cathode, the intensity of the oxygenated groups absorption bands considerably increases (stretching vibrations of the C=O group in –C–(O)–F or perfluoroketone groups; absorption band at 1885 cm^{-1}), probably as a result of oxidation by plasma. The absorption band at 1775 cm^{-1} is presumably due to the C=C double bonds formed as a result of fluorine atom abstraction from the polymer chain. Most likely the reason for the decrease in the amount of CF_2 groups is the detachment of fluorine resulting in formation of double bonds in the chain. As a confirmation of this reasoning is the appearance of the absorption band at 1775 cm^{-1}. The intensity of the absorption bands at 3500–3600 cm^{-1} reveals increase of water adsorption as well.

These results are confirmed by ESCA data. The spectrum of the initial PTFE film consists of a C_{1S} peak at binding energy of 292.2 eV with width of 1.3 eV and a small peak at 286.0 eV, which is due to the presence of a certain amount of hydrocarbon impurities in the polymer [12]. The initial F_{1S} spectrum consists of one peak at 690.2 eV with a width of 1.7 eV. After the discharge treatment of the film, a C_{1S} peak at 287.0 eV corresponding to oxidized carbon (C=O) formations appears in the spectrum while the peak at 286.0 eV remains unchanged. The width of the F_{1S} peak increases from 1.7 to 2.0 eV, and its shape somewhat changes. In addition, it was found that polymer contains $< 4\%$ oxygen. These results are close to those, obtained by Wade et al [12], who studies the effect of treatment of PTFE films by oxygen RF (13.56 MHz) discharge.

The adhesion characteristics were examined for the initial PTFE film and the one modified on the anode or cathode of a DC discharge. The discharge conditions are: air pressure – 10 Pa, discharge current – 50 mA and treating time – 10 and 60 s. The data in table 2, reveal that the initial PTFE film exhibits a low with respect to the deposited Al layer peel resistance (**A**) but after the modification of the film on the anode and/or cathode **A** substantially increases.

Table 2. Peel strength (**A**) for the initial PTFE film and the film modified in DC discharge (50 mA) with a physical-vapor deposited Al layer (Scotch® 810/Al)

Sample		**A** (N/m)
Initial		78±18
Treated on the cathode	10 s	188±6
	60 s	181±5
Treated on the anode	10 s	189±13
	60 s	169±15
Scotch® 810/Scotch® 810		198±5

Table 3. Peel strength (**A**) for the initial PTFE film and the film modified in DC discharge (50 mA) (Scotch® 810/PTFE)

Sample		**A**, N/m
Initial		30±6
Treated on the cathode	10 s	76±3
	60 s	134±4
Treated on the anode	10 s	250±20
	60 s	200±10
Scotch® 810/Scotch® 810		198±5

Table 3 shows that the anode-treated PTFE film exhibits higher **A** values for the Scotch® 810/PTFE joint.

These results are illustrated in figure 3 by photographs of specimens of the initial and the anode-modified PTFE films after T-peel testing.

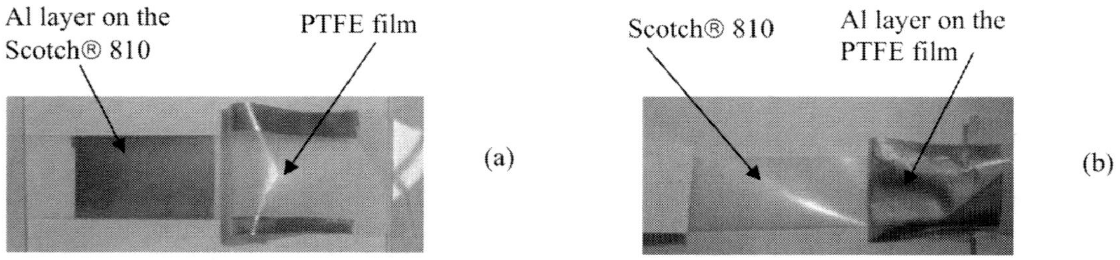

Figure 3. Photographs of the specimens of the (a) initial and (b) anode-modified PTFE films after T-peel testing.

It is seen at figure 3(a) that the Scotch® 810 tape pulls off completely the aluminum layer from the untreated PTFE film but it is fully retained (figure 3(b)) on the anode-modified film.

The results obtained in this work show that DC discharge treatment of PTFE films leads to substantial enhancement of the contact properties of the polymer surface. It also alters considerably the adhesion properties of PTFE film surface. The proposed procedure can be successfully used for quantitative evaluation of the adhesion characteristics of the surface of plasma-modified thin polymer films.

Acknowledgments

This work was supported by the Presidium of Russian Academy of Sciences under the program "Development of the methods for the manufacture of chemicals and design of new materials", grant 7P; the President of the Russian Federation under the program for assistance to Leading National Scientific Schools, grant NSh-247.2008.3, and Government contract № 02.740.11.0143.

References

[1] *Encyclopaedia of Polymer Science and Technology* 2007 ed H Mark (Hoboken: Wiley)
[2] *Plasma Deposition, Treatment and Etching of Polymers* 1990 ed R D'Agostino (N.Y.: Academic Press)
[3] *Encyclopedia of Low-Temperature Plasma* (Introductory Volume IV) 2000 ed V E Fortov (Moscow: Nauka)
[4] Lunkwitz K, Lappan U and Lehman D 2000 *Radiat. Phys. & Chem.* **57** 373
[5] Koh S-K, Park S-C, Kim S-R, Choi W-K, Jung H-J and Pae K D 1997 *J. Appl. Polym. Sci.* **64**

1913
[6] Kim S R 2000 *J. Appl. Polym. Sci.* **77** 1913
[7] Chen J-R and Wakida T 1997 *J. Appl. Polym. Sci.* **64** 1733
[8] Yamada Y, Yamada T, Tasaka S and Inagaki N 1996 *Macromolecules* **29** 4331
[9] Badey J P, Ubraczewski–Espuche E, Junget Y, Sage D, Duc T M and Chabret B 1994 *Polymer* **35** 2472
[10] Tu Ch-Y, Liu Y-L, Lee K-R and Lai J-Y 2005 *Polymer* **46** 6976
[11] Kumar R, Singh R, Kumar M and Barthwal S 2007 *J. Appl. Polym. Sci.* **104** 767
[12] Wade W L, Mammone R J and Binder M 1991 *J. Appl. Polym. Sci.* **43** 1589
[13] Park Y W, Tasaka S and Inagaki N 2002 *J. Appl. Polym. Sci.* **83** 1258
[14] Noeske M, Degenhardt J, Strudthoff S and Lommatzsch U 2004 *Intern. J. of Adhesion & Adhesives* **24** 171
[15] Lapeikova B, Lapeik L Jr, Smolka P, Dlabaja R and Hui D 2006 *J. Appl. Polym. Sci.* **102** 1827
[16] Gil'man A, Drachev A, Kuznetsov A, Lopukhova G and Potapov V 1997 *High Energy Chemistry* **31** 121
[17] Garbassi F, Morra M and Occhiello E 1994 *Polymer Surfaces from Physic to Technology* (Chichester: Wiley) p 166
[18] Dechant J, Danz R, Kimmer W and Schmolke R 1972 *Ultrarotspektroskopische Untersuchungen an Polymeren* (Berlin: Akademie)
[19] http://www.srdata.nist.gov

Journal of Physics: Conference Series **516** (2014) 012013 doi:10.1088/1742-6596/516/1/012013

Landau Damping in Space Plasmas with Two Electron Temperature Non-Maxwellian Distribution Functions

M. N. S. Qureshi, S. Sehar and H. A. Shah
Department of Physics, GC University, Katchery Road, Lahore 54000, Pakistan

E-mail: nouman_sarwar25@hotmail.com

Abstract. Space plasmas generally posses distribution functions that exhibit high or super thermal energy tails in velocity space which can have different temperatures, a dense cold population and a hot population. Moreover, in laboratory plasmas when a laser or electron beam is passed through a dense plasma, hot low density electron populations can be generated. Presence of such low density electron distributions can act to increase the magnitude of the wave damping rate. In this paper we employ non-Maxwellian distribution function such as the generalized (r , q) distribution function with two electron temperatures to study the Landau damping of electrostatic waves. The results show that the Landau damping increases significantly when the percentage of high energy particles increases and with the increase of the high energy tail and less pronounced shoulders in the profile of the distribution function.

1. Introduction

Electron velocity distribution functions (VDFs) with pronounced superthermal tails and shoulders are frequently observed in space plasmas [1-5], which are often modeled by κ-distributions or generalized (r , q) distribution functions [3,6-8]. Observations of electron VDFs from solar wind shown significant deviations from Maxwellian distribution function and; a dense thermal core and a hot superthermal population 'halo' can be distinguished in slow solar wind [9-12]; whereas in a fast solar wind a core and two hot superthermal populations 'halo' and 'strahl' are generally observed [13-15]. Such electron VDFs cannot be modelled by simple one electron population κ-distributions.

In laboratory plasmas a small population of electrons possessing much higher energies than the original laser beam can be produced [16-18]. Simulation results of electron or laser beam propagation in dense plasmas often show electron distributions that are characterized by power-law tails of hot electrons superposed on an approximately Maxwellian bulk distribution [19, 20]. The presence of such low density electron distributions can act to increase the wave damping rate. In this paper, we studied Landau damping with generalized generalized (r, q) distribution function consisting of two populations, a hot population and a cold bulk population. The underlined theory can be used to understand the physical picture in the laboratory and space plasmas.

The generalized (r , q) distribution functions is the sum of a fractional 'F' hot and a cold bulk electron distributions, which is written as

Journal of Physics: Conference Series **516** (2014) 012013 doi:10.1088/1742-6596/516/1/012013

$$f_{rq}(v) = \frac{(1-F)\,C}{\pi}\left(\frac{a}{D\,T_c}\right)^{3/2}\left[1+\frac{1}{q-1}\left(\frac{a}{D^2}\frac{v^2}{T_c}\right)^{r+1}\right]^{-(q)} + \frac{(F)\,C}{\pi}\left(\frac{a}{D\,T_h}\right)^{3/2}\left[1+\frac{1}{q-1}\left(\frac{a}{D^2}\frac{v^2}{T_h}\right)^{r+1}\right]^{-(q)} \tag{1}$$

where,

$$C = \frac{3\,(q-1)^{-3/2(1+r)}\,\Gamma(q)}{\Gamma\left(q-\frac{3}{2(1+r)}\right)\Gamma\left(1+\frac{3}{2(1+r)}\right)} \tag{2}$$

$$D = \frac{3\,(q-1)^{-1/(1+r)}\,\Gamma\left(q-\frac{3}{2(1+r)}\right)\Gamma\left(\frac{3}{2(1+r)}\right)}{\Gamma\left(q-\frac{5}{2(1+r)}\right)\Gamma\left(\frac{5}{2(1+r)}\right)} \tag{3}$$

Here, Γ is the usual Gamma function; T_c and T_h are the cold and hot electron temperatures and $a = m/2k_B$, where m is the electron mass and k_B is the Boltzmann constant. The total electron density is the sum of the hot and cold electron densities. In the limit $r = 0$ and $q = \kappa + 1$, the equation (1) reduces to κ-kappa distribution function and in the limiting case when $r = 0$ and $q \to \infty$, the equation (1) reduces to the well known Maxwellian distribution function. We note that $q > 1$ and $q(1+r) > 5/2$ are the conditions which arise from the normalization and definition of temperature for the distribution function given in equation (1).

Figures 1 and 2 show the profiles of generalized (r , q) distribution function for 1% hot population at 10 eV added to cold dense population at 1 eV for different values of r and q. Figure 1 is plotted for different values of spectral index q when r = 1 (upper panel) and r = 2 (lower panel). From the upper panel when r = 1, we note that as the value of q increases the high energy tail decreases. From the lower panel when r = 2, we note that as the value of q increases the high energy tail decreases similar to the upper panel and shoulders in the profile of the distribution function become more prominent with the increase of r. Figure 2 is plotted for different values of spectral index r when q = 2 (upper panel) and q = 5 (lower panel). For q = 2 in the upper panel, we note that as r increases the shoulders in the profile of distribution function tend to increase and the high energy tail decreases. For q = 5 in the lower panel, we note that as r increases the shoulders in the profile of distribution function tend to increase similar to the upper panel and the high energy tail decrease as compared to the upper panel. Therefore, from the Figures 1 and 2 we can note that when q increases the high energy tail decreases and when r increases shoulders in the profile of distribution becomes more prominent.

2. Dispersion Relation

We follow the general formulism of kinetic theory to derive the dispersion relation for the electrostatic waves with complex frequency $\omega = \omega_r + i\omega_i$ and real wave number k. This procedure yields the propagation and damping characteristics of electron plasma waves in the limiting case when $k\lambda_D << 1$, where

$$\lambda_D = \left(\frac{T_\alpha}{4\pi N e^2}\right) = \left(\frac{(q-1)^{1/(1+r)}\,\Gamma\left(q-\frac{5}{2(1+r)}\right)\Gamma\left(\frac{5}{2(1+r)}\right)}{3\,\Gamma\left(q-\frac{3}{2(1+r)}\right)\Gamma\left(\frac{3}{2(1+r)}\right)}\right)^{1/2}\left(\frac{v_{th\alpha}}{\omega_p}\right) \tag{4}$$

is the electron Debye length, v_{th} is the thermal velocity, ω_p is the electron plasma frequency, T is the electron temperature, $\alpha = c, h$ (for cold or hot population) and N is the total number density. The general dispersion relation of electrostatic waves is

$$1 - \frac{\omega_p^2}{k^2}\int_{-\infty}^{\infty}\frac{\partial f/\partial v}{(v-\omega/k)}\,dv = 0 \tag{5}$$

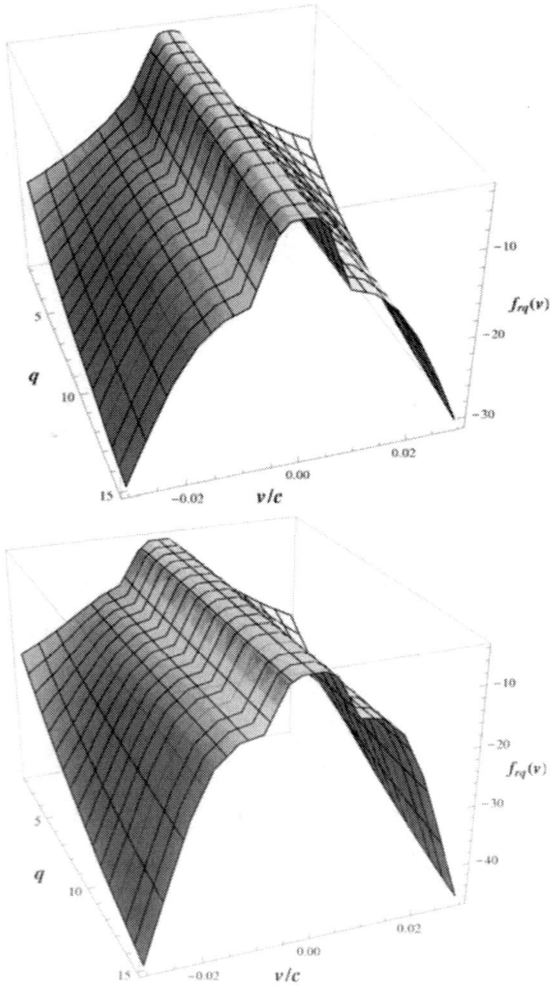

Figure 1. Profiles of generalized (r, q) distribution function for different values of q when $r = 1$ (upper panel), $r = 2$ (lower panel), and $F = 0.01$, $T_c = 1$ eV and $T_h = 10$ eV.

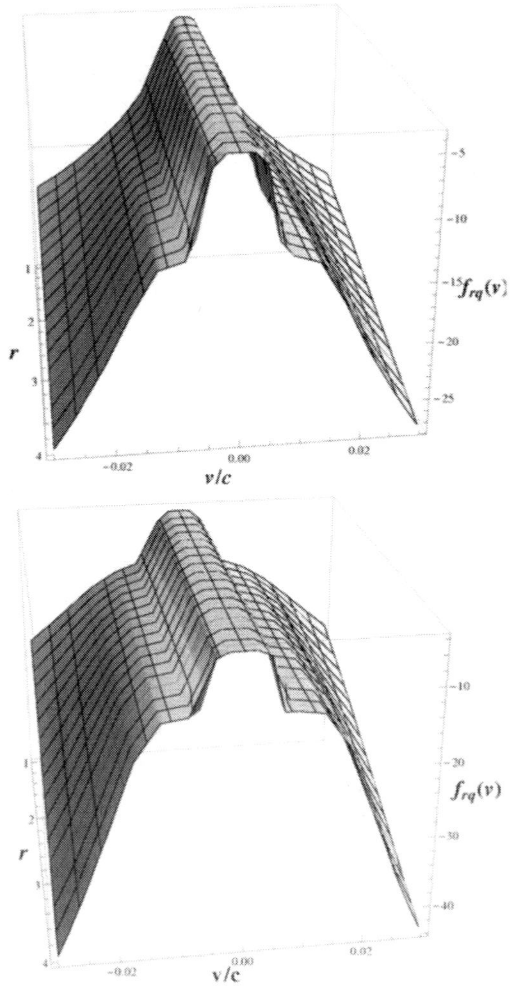

Figure 2. Profiles of generalized (r, q) distribution function for different values of r when $q = 2$, (upper panel), $q = 5$ (lower panel), and $F = 0.01$, $T_c = 1$ eV and $T_h = 10$ eV.

Using equation (1) in above equation, we get the general dispersion relation as

$$1+\frac{2\,\omega_p^2}{k^2}\left[\frac{(1-F)}{v_{thc}^2}\left\{H+\xi_c\,Z_1^{(r,q)}(\xi_c)\right\}+\frac{F}{v_{thh}^2}\left\{H+\xi_h\,Z_1^{(r,q)}(\xi_h)\right\}\right]=0 \tag{6}$$

where

$$H=\frac{3\,(q-1)^{-1/(1+r)}\;\Gamma\left(q-\frac{1}{2(1+r)}\right)\Gamma\left(1+\frac{1}{2(1+r)}\right)}{2\,\Gamma\left(q-\frac{3}{2(1+r)}\right)\Gamma\left(1+\frac{3}{2(1+r)}\right)} \tag{7}$$

and

$$Z_1^{(r,q)}(\xi_\alpha)=\frac{3\,(q-1)^{-3/2(1+r)}\;\Gamma(q)}{4\Gamma(q-3/2(1+r))\;\Gamma(1+3/2(1+r))}\int_{-\infty}^{\infty}\frac{ds}{(s_\alpha-\xi_\alpha)}\left(1+\frac{1}{q-1}\,s_{c,h}^{2(r+1)}\right)^{-q} \tag{8}$$

is the generalized plasma dispersion function [6]. Here $\xi_\alpha=\omega/kv_{th\alpha}$, $s_\alpha=v/v_{th\alpha}$ and again $\alpha=c,h$.

For the Langmuir waves propagating in unmagnetized plasma it is justified to consider the ions as immobile giving uniform background that simply maintains the charge neutrality. Therefore, neglecting the ion terms in equation (6) and using the appropriate limiting form $\xi_\alpha>>1$ of the generalized plasma dispersion function (8), the dispersion relation for hot and cold species can be written as

$$D_r(\Omega)+i\,D_i(\Omega)=0 \tag{9}$$

where the real and imaginary terms are

$$D_r(\Omega)=1-\frac{1}{\Omega^2}\left[(1-F)+F\right]-\frac{2\,K\,D\,k^2\lambda_D^2}{\Omega^4}\left[(1-F)+\frac{T_h}{T_c}F\right] \tag{10}$$

and

$$D_i(\Omega)=\frac{2\pi\Omega C}{D^{3/2}\,k^2\lambda_D^2}\left[\begin{array}{l}(1-F)\left(1+\dfrac{1}{q-1}\left\{\dfrac{\Omega^2}{D\,k^2\lambda_D^2}\right\}^{r+1}\right)^{-(q)}\\ +F\left(\dfrac{T_c}{T_h}\right)^{3/2}\left(1+\dfrac{1}{q-1}\left\{\dfrac{\Omega^2}{D\,k^2\lambda_D^2}\dfrac{T_c}{T_h}\right\}^{r+1}\right)^{-(q)}\end{array}\right] \tag{11}$$

respectively. Here $\Omega=\omega/\omega_p$ and

$$K=\frac{3\,(q-1)^{1/(1+r)}\;\Gamma\left(q-\frac{5}{2(1+r)}\right)\Gamma\left(1+\frac{5}{2(1+r)}\right)}{10\,\Gamma\left(q-\frac{3}{2(1+r)}\right)\Gamma\left(1+\frac{3}{2(1+r)}\right)} \tag{12}$$

The solution of $D_r(\Omega_r)=0$ gives the real frequency Ω_r of the wave and the solution is then given as (for positive root)

$$\Omega_r\simeq\sqrt{\frac{1}{2}+\frac{1}{2}\left(1+8\,D\,K\,k^2\lambda_D^2\left(1-F+\frac{T_h}{T_c}F\right)\right)^{1/2}} \tag{13}$$

Journal of Physics: Conference Series **516** (2014) 012013 doi:10.1088/1742-6596/516/1/012013

And the damping rate of the wave is determined by

$$D_i \simeq -\left.\frac{D_i}{\partial D_r / \partial \Omega}\right|_{\Omega_r} \tag{14}$$

Using equation (13), in evaluating the above equation, gives

$$\Omega_i = -\frac{\pi C}{D^{3/2}}\frac{\Omega^8}{k^3\lambda_D^3}\left(\Omega^4 + 4KD\Omega^2 k^2\lambda_D^2\left(1 - F + \frac{T_h}{T_c}F\right)\right)^{-1}$$
$$\left[(1-F)\left(1+\frac{1}{q-1}\left\{\frac{\Omega^2}{Dk^2\lambda_D^2}\right\}^{r+1}\right)^{-(q)} + F\left(\frac{T_c}{T_h}\right)^{3/2}\left(1+\frac{1}{q-1}\left\{\frac{\Omega^2}{Dk^2\lambda_D^2}\frac{T_c}{T_h}\right\}^{r+1}\right)^{-(q)}\right] \tag{15}$$

In the limit $8\,K\,D\,k^2\lambda_D^2(1-F+F(T_c/T_h)) << 1$, $\Omega_r = \sqrt{1+2KDk^2\lambda_D^2(1-F+F(T_c/T_h))}$ and the equation (15) can be further reduced to

$$\Omega_i = -\frac{\pi C}{D^{3/2}}\frac{1}{k^3\lambda_D^3}$$
$$\left[\begin{array}{l}(1-F)\left(1+\frac{1}{q-1}\left\{\frac{1}{Dk^2\lambda_D^2} + 2K\,(1-F+F(T_h/T_c))\right\}^{r+1}\right)^{-(q)} \\ +F\left(\frac{T_c}{T_h}\right)^{3/2}\left(1+\frac{1}{q-1}\left\{\frac{1}{Dk^2\lambda_D^2} + 2K\,(1-F+F(T_h/T_c))\left(\frac{T_c}{T}\right)\right\}^{r+1}\right)^{-(q)}\end{array}\right] \tag{16}$$

3. Numerical Solution

The numerical solution of equation (16) is shown in Figures 3 to 5 for different values of r and q. Figure 3 depicts the magnitude of damping rates for 1% hot population at 10 eV added to cold dense population at 1 eV in the limit $k\lambda_D << 1$ in the appropriate limiting form $\xi_\alpha >> 1$ of the generalized plasma dispersion function (8) for different values of q when r is fixed. From Figure 3 upper panel, we can see that the damping rate significantly increases when q decreases from 15 to 2 when r = 1. This is due to the fact that as the high energy tail increases in the profile of distribution function, the damping rate increases for the low values of q which can be seen in Figure 1 (upper panel). From the lower panel of Figure 3, we can note that the damping rate increases when q decreases from 15 to 2 when r = 2 similar to the damping rate as shown in upper panel but the comparison of upper and lower panels shows that the damping rate decreases as r increases. This is due to the fact that as r increase, the shoulder increases in the profile of the distribution function and high energy tail decreases which correspond to the Figure 1 (lower panel). Thus the damping rate decreases when either r or q increases.

Figure 4 depicts the magnitude of damping rates for 1% hot population at 10 eV added to cold dense population at 1 eV in the limit $k\lambda_D << 1$ in the appropriate limiting form $\xi_\alpha >> 1$ of the generalized plasma dispersion function (8) but for different values of r when q is fixed. From Figure 4 (upper panel), we can see that the damping rate significantly decreases when r increases from 1 to 4 when q = 2. This is due to the fact that as the shoulders increase in the profile of distribution function the high energy tail decreases which can be seen in Figure 2 (upper panel) and hence the damping rate decreases when r increases. From the lower panel of Figure 4, we can note that the damping rate decreases when r increases from 1 to 4 when q = 5 similar to the damping rate as shown in upper panel. But the comparison of upper and lower panels shows that the damping rate decreases

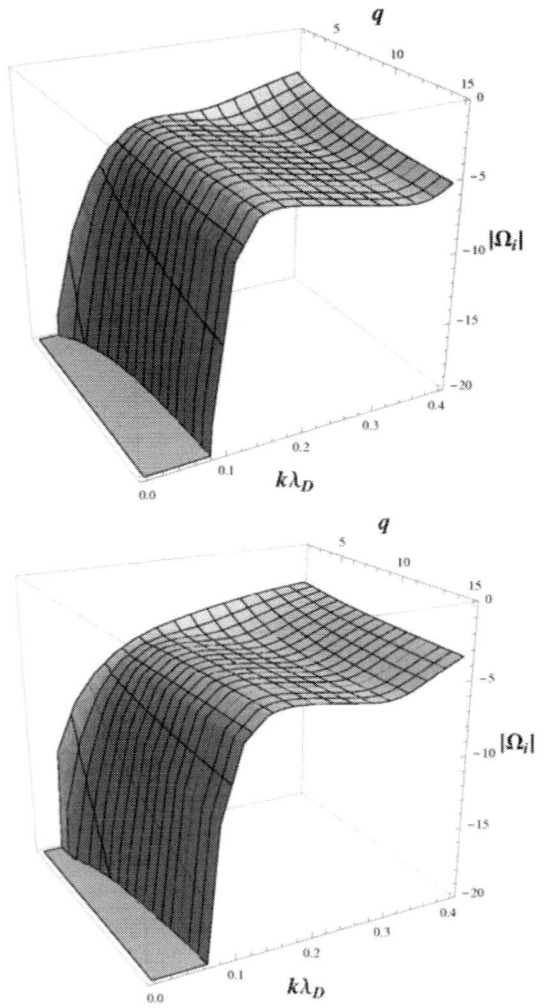

Figure 3. Magnitude of Landau damping for different values of q when $r = 1$ (upper panel), $r = 2$ (lower panel), and for $F = 0.01$, $T_c = 1$ eV and $T_h = 10$ eV.

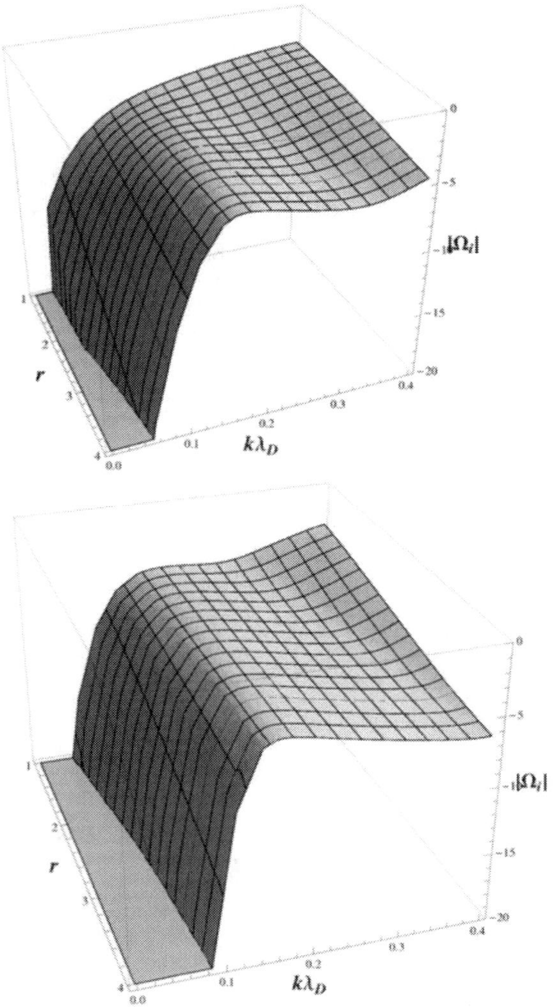

Figure 4. Magnitude of Landau damping for different values of r when $q = 2$ (upper panel), $q = 5$ (lower panel), and for $F = 0.01$, $T_c = 1$ eV and $T_h = 10$ eV.

significantly as q increases. This is due to the fact that as q increase, the high energy tail decreases in the profile of the distribution function which correspond to the Figure 2 (lower panel). For fixed value of $k\lambda_D$, the comparison of Figure 3 and 4 shows that the damping rate significantly increases when either the value of r or q decreases in the range when $k\lambda_D < 0.15$.

Figure 5 shows the magnitude of damping rates for the larger fraction of the hot population (10%) at 10 eV added to cold dense population at 1 eV in the same limiting cases as for Figures 3 and 4 for different values of r and q. From Figure 5 upper panel, we can note that the damping rate enhances significantly when the percentage of hot population increases as compared to the lower percentage of hot population Figure 3 (upper panel) for r = 1. Similarly from Figure 5 lower panel, we can note that the damping rate enhances significantly when the percentage of hot population increases as compared to the lower percentage of hot population Figure 3 (lower panel) for q = 2. Therefore, for the larger fraction of the hot population (10%), the Landau damping enhances significantly for all values of r and q and in the range $0.0 < k\lambda_D < 0.4$.

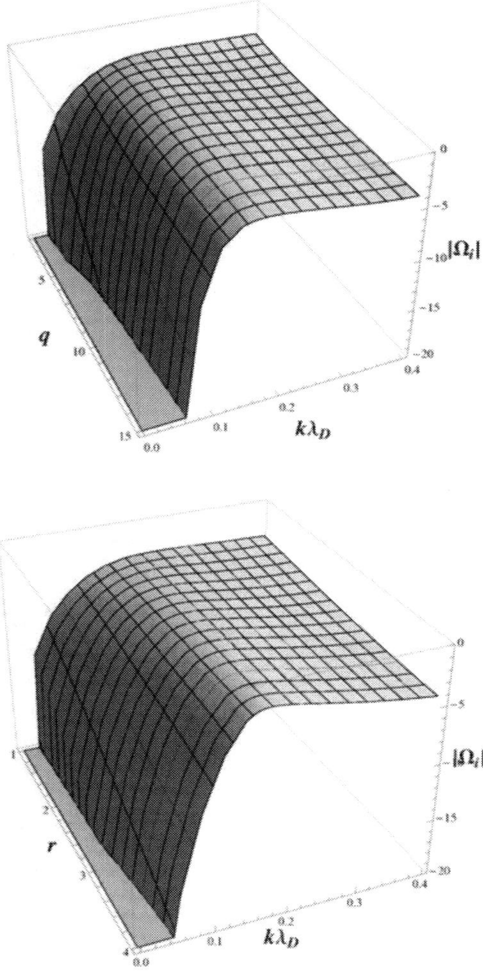

Figure 5. Magnitude of Landau damping when the hot population is 10% of the bulk population, i.e. $F = 0.1$, for different values of q when $r = 1$ (upper panel), and for different values of r when $q = 2$ (lower panel). Other parameters are the same as that of Figures 3 and 4.

4. Conclusion

In this paper, the effect of enhanced Landau damping of electrostatic waves is studied in the presence of low density superthermal electron distribution when added to dense bulk population of cold plasmas. We followed the standard kinetic approach to evaluate the damping rate of Langmuir waves modelled generalized (r, q) distribution function. We have shown that Landau damping increases as the q decreases from 15 to 2 or r decreases from 4 to 1 when we take hot population as the 1% of the bulk cold population. The same trend has also been observed when we consider 10% hot population. But when we consider the increased hot population, the Landau damping significantly increases as compared to the case when we take lower percentage of hot population.

5. References

[1] Qureshi M N S, Pallocchia G, Bruno R, Cattaneo M B, Formisano V, et al. 2003 Solar Wind-10 AIP
[2] Vasyliunas V M 1990 J. Geophys. Res. 73, 2839
[3] Gosling J T, Ashbridge J R, Bame S J, Feldman W C and Zwickl R D 1981 J. Geophys. Res. 86, 547
[4] Lin R P, Levedahi W K, Lotko W, Gurnett D A and Scarf F L 1986 Astrophys. J. 308, 954
[5] Fitzenreiter R J, Ogilivie K W, Chornay D J and Keller J 1998 Geophys. Res. Lett. 25, 249
[6] Qureshi M N S, Shi J K and Ma S Z 2005 Phys. Plasmas 12, 122902
[7] Mace R L and Hellberg M A 1995 Phys. Plasmas 2, 2098
[8] Maksimovic M, Pierrard V and Riley P 1997 Geophys. Res. Lett. 24, 1151
[9] Lin R P 1980 Sol. Phys. 67, 393
[10] Lin R P 1998 Space Sci. Rev. 86, 61
[11] Pagel C, Crooker N U, Larson D E, Kahler S W and Qwens M J 2005 J. Geophys. Res. 110, A01103
[12] Gaelzer R, Ziebell L F, Vinas A F, Yoon P H and Ryu C –M 2008 Astrophys. J. 677, 676
[13] Pierrard V, Maksimovic M and Lemaire J 2001 Astrophys. Space Sci. 277, 195
[14] Pilipp W G, Miggenrieder H and Montgomery M D 1987 J. Geophys. Res. 92, 1075
[15] Hammond C M, Feldman W C, McComas D J, Phillips J L and Forsyth R J 1996 Astron. Astrophys. 316, 350
[16] Freund H P, Smith R A and Papadopoulos K 1981 Phys. Fluids 24, 442
[17] Fourkal E, Bychenkov V Y, Rozmus W, Sydora R, Kirkby C, Capjack C E, Glenzer S H and Baldis H A 2001 Phys. Plasmas 8, 550
[18] Yoon P H, Rhee T and Ryu C –M 2005 Phys. Rev. Lett. 95, 215003
[19] Tanaka M and Papadopoulos K 1983 Phys. Fluids. 26, 1697
[20] Wang J G, Newman D L and Goldman M V 1997 J. Atmos. Sol. Phys. 59, 2461

Journal of Physics: Conference Series **516** (2014) 012014 doi:10.1088/1742-6596/516/1/012014

A review on Lamb´s atmospheric oscillations using initial value problem approach

Ángel De Andrea González
Departamento de Física. Escuela Politécnica Superior.
Universidad Carlos III de Madrid. Av. Universidad, 30. 28911 Leganés. Spain.

E-mail: aandrea@fis.uc3m.es

Abstract. Waves at a surface of discontinuity in the atmosphere were analysed in 1910 by Lamb, who derived, using normal mode approach, an analytical dispersion relation for a discrete mode (surface mode). Lamb examined the case of waves propagated along a horizontal plane where the equilibrium temperature is discontinuous. For simplicity, the upper and the lower regions are considered incompressible. The oscillations are treated in the ideal (dissipationless) limit and the uniform gravitational acceleration is taken to be co-aligned with the prevailing temperature gradient. In this work, in order to show how the modes appear in the response of a surface discontinuity to an initial perturbation, we consider the initial value problem (IPV). The main difference from the standard analysis is that solutions to the linearized equations of motion which satisfy general conditions are obtained in terms of Fourier-Laplace transform of the hydrodynamics variables. These transforms can be inverted explicitly to express the fluid variables as integrals of Green´s functions multiplied by initial data. In addition to discrete mode (surface mode), sets of continuum modes due to branch cuts in the complex plane, not treated explicitly in the literature, appears.

1. Introduction

Several authors have purposed solutions of the initial value problem through the use of a Laplace Transform in time [2-4]. The principal finding of the initial value problem approach is that, in addition to the discrete eigenvalues linked to the normal modes, there exists a continuous spectrum of eigenvalues. Thus, the modal approach cannot provide a complete solution.

2. Linear stability analysis

We consider two infinitely incompressible plasmas in a steady state which are separated by a plane interface at y=0 and subjected to a gravitational field $\vec{g}$ directed perpendicular to the interface, where q and q^* are both the inverse density-gradient scale length at each region.

By following the approach developed by Lamb (1910) [1], it is possible to reduce the linearized equations of continuity, momentum and energy for a polytrope to a set of partial differential equations. These equations can be easily combined into a stream-function equation:

$$\frac{\partial}{\partial \tilde{t}^2}\left(\frac{\partial^2 \tilde{\psi}}{\partial \tilde{y}^2}-\tilde{\psi}\right)+\frac{1}{\tilde{k}}\left(\frac{\partial^3 \tilde{\psi}}{\partial \tilde{y}\partial \tilde{t}^2}-\tilde{\psi}\right)=0 \tag{1}$$

Where the dimensionless normalized parameters are: $\tilde{t}=t\sqrt{kg}$, $\tilde{k}=\frac{k}{q}$, $\tilde{t}=t\sqrt{kg}$. We define the Laplace transform $\tilde{\psi}_{\omega}(\tilde{y},\tilde{\omega})$ in the complex $\tilde{\omega}$-plane:

$$\tilde{\psi}_{\tilde{\omega}}(\tilde{y},\tilde{\omega})=\frac{1}{2\pi}\int_{-i\infty+\sigma}^{+i\infty+\sigma}\tilde{\psi}(\tilde{y},\tilde{t})e^{-i\tilde{\omega}\tilde{t}}d\tilde{\omega} \tag{2}$$

Then, we consider the initial value problem for equation (1), and introduce the Laplace transform. So, we obtain the following equation:

$$\frac{\partial^2 \tilde{\psi}_{\tilde{\omega}}}{\partial \tilde{y}^2}+\frac{1}{\tilde{k}}\frac{\partial \tilde{\psi}_{\tilde{\omega}}}{\partial \tilde{y}}-\left(1-\frac{1}{\tilde{k}\tilde{\omega}^2}\right)\tilde{\psi}_{\tilde{\omega}}=\Gamma(\tilde{y}) \tag{3}$$

being:

$$\Gamma(\tilde{y})=\frac{1}{\tilde{\omega}^2}\left[\frac{\partial^2\left(i\tilde{\omega}\tilde{F}_1-\tilde{F}_2\right)}{\partial \tilde{y}^2}+\frac{\partial\left(i\tilde{\omega}\tilde{F}_1-\tilde{F}_2\right)}{\tilde{k}\partial \tilde{y}}-\left(i\tilde{\omega}\tilde{F}_1-\tilde{F}_2\right)\right] \tag{4}$$

where the Laplace transform function $\tilde{\psi}(\tilde{y},\tilde{t})$ is:

$$L\left[\frac{\partial^2\tilde{\psi}(\tilde{y},\tilde{t})}{\partial \tilde{t}^2}\right]=-\tilde{\omega}^2\tilde{\psi}_s+i\tilde{\omega}\tilde{F}_1-\tilde{F}_2 \tag{5}$$

With the following initial conditions:

$$\tilde{F}_1=\tilde{\psi}(\tilde{y},\tilde{t})\Big|_{\tilde{t}=0^+};\ \tilde{F}_2=\frac{\partial\tilde{\psi}(\tilde{y},\tilde{t})}{\partial \tilde{y}}\Bigg|_{\tilde{t}=0^+} \tag{6}$$

With Laplace contour σ chosen that $\tilde{\psi}_{\tilde{\omega}}$ has no singularities in $\mathrm{Re}(\tilde{\omega})\geq\sigma$. Equation is subjected to the conditions at $|\tilde{y}|\to\infty$ that $\tilde{\psi}_{\tilde{\omega}}\to 0$, the continuity of $\tilde{\psi}_{\tilde{\omega}}$ at $\tilde{y}=0$ and the jump conditions on $\frac{\partial\tilde{\psi}_s}{\partial\tilde{y}}$ at $\tilde{y}=0$.

We consider the Green´s function associated with equation (3). So, the solution of equation (1) is:

$$\tilde{\psi}_s(\tilde{y},\tilde{\omega}) = \int_0^\infty G(\tilde{y},\tilde{\omega},\tilde{y}_o)\Gamma(\tilde{y})\, d\tilde{y}_o \tag{7}$$

2.1. *Results and discussion*

The Green´s function for $\tilde{y} > 0$:

$$G(\tilde{y},\tilde{\omega},\tilde{y}_o) = \frac{\tilde{\omega}^2 e^{-\tilde{y}\left(\beta+\frac{1}{2\tilde{k}}\right)-\tilde{y}_o\left(\alpha+\frac{1}{2\tilde{k}r}\right)}}{\left[-1+r-\tilde{\omega}^2\left(\alpha-\frac{1}{2\tilde{k}r}\right)-r\tilde{\omega}^2\left(\beta+\frac{1}{2\tilde{k}}\right)\right]} \tag{8}$$

In accord with the boundary conditions $G(\pm\infty,\tilde{\omega},\tilde{y}_o)=0$, the square roots are defined so that Re(α)>0 y Re(β)>0 for Re ($\tilde{\omega}$)>σ. It is evident that Green´s function has a pole (discrete spectra). Expanding Green´s function about α=0, and β=0, this function does not turn out to be an even function of α and β. So, the Green´s function has branch points at $\tilde{\omega}^2 = \gamma_b^2$ (associated with continuous spectra neglected by Lamb):

$$\gamma_{b1}^2 = \frac{4\tilde{k}}{1+4\tilde{k}^2} \quad \text{y en} \quad \gamma_{b2}^2 = \frac{4\tilde{k}r}{1+4\tilde{k}^2r^2} \quad (\text{i.e., } \alpha=0 \text{ y } \beta=0) \tag{9}$$

It is interesting to follow the migration of the pole as $\tilde{k}$ is varied. For $\tilde{k} > \tilde{k}_{cr}$, $\gamma_p > \gamma_{b1}$. If $\tilde{k}$ is diminished, then the γ_p pole moves toward γ_{b1}. For $\tilde{k} = \tilde{k}_{cr}$, then $\gamma_p = \gamma_{b1}$. For $\tilde{k} < \tilde{k}_{cr}$ the pole drops on to the lower sheet of β and the eigenmode disappears. The disappearance of the γ_p pole was neglected by Lamb too. Thus, there will be some critical wave number $\tilde{k}_{cr}$ such for $\tilde{k} > \tilde{k}_{cr}$ ($\tilde{k} < \tilde{k}_{cr}$) eigenmodes exist (do not exist). Lamb obtained the following dispersion relation [1]:

$$\left(\tilde{\omega}^2\right)^2 - \frac{2\tilde{\omega}^2}{\left[\tilde{k}(1+r)\right]} - \left(\frac{r-1}{r+1}\right)^2 = 0 \tag{10}$$

Being, $r=\frac{q}{q^\bullet}$ .We show that Lamb´s dispersion relation, the critical wave number $\tilde{k}_{cr}$ is:

$$\tilde{k}_{cr} = \frac{1}{2}\sqrt{\frac{3+r}{r-1}} \quad \text{for } r \neq 1 \tag{11}$$

Journal of Physics: Conference Series **516** (2014) 012014 doi:10.1088/1742-6596/516/1/012014

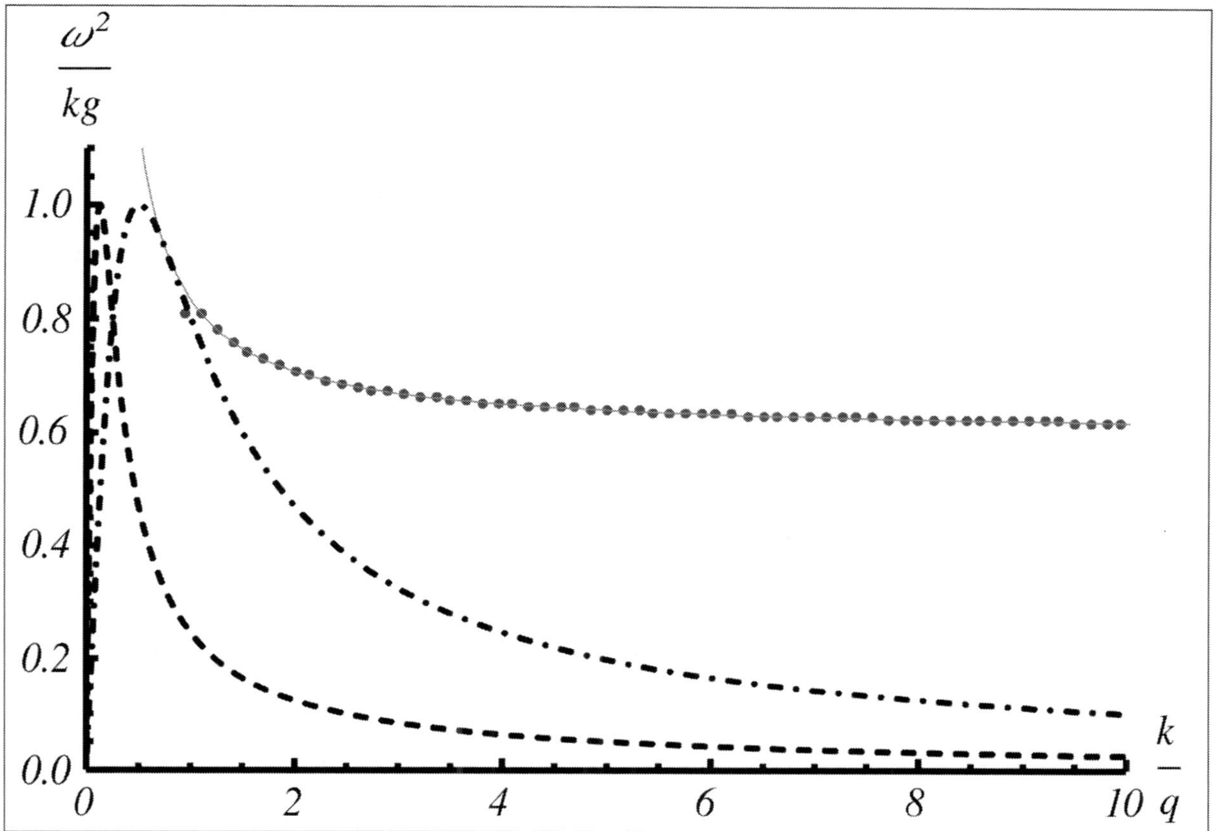

Figure 1. Dimensionless normalized growth rate versus dimensionless normalized wave number for r=4. Solid line corresponds to the Lamb´s results; dots correspond to our numerical results. There is an excellent agreement with Lamb´s results for $\tilde{k} > \tilde{k}_{cr}$. Dashed line and dot dashed line correspond to branch points.

3. Conclusion

Waves at a surface of discontinuity in the atmosphere are analysed using initial value problem (IPV). So that, sets of continuum modes due to branch cuts in the complex plane, not treated explicitly in the literature, explain the disappearance of the dispersion relation. Thus, there will be some critical wave number $\tilde{k}_{cr}$ such that for $\tilde{k} > \tilde{k}_{cr}$ ($\tilde{k} < \tilde{k}_{cr}$) eigenmodes exist (do not exist).

4. References

[1] Lamb H 1910 *Proc. Lond. Math. Soc.* A **84** 551-72
[2] Ott E and Russell D A 1978 *Physical Review Letters* **41** 1048-51
[3] Russell D A and Ott E 1979 *Journal of Geophysical Research* **84** 6573-79.
[4] Bogdan T J and Cally P S 1997 *Proc. R. Soc. Lond. A, Math. Phys. Eng. Sci.*, **453** 943 - 61

4th International Workshop & Summer School on Plasma Physics 2010 IOP Publishing
Journal of Physics: Conference Series **516** (2014) 012015 doi:10.1088/1742-6596/516/1/012015

The influence of a semi-infinite atmosphere on solar oscillations

Ángel De Andrea González
Departamento de Física. Escuela Politécnica Superior.
Universidad Carlos III de Madrid. Av. Universidad, 30. 28911 Leganés. Spain.

E-mail: aandrea@fis.uc3m.es

Abstract. The influence of a semi-infinitive atmosphere on solar investigations is investigated using a model in which the corona is represented by a gravitationally stratified fluid. The solar corona can be modeled as a semi-infinitive region of plasma that occupies the space above the xy- plane in Cartesian coordinates with the z-axis taken along the gravitational acceleration $\vec{g} = g\vec{u}_z$. This assumption is reasonable, as the plasma density of the atmosphere is much lower than the density of the photosphere. So, we consider the phostosphere as a solid and immobile boundary for the atmosphere. The standard mathematical procedures in Helioseismology field are based on normal mode approach for various solar models. The region is assumed quasi-isothermal and without magnetic fields. In this work, in order to show how the modes appear in the response to an initial perturbation, we consider the initial value problem (IPV). The p-modes and g-modes possess only continuous spectra –as opposed to discrete spectrum like that previous investigators found for this problem- and the solution to the initial value problem is obtained through an appropiate Green´s function.

1. Introduction

Several authors have purposed solutions of the initial value problem through the use of a Laplace Transform in time [5-6]. The principal finding of the initial value problem approach is that, in addition to the discrete eigenvalues linked to the normal modes, there exists a continuous spectrum of eigenvalues. Thus, the modal approach cannot provide a complete solution.

We use the Cartesian coordinates with the z-axis pointing towards the interior of the Sun and parallel to the constant gravity $\vec{g}$. As to the applicability of the Cartesian geometry in our analysis, it is valid for perturbations of a sufficiently small wavelength when the effects of stellar sphericity are unimportant. This then places the centre of the Sun into $z/H \to -\infty$. Our model assumes an isothermal atmosphere at $z < 0$ above the solar surface. The medium is considered a perfect gas. In this paper we exclude macroscopic motions and magnetic fields [1-3].

Journal of Physics: Conference Series **516** (2014) 012015 doi:10.1088/1742-6596/516/1/012015

2. Linear stability analysis and the initial problem value

We start from the standard set of hydrodynamic equations describing the dynamics processes in a fully ionized hydrogen plasma in presence of gravity with constant acceleration [4].

$$\frac{\partial\rho}{\partial t}+\vec{\nabla}\cdot(\rho\vec{v})=0 \tag{1}$$

$$\rho\frac{\partial\vec{v}}{\partial t}+\rho\ \vec{v}\cdot\vec{\nabla}\cdot(\rho\vec{v})=-\vec{\nabla}p-\rho\vec{g} \tag{2}$$

$$\frac{\partial\rho}{\partial t}+\vec{v}\cdot\vec{\nabla}p=\gamma\frac{p}{\rho}\left(\frac{\partial\rho}{\partial t}+\vec{v}\cdot\vec{\nabla}\rho\right) \tag{3}$$

Considering the displacement vector $\vec{\xi}$, where $\vec{v}_1=\frac{\partial\vec{\xi}}{\partial t}$, linearization leads to:

$$\rho\frac{\partial^2\vec{\xi}}{\partial t^2}-\vec{\nabla}\left(\gamma\ p\vec{\nabla}\cdot\vec{\xi}\right)+\rho\ \vec{\nabla}\left(\vec{g}\cdot\vec{\xi}\right)-\rho\ \vec{g}\vec{\nabla}\cdot\vec{\xi}=0 \tag{4}$$

2.1. *Results and discussion*

We define the Laplace transform $\xi_z(z,\omega)$ in the complex - ω plane:

$$\xi_z(z,\omega)=\frac{1}{2\pi}\int_{-i\infty+\sigma}^{+i\infty+\sigma}\xi_z(z,\omega)e^{-i\omega t}d\omega \tag{5}$$

Then, we consider the initial value problem and introduce the Laplace transform. So, we obtain the following ODE:

$$\frac{d^2\xi_z}{dz^2}+\lambda\frac{d^2\xi_z}{dz^2}+\frac{\omega^4-k^2v_s^2\omega^2+k^2v_s^2N^2}{v_s^2\omega^2}\xi_z=f\left(z,\omega,k,F_1,F_2\right) \tag{6}$$

being:

$N^2=\frac{(\gamma-1)g^2}{v_s^2}$ the Brunt-Väisäläa frequency; $\lambda=\frac{1}{H}$; F_1 and F_2 are initial conditions:

$$F_1=\xi_z(z,t)\big|_{t=0^+}\ ;\ F_2=\left.\frac{\partial\xi_z(z,t)}{\partial z}\right|_{t=0^+} \tag{7}$$

We can consider the Green´s function for the former ODE, choosing the boundary conditions $G(-\infty,\omega,z_o)=0$ and $G(0,\omega,z_o)=0$:

$$G(z,\omega,z_o)=\begin{cases} Ae^{\frac{z(-1+\alpha)}{2H}} & z<-z_o \\ Be^{\frac{z(-1+\alpha)}{2H}}+Ce^{\frac{z(-1-\alpha)}{2H}} & -z_o<z<0 \end{cases} \tag{8}$$

On the other hand:

$$\alpha=\sqrt{1-\frac{4H^2\left[g^2k^2(\gamma-1)+\omega^2\left(\omega^2-v_{sc}^2k^2\right)\right]}{v_{sc}^2\omega^2}} \tag{9}$$

In accord with the boundary condition $G(-\infty,\omega,z_o)=0$, the square root are defined so that Re(α)>0 for Re (ω)$>\sigma$. So, the solution is:

$$\xi_z(z,\omega)=\int_0^{\infty}G(z,\omega,z_o)f(z,\omega,k,F_1,F_2)\,dz_o \tag{10}$$

On the other hand, the continuity of $G(z,\omega,z_o)$ at $z=-z_o$, and the jump condition on $\frac{dG(z,\omega,z_o)}{dz}$ at $z=-z_o$, yields:

$$G(z,\omega,z_o)=\begin{cases} \frac{1}{\alpha}e^{\frac{z+z_o-z\alpha+z_o\alpha}{2H}}\left(e^{\frac{z_o\alpha}{H}}-1\right)H & z<-z_o \\ \frac{1}{\alpha}e^{\frac{(z+z_o)(1+\alpha)}{2H}}\left(e^{\frac{z_o\alpha}{H}}-1\right)H & -z_o<z<0 \end{cases} \tag{11}$$

Expanding Green´s function about α=0, this function does not turn out to be an even function of α. So, the Green´s function has branch points at $\omega^2=\gamma_b^{\,2}$ (associated with continuous spectra for p-modes and g-modes). So, the infinite thickness causes the degenerate g-modes and p-modes to be distended into a continuous spectrum.

The following figure 1 shows branch points and domains for p-modes continuous spectra and g-modes continuous spectra. The temperature for solar corona is 10^6 K. Gravitational acceleration is assumed constant within the domains of studied harmonic perturbations and we take g= 274 m/s^2 as it is on the solar surface.

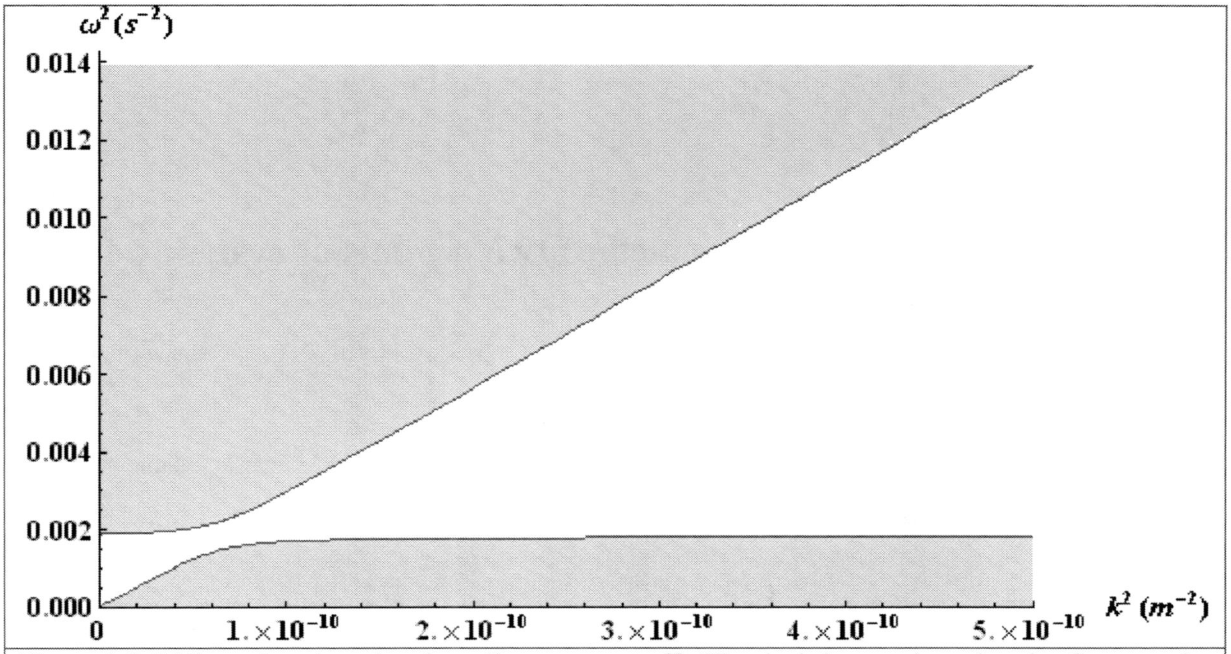

Figure 1. Solid lines show branch points (frecuency cut-off) .The shaded areas show domains for p-modes continuous spectra (top) and g-modes continuous spectra (bottom). Discrete modes do not exist.

3. Conclusion

Oscillations in solar corona is analyzed using initial value problem (IPV), so that a set of continuum p- modes and g-modes due to branch cuts in the complex plane, not treated explicitly in the literature, appears. On the other hand, discrete modes do not exist. The main consequence of a semi-infinitive atmosphere is a set of continuum p-modes and g-modes.

4. References

[1] Vanlommel E and Cadez V M 1998 *Solar Physics* **182** 263-81
[2] Schmitz F and Steffens S 1999 *Astron. Astrophys.* **334** 973-80
[3] Cadez V M and Jovanovic G 2008 *Publ. Astron. Obs. Belgrade* **84** 495-98
[4] Goedbloed H and Poeds S 2004 *Principles of Magnetohydrodynamics* (Cambridge: University Press)
[5] Ott E and Russell D A 1978 *Physical Review Letters* **41** 1048-51
[6] Russell D A and Ott E 1979 *Journal of Geophysical Research* **84** 6573-79

Journal of Physics: Conference Series **516** (2014) 012016 doi:10.1088/1742-6596/516/1/012016

Initial value problem approach of two regions of system solar: the continuous spectrum

Ángel De Andrea González
Departamento de Física. Escuela Politécnica Superior.
Universidad Carlos III de Madrid. Av. Universidad, 30. 28911 Leganés. Spain.

E-mail: aandrea@fis.uc3m.es

Abstract. The standard mathematical procedures in Helioseismology field are based on normal mode approach for various models of solar interior and atmosphere. We consider a two region model of a system solar interior and solar atmosphere. For simplicity, the two different regions are assumed quasi-isothermal, semi-infinite, without magnetic fields and separated by a boundary z=0. It was found useful to phrase of stability as initial value problem (IPV) in order to ensure the inclusion of certain continuum modes otherwise neglected. In addition to discrete mode (f-mode), sets of continuum modes due to a branch cuts in the complex plane, not treated explicitly in the literature, appears. It will be seen that an ambiguity of the usual normal mode method is avoided.

1. Introduction

Since Helioseismology actually depends on surface oscillations for detection, the extension with free boundary modes is an important step towards reality of a solar model. Several authors have purposed solutions of the initial value problem through the use of a Laplace Transform in time [2-4]. The principal finding of the initial value problem approach is that, in addition to the discrete eigenvalues linked to the normal modes, there exists a continuous spectrum of eigenvalues. Thus, the modal approach cannot provide a complete solution.

We use the Cartesian coordinates with the z-axis pointing towards the interior of the Sun and parallel to the constant gravity $\vec{g}$. As to the applicability of the Cartesian geometry in our analysis, it is valid for perturbations of a sufficiently small wave length when the effects of stellar sphericity are unimportant. This then places the centre of the Sun into $z/H \to \infty$. Our model assumes an isothermal atmosphere at $z<0$ above the solar surface and the isothermal solar interior (special case) with at $z>0$. This model would permit the investigation of a surface mode (f-mode), which is a free boundary oscillation of the solar surface involving both the interior and the atmosphere of the Sun The medium is considered a perfect gas. In this paper we exclude macroscopic motions and magnetic fields.

Journal of Physics: Conference Series **516** (2014) 012016 doi:10.1088/1742-6596/516/1/012016

2. Linear stability analysis and the initial problem value

We start from the standard set of hydrodynamic equations describing the dynamics processes in a fully ionized hydrogen plasma in presence of gravity with constant acceleration [1-2]:

$$\frac{\partial \rho}{\partial t} + \vec{\nabla}\cdot(\rho\vec{v}) = 0 \tag{1}$$

$$\rho\frac{\partial \vec{v}}{\partial t} + \rho\ \vec{v}\cdot\vec{\nabla}\cdot(\rho\vec{v}) = -\vec{\nabla}p - \rho\vec{g} \tag{2}$$

$$\frac{\partial \rho}{\partial t} + \vec{v}\cdot\vec{\nabla}p = \gamma\frac{p}{\rho}\left(\frac{\partial \rho}{\partial t} + \vec{v}\cdot\vec{\nabla}\rho\right) \tag{3}$$

Considering the displacement vector $\vec{\xi}$, where $\vec{v}_1 = \frac{\partial \vec{\xi}}{\partial t}$ linearization leads to:

$$\rho\frac{\partial^2\vec{\xi}}{\partial t^2} - \vec{\nabla}\left(\gamma\ p\vec{\nabla}\cdot\vec{\xi}\right) + \rho\ \vec{\nabla}\left(\vec{g}\cdot\vec{\xi}\right) - \rho\ \vec{g}\vec{\nabla}\cdot\vec{\xi} = 0 \tag{4}$$

2.1. *Results and discussion*

We define the Laplace transform $\xi_z(z,\omega)$ in the complex - ω plane:

$$\xi_z(z,\omega) = \frac{1}{2\pi}\int_{-i\infty+\sigma}^{+i\infty+\sigma}\xi_z(z,\omega)e^{-i\omega t}d\omega \tag{5}$$

Then, we consider the initial value problem and introduce the Laplace transform. So, we obtain the following ODE:

$$\frac{d^2\xi_z}{dz^2} + \lambda\frac{d^2\xi_z}{dz^2} + \frac{\omega^4 - k^2v_s^2\omega^2 + k^2v_s^2N^2}{v_s^2\omega^2}\xi_z = f(z,\omega,k,F_1,F_2) \tag{6}$$

Being $N^2 = \frac{(\gamma-1)g^2}{v_s^2}$ the Brunt-Väisäläa frequency; $\lambda = \frac{1}{\tau H}$ for (z<0) and $\lambda = \frac{1}{H}$ (z>0). On the other hand, F_1 and F_2 are initial conditions:

$$F_1 = \xi_z(z,t)\big|_{t=0^+}\ ;\ F_2 = \left.\frac{\partial\xi_z(z,t)}{\partial z}\right|_{t=0^+} \tag{7}$$

We can find the Green´s function for the former ODE subject to conditions at $|z| \to \infty$. The continuity of $G(z,\omega,z_o)$ at $z=0$ and $z=-z_o$, and the jump conditions on $\frac{dG(z,\omega,z_o)}{dz}$ at $z=0$ and $z=-z_o$. So we have:

$$G(z,\omega,z_o)=\begin{cases} Be^{\frac{z(-1+\alpha)}{2H\tau}} & z<-z_o \\ Ce^{\frac{z(-1+\alpha)}{2H\tau}}+De^{\frac{z(-1-\alpha)}{2H\tau}} & -z_o<z<0 \\ Ae^{\frac{z(-1-\beta)}{2H}} & z>0 \end{cases} \tag{8}$$

On the other hand:

$$\alpha=\sqrt{1-\frac{4H^2\tau\left[g^2k^2(\gamma-1)+\tau\,\omega^2\left(\omega^2-v_{sc}^2k^2\right)\right]}{v_{sc}^2\omega^2}};\quad \beta=\sqrt{1-\frac{4H^2\left[g^2k^2(\gamma-1)+\omega^2\left(\omega^2-v_{sc}^2k^2\right)\right]}{v_{sc}^2\omega^2}} \tag{9}$$

So, the solution is:

$$\xi_z(z,\omega)=\int_0^{\infty}G(z,\omega,z_o)\,f(z,\omega,k,F_1,F_2)\,dz_o \tag{10}$$

The Green´s function for $z>0$:

$$G(z,\omega,z_o)=\frac{-2H\tau\,\omega^2e^{\frac{1+\beta}{2H}z+\frac{\tau(1+\beta)z-(z+z_o)(1+\alpha)}{2H\tau}}}{2gHk^2(\tau-1)-(\alpha+\beta)\omega^2}e^{\frac{1+\beta}{2H}z} \tag{11}$$

In accord with the boundary conditions $G(\pm\infty,s,z_o)=0$, the square root are defined so that Re(α)>0 y Re(β)>0 for Re (ω)$>\sigma$. It is evident that Green´s function has a pole (f-mode).

Expanding Green´s function about α=0, and β=0, this function does not turn out to be an even function of α and β .So, the Green´s function has branch points at $\omega^2=\gamma_b^2$ (associated with continuous spectra).

It is of interest to follow the migration of the pole as k is varied. For $k>k_{cr}$, $\gamma_p>\gamma_{b1}$. If k is diminished, then the γ_p pole moves toward γ_{b1}. For $k=k_{cr}$, then $\gamma_p=\gamma_{b1}$. For $k<k_{cr}$ the pole drops on to the lower sheet of β y the f-mode disappears. The disappearance of the γ_p pole (f- mode) is important. Thus, there will be some critical wave number k_{cr} such for $k>k_{cr}$ ($k<k_{cr}$) f-mode exist (do not exist).

The following figure 1 shows dispersion curve. The temperature for the solar interior is 10^4 K, and for the corona 10^6 K, being the discontinuous drop $\tau = 100$. Gravitational acceleration is assumed constant within the domains of studied harmonic perturbations and we take g= 274 m/s^2 as it is on the solar surface.

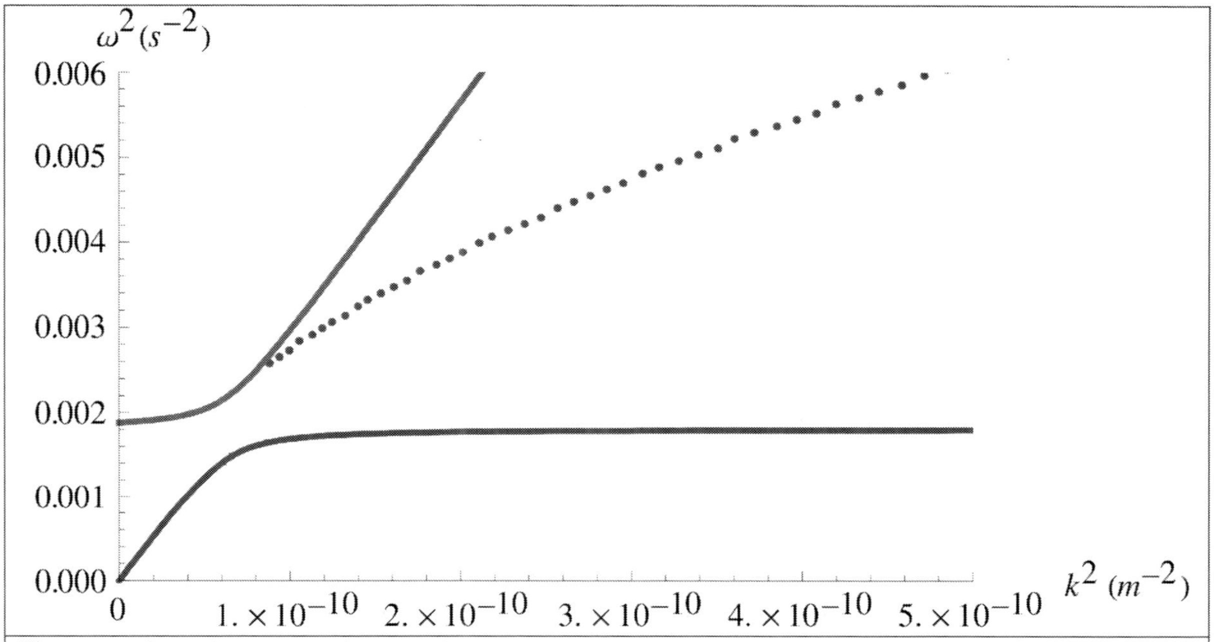

Figure 1. Dispersion curve (f-mode) for $\tau = 100$. Dots correspond to our numerical results. Solid lines corresponds to branch points (frecuency cut-off) . For $k^2 < k^2_{cr}$ f-mode does not exist.

Here, the critical wave number is $k_{cr} \approx 9 \cdot 10^{-6} m^{-1}$.

3. Conclusion

Surface mode (f-mode) is analysed in a two regions solar model using initial value problem (IPV). In the special case, where the solar interior is isothermal, there will be some critical wave number k_{cr} such that for $k > k_{cr}$ ($k < k_{cr}$) f-mode exist (do not exist). On the other hand, a set of continuum p-modes and g-modes due to branch cuts in the complex plane, not treated explicitly in the literature, explain the disappearance of the f-mode.

4. References

[1] Vanlommel E and Cadez V M 1998 *Solar Physics* **182**, 263-81
[2] Schmitz F and Steffens S 1999 *Astron. Astrophys* **334** 973-80
[3] Ott E, Russell D A 1978 *Physical Review Letters* **41** 1048-51
[4] Russell D A and Ott E 1979 *Journal of Geophysical Research* **84** 6573-79

Journal of Physics: Conference Series **516** (2014) 012017 doi:10.1088/1742-6596/516/1/012017

Differential cross section in PIC-MCC simulation of plasma-solid interaction

T Ibehej and R Hrach

Faculty of Mathematics and Physics, Charles University, V Holešovičkách 2, Prague 8, Czech Republic

E-mail: tomas.ibehej@gmail.com

Abstract. This paper focuses on the particle modelling of the interaction of low-temperature plasma with immersed cylindrical Langmuir probe. Particle modelling gives comprehensive information about the system, which is being analyzed, but the simulations are highly time-consuming. Especially in the high-pressure plasma the interactions between charged and neutral particles play an important role and they are often simulated by the Monte Carlo method. For this paper we prepared a simulation, which contains these interactions in as accurate way as possible, including the angular dependence of the electron-neutral elastic collision. We showed the effect of neglecting this angular dependence as well as the energy dependences of the cross sections of all included interactions.

1. Introduction

To understand plasma-solid interaction is essential in modern plasma-assisted technologies as well as for the interpretation of plasma probe diagnostics. The theoretical analyses of processes taking place in low-temperature plasmas are complicated especially due to the collisions. Hence, computer simulations present a good possibility to describe the examined systems.

The particle simulations of plasma provide results in both macroscopic and microscopic scales. Their time requirements are, however, very high and many techniques are being developed to speed up the calculations. The simulations based on Particle in Cell technique with Monte Carlo collisions (PIC-MCC) are widely used. So far, the PIC-MCC simulations have included collisions with either constant or energy-dependent cross sections. Recently, the Monte Carlo collision model has been improved in accuracy and speed [1] and the effect of further improvement of this model in sense of adding the angular dependencies of the cross sections is presented in this contribution.

Three simulations of interactions between the charged particles and the neutral ones with different levels of accuracy are compared and presented as a part of the PIC-MCC simulation. The simulated system is electropositive argon plasma in DC glow discharge with immersed Langmuir probe.

2. Description of the simulation

The simulation is based on the combination of molecular dynamics (MD) and Monte Carlo (MC) techniques. The trajectories of the charged particles (electrons and argon ions) are determined by the numerical solution of the Newton's second law of motion, which includes the electrostatic interactions between the charged particles and with the probe. The interactions with neutral particles are included as random events using the Monte Carlo technique.

Journal of Physics: Conference Series **516** (2014) 012017 doi:10.1088/1742-6596/516/1/012017

2.1. The molecular dynamics simulation

The two-dimensional working area is square-shaped with the side of 2 cm. In the centre of the area there is a cylindrical probe with positive bias and in the undisturbed plasma outside the working area we assumed Maxwell-Boltzmann distributions of argon ions and electrons.

For the integration of Newton's equations of motion we used the velocity Verlet algorithm [2]. In order to speed up the relaxation time we implemented different time steps for electrons 1×10^{-12} s and for argon ions 1×10^{-9} s. The force acting on the charged particles is obtained in each time step by solving Poisson's equation on the grid of 200×200 nodes. The algorithm we used is usually called Particle in Cell - Nearest Grid Point (PIC-NGP) algorithm and it is described in details in [3]. For numerical solving of the Poisson's equation the C library UMFPACK [4] was used.

The main input parameters of the undisturbed argon plasma and infinite cylindrical Langmuir probe are shown in table 1.

Table 1. The parameters of the MD part of the simulation.

Parameter	Value
probe radius	100 μm
probe bias	+10 V
plasma number density	3.22×10^{22} m^{-3}
electron number density	1×10^{15} m^{-3}
electron mean energy	3 eV
argon ion mean energy	39 meV

2.2. The Monte Carlo simulation

The Monte Carlo part of the simulation is responsible for the interactions between the charged particles and the neutral argon atoms. For electrons we chose elastic collision, excitation to the first excited level (11.55 eV) and ionization (15.755 eV). For argon ions it is elastic collision and charge transfer. The cross sections of the electron interactions were obtained from [5] and for the argon ions from [6]. The angular dependency of the electron-argon elastic collision was taken from [7].

The most accurate version of simulation contains all dependencies that have been mentioned including the angular dependency and in the resulting graphs it is called "angle". For the generation of electron random free path ξ we could use the simplified formula

$$\xi = -\lambda \ln \gamma \qquad (1)$$

where λ is the mean free path of electrons and γ is uniformly distributed random number from the interval (0;1). The equation (1) can be used only if the velocity of scattering centers can be neglected and the mean free path is independent of energy. The second requirement we fulfilled using the "null collision" method [8] by adding a virtual collision. During the simulation, if the elastic collision was chosen, we randomly chose the scattering angle with the probability distribution, which corresponds to the angular cross section dependency for particular electron energy.

For the ion interactions the velocity of the neutral argon atoms cannot be neglected. Therefore, we used the "advanced null collision method" [9]. The null collision cross section is chosen in such a way that the expression $v\sigma(v)$ is constant for all relative velocities v that might occur in the system. The probability that during the time step Δt an electron undergoes collision is then given by the formula

$$P = 1 - \exp(-nv\sigma\Delta t). \qquad (2)$$

Journal of Physics: Conference Series **516** (2014) 012017 doi:10.1088/1742-6596/516/1/012017

The simulation "energy" shows the effect of neglecting angular dependency of the electron-argon elastic cross section and in the simulation "constant" the cross sections of all interactions are constant. Using numerical integration we computed mean values of the cross sections of each interaction. In this case for electrons the equation (1) could be used without applying the null collision method.

3. Results and discussion

In order to show the effect of the described simplifications we chose several output parameters. Through these parameters we compared the simulations "angle", "energy" and "constant".

3.1. Electrostatic potential and electron number density

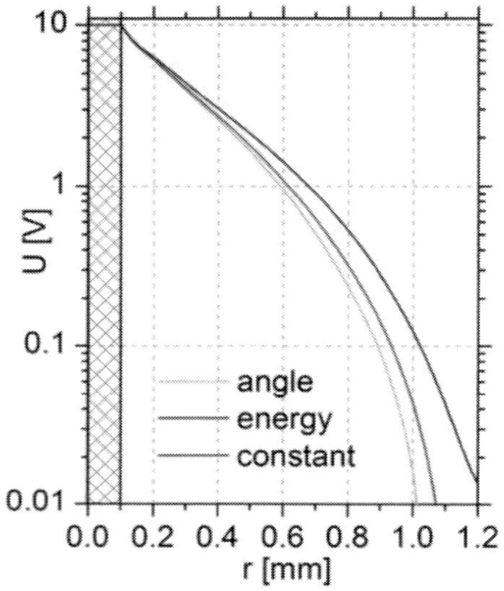

Figure 1. Radial distribution of potential near the probe. The shaded area shows the location of the probe.

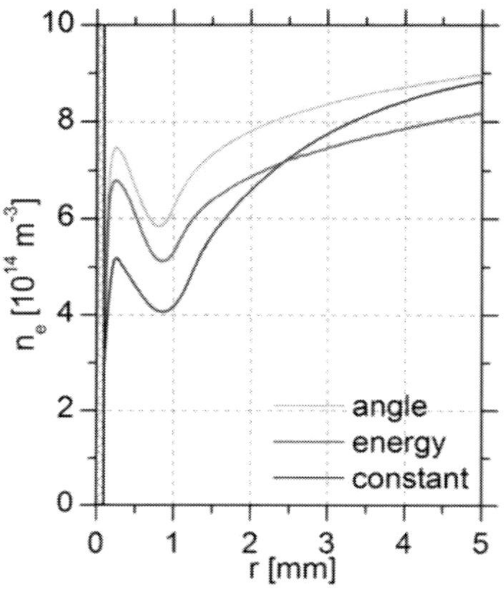

Figure 2. Radial distribution of electron number density. The shaded area shows the location of the probe.

The figures 1 and 2 show the radial distributions of the electrostatic potential and electron number density near the probe. The results are similar for simulations "energy" and "angle". The simulation "constant" leads to a broader sheath region. The distributions of the electron number density have a similar tendency in the simulations "angle" and "energy". The simulation "constant" leads again to biased results – the maximum near the probe is lower and it is followed by a faster radial increase. The local maximum near the probe is the result of cylindrical geometry and it was already presented as a result of other particle and fluid simulations, e.g. [10].

3.2. Electron angular distribution function

The figures 3 and 4 compare the angular distributions of electrons. The angle 0° represents the direction towards the probe. We chose two different locations for comparison – the first one (figure 3) is located directly at the probe, thus in the sheath region and the second one (figure 4) is located 2 mm from the centre of the working area in the pre-sheath region. The results of simulation "angle" are in these comparisons significantly different from the results of the other simulations, especially in the direction near 0°. As we can see on the graphs, there are fewer electrons heading towards this probe and more ones heading in the perpendicular directions. In the direction from the probe, which is not as important for the applications, the simulation "energy" gives the most biased results.

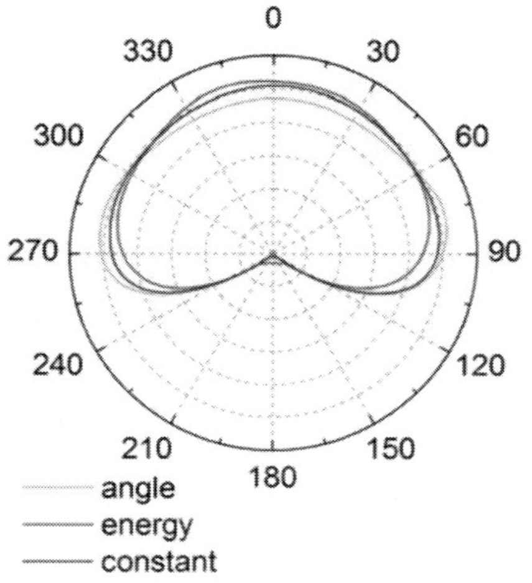

Figure 3. The angular distribution of electrons from the annulus with radius 0.1 – 0.13 mm.

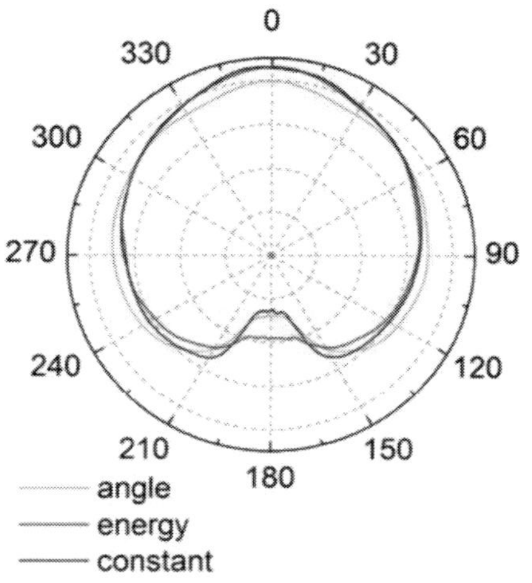

Figure 4. Same as in figure 3. The mean radius of the annulus is 2 mm and width 0.03 mm.

4. Conclusion

The results of the three simulations with different inclusion of collision processes were compared. We compared radial distributions of electrostatic potential, electron number densities and distributions of the velocity direction. As expected, the angular distribution was quite strongly influenced by the use of the angular dependence of electron-argon elastic collision.

In the other comparisons, the angular dependency does not play such an important role and the most biased results are produced by the simulation, which assumed the constant cross sections. On the example of the argon plasma we have shown that using differential cross sections in the particle simulations of plasma-solid interaction is essential if the angle-dependent results are needed, otherwise the previously used Monte Carlo simulation, as used e.g. in [1], are sufficient. The simplification of constant cross sections is in all cases unsatisfactory.

Acknowledgements

The work is a part of the research plan MSM0021620834 financed by the Ministry of Education of the Czech Republic. The authors acknowledge the support of the project SVV-2010-261302, of the Grant Agency of Czech Republic – grant No GACR P205/10/0979 and of the Grant Agency of Charles University – grant No GAUK-46310/2010.

References

[1] Hrach R, Roučka Š, Hrachová V and Schmiedt L 2010 *Vacuum* **84** 94
[2] Swope W C, Andersen H C, Berens P H and Wilson K R 1982 *J. Chem. Phys.* **76** 637
[3] Birdsall C K and Langdon A B 1991 *Plasma Physics via Computer Simulation* (Bristol: IOP Publishing Ltd) pp 16-22
[4] Davis T A 2004 *ACM T Math Software* **30** 196
[5] Phelps A V (unpublished) ftp://jila.colorado.edu/collision_data/
[6] Horváth M 2002 *Ph.D. Thesis* (Prague: MFF UK)
[7] Panajotovic R, Filipovic D, Marinkovic B, Pejcev V, Kurepa M and Vuskovic L 1997 *J. Phys.* B **30** 5877
[8] Skullerud H R 1968 *J. Phys.* D **1** 1567
[9] Roučka Š 2008 *Thesis* (Prague: MFF UK)
[10] Havlíčková E, Bartoš P and Hrach R 2007 *J. Phys. Conf. Ser.* **63** 012019

4th International Workshop & Summer School on Plasma Physics 2010 IOP Publishing
Journal of Physics: Conference Series **516** (2014) 012018 doi:10.1088/1742-6596/516/1/012018

Comparison of several algorithms of the electric force calculation in particle plasma models

J Lachnitt and R Hrach

Department of Surface and Plasma Science, Faculty of Mathematics and Physics, Charles University, V Holešovičkách 2, 180 00 Prague 8, Czech Republic

E-mail: pepalogik@seznam.cz

Abstract. This work is devoted to plasma modelling using the technique of molecular dynamics. The crucial problem of most such models is the efficient calculation of electric force. This is usually solved by using the particle-in-cell (PIC) algorithm. However, PIC is an approximative algorithm as it underestimates the short-range interactions of charged particles. We propose a hybrid algorithm which adds these interactions to PIC. Then we include this algorithm in a set of algorithms which we test against each other in a two-dimensional collisionless magnetized plasma model. Besides our hybrid algorithm, this set includes two variants of pure PIC and the direct application of Coulomb's law. We compare particle forces, particle trajectories, total energy conservation and the speed of the algorithms. We find out that the hybrid algorithm can be a good replacement of direct Coulomb's law application (quite accurate and much faster). It is however probably unnecessary to use it in practical 2D models.

1. Introduction

Computer modelling is a very important approach in plasma physics. This work is focussed on the molecular dynamics, i.e. deterministic particle modelling, which is used for the detailed study of plasma behaviour in cases like plasma-solid interaction. In molecular dynamics, the motion of individual particles is modelled using Newton's second law. The crucial point is the calculation of forces. In self-consistent models, the forces are updated in each step according to the actual particle positions (and velocities if we include magnetic field). This usually takes lots of computation time and many optimizations have to be done to make the model run reasonably fast (say, days instead of years of computation time).

We focus on high-temperature plasma so we have to include the electric and magnetic forces in our model. In low-temperature plasma, we would have to add the collisions with neutral particles.

Apparently the most widely used method for the calculation of electric force is the particle-in-cell (PIC) technique. It is relatively fast, relatively simple and it allows easy application of most types of boundary conditions. The main drawback of particle-in-cell is that it greatly underestimates the short-range interactions between charged particles as it sums their charges in the cells of the grid. The question is how much this inaccuracy affects practical plasma models.

According to [1], it seems that the short-range interactions (alias Coulomb collisions) are quite important in the case of magnetized plasma, as shown in figure 1. However, it has been found that the effect is probably exaggerated in the right picture and the truth lies somewhere in the middle. This is one of the reasons why we study this problem.

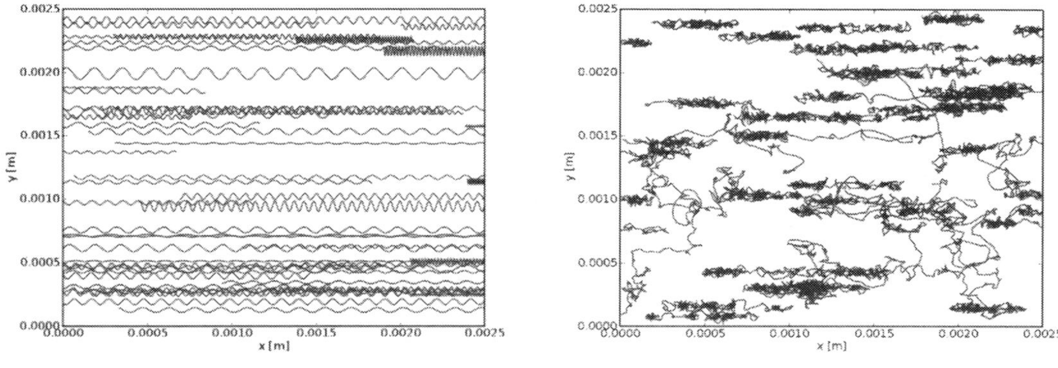

Figure 1. Electron trajectories in a two-dimensional magnetized plasma model. The electric force was calculated using PIC (left) and a tree algorithm (right). Reprinted from [1].

2. Our model

Our model is two-dimensional and self-consistent and represents a region in a collisionless magnetized plasma. For the comparison of different electric force calculation algorithms, it is necessary to use the same boundary conditions for all the algorithms. We found periodic boundary conditions to be the only boundary conditions that satisfy this requirement. Hence our model is fully periodic.

We include the magnetic field in the simplest way – as a constant field along the x axis. We use the algorithm referred to as half acceleration, rotation, half acceleration [2].

3. The methods

Here we describe the electric force calculation methods which we have used.

3.1. Particle-in-cell

This technique consists in solving the discretized field equations

$$\Delta\varphi = -\frac{\rho}{\varepsilon_0}, \tag{1}$$

$$\boldsymbol{E} = -\nabla\varphi \tag{2}$$

on a grid covering the modelled region. There exist several variants of PIC which differ in the way how the charges of the particles are counted (or weighted) to the cells and, after solving (1) and (2), how the electric field (evaluated at the grid points) is weighted back to the particles (both weightings should be the same) [2]. We use these two variants: PIC with nearest-grid-point (zero order) weighting, which we will refer to as PIC-NGP, and PIC with linear (first order) weighting, so called cloud-in-cell, which we will refer to as PIC-CIC.

3.2. Coulomb's law

This is the most direct and theoretically the most accurate method of calculating the electric force. In our model, Coulomb's law does not have the well-known form. In a two-dimensional model, the particles must be regarded as infinite charged rods, so Coulomb's law takes the form

$$\boldsymbol{E} = \frac{\lambda\boldsymbol{r}}{4\varepsilon_0 r^2} \tag{3}$$

Journal of Physics: Conference Series **516** (2014) 012018 doi:10.1088/1742-6596/516/1/012018

where λ is the linear charge density. Furthermore, our model is periodic so when we calculate the force at particle 1 caused by particle 2, we have to count in all the images of particle 2 in the whole 2D space. So we use a double infinite sum of (3) in our program.

3.3. Hybrid algorithm

The aim of our hybrid algorithm is to be more accurate than PIC and faster than Coulomb's law. Here we will give a brief description of the algorithm.

First, PIC-NGP is performed so we have the electric field calculated at the cell centres. Then, for each particle, we subtract the Coulomb forces (3) of the eight neighbouring cells (see figure 2), using the coordinates of the cell centres. Finally, we add Coulomb forces of all the particles inside the grey region in figure 2 (except for the target particle, of course).

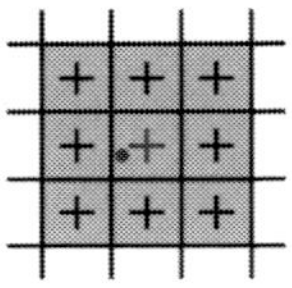

Figure 2. Part of the grid for the explanation of the hybrid algorithm. The blue spot represents the target particle (i.e. the particle for which we calculate the electric field) and the crosses represent cell centres.

4. Non-physical heating

Non-physical heating is a common problem in molecular dynamics. It is a result of inaccurate time integration due to the finite time step. This problem arises especially in the Coulomb and hybrid methods. In Coulomb collisions, the particles have large acceleration and large trajectory curvature. In the model, they usually get closer together than they would in reality, so they gain additional energy, producing non-physical heating.

We have devised and tested the following method of reducing the non-physical heating. At the beginning of the program, we estimate the maximal particle velocity, according to Maxwell--Boltzmann distribution and the number of particles. Then, during the model run, we check the velocity of each particle each time it is updated. If the velocity is larger than the precomputed limit, we change it to this limit, preserving the original direction.

5. Results

Table 1 summarizes the input parameters we have used.

Table 1. Numerical parameters.

Parameter	Value
Gas	Deuterium
Plasma density	1×10^{18} m^{-3}
Plasma temperature	1×10^{6} K
Magnetic field	1 T
Number of particles	Various
Modelled region size	2 cm × 2 cm
Grid	100 × 100 cells
Time step	1×10^{-12} s

5.1. Force comparison

We have made histograms which show the correlation between the electric field calculated by the given method and the electric field calculated by the periodic Coulomb's law (taken as the reference). These histograms were created using 40 000 particles and they are shown in figure 3.

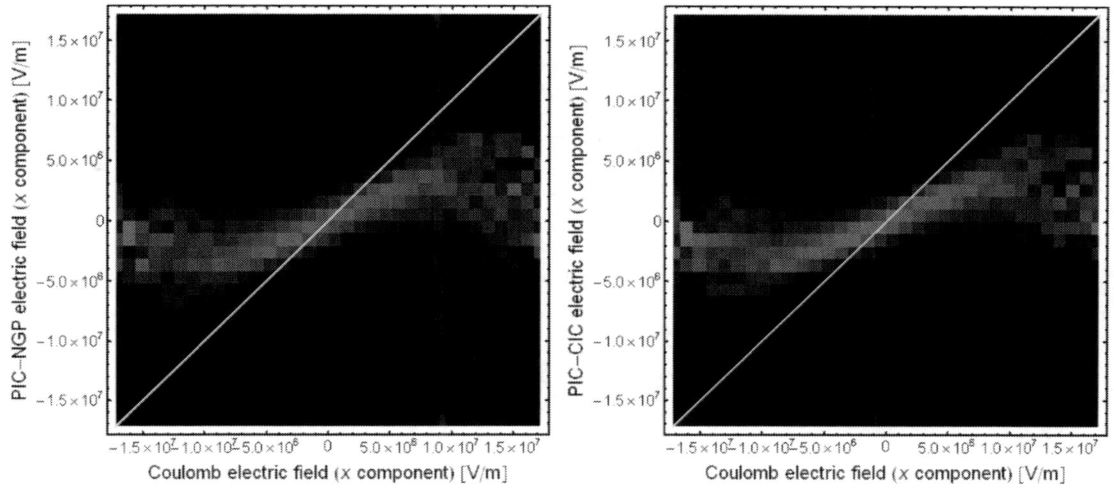

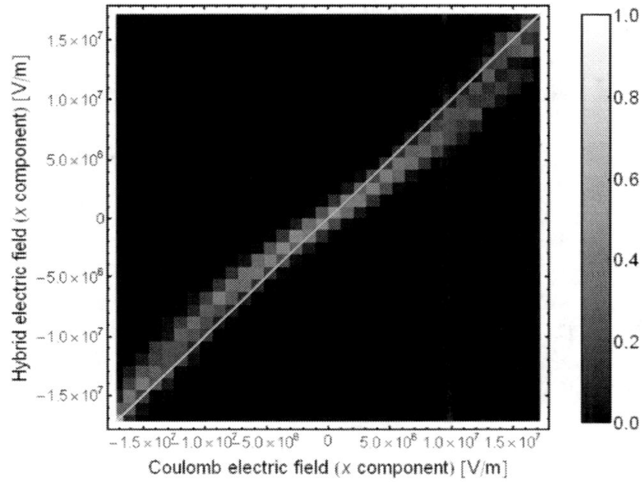

Figure 3. Force correlation histograms. The particles were divided into columns according to the electric field they experience (calculated exactly by the periodic Coulomb algorithm). In each column, there is a histogram of the electric field calculated by the given algorithm (see vertical labels) at the particles. The colour scale is common to all the pictures. The green line represents the ideal profile.

5.2. Particle trajectories

We originally performed the computations with only 2 000 particles (1 000 ions and 1 000 electrons) because of the very slow periodic Coulomb calculations; but there was a huge non-physical heating and other problems. As our hybrid algorithm seems to be a good replacement of the direct Coulomb calculation (see figure 3), we decided to exclude the Coulomb algorithm from testing and repeat the computations with 200 000 particles. The resulting trajectories are shown in figure 4.

5.3. Energy conservation

Because there are no physical energy sources or leaks in our model, the total energy should remain constant. Figure 5 shows the total energy evolution for the tested algorithms. These results come from the same computations as the trajectories in figure 4 (i.e. with 200 000 particles).

5.4. Speed

Our computations have been performed on an ordinary PC (Athlon64 @ 1.8 GHz). The computation times are summarized in table 2.

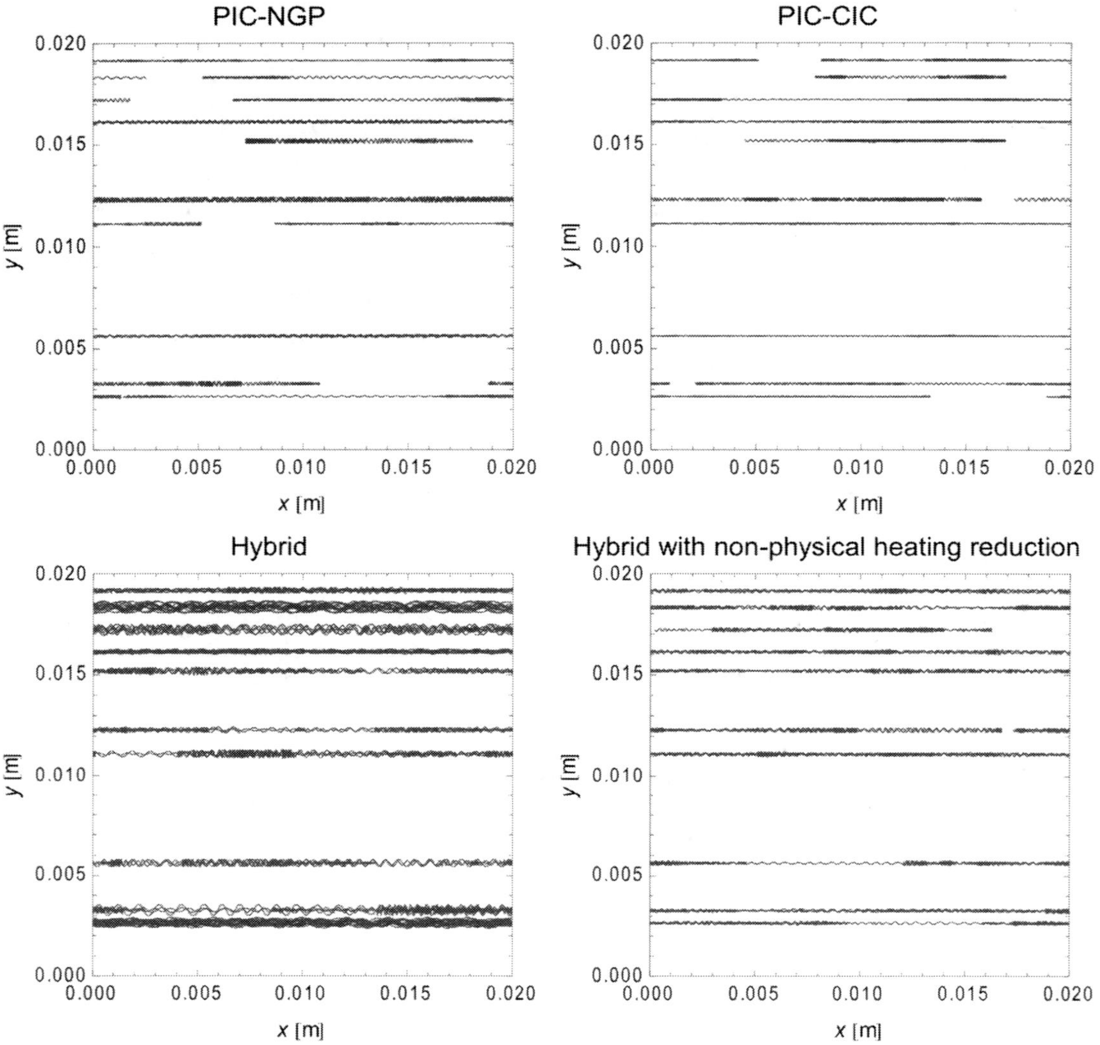

Figure 4. Trajectories of 10 almost randomly chosen electrons during the time of 1×10^{-8} s. All the pictures show the trajectories of the same electrons for the same initial conditions.

6. Discussion and conclusions

The proposed hybrid algorithm really does what it is supposed to do – it calculates the forces very close to the results of exact Coulomb's law calculations while it is much faster. On the other side, it suffers from the non-physical heating markedly just like the direct Coulomb's law calculations.

The proposed method of reducing the non-physical heating does really reduce the non-physical heating but it obviously cannot do wonders. It also distorts the energy distribution function (however, the non-physical heating distorts it in another way).

The most important result of our contribution can be found in figure 4 – particle trajectories. For given parameters (typical edge plasma), we have found no transitions between magnetic field lines even if we use Coulomb's law for close particle encounters (the hybrid algorithm). Moreover, we have found no such transitions even in the case of 2 000 particles and using the periodic Coulomb's law directly (picture not presented). We performed the computations also for the parameters of low--temperature plasma (though physically impossible without neutral particles) and the results were

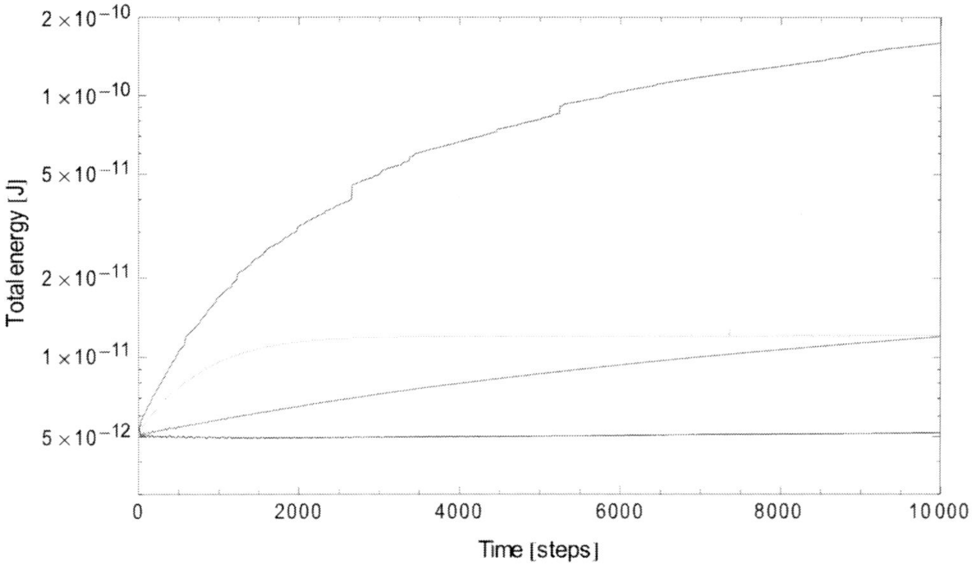

Figure 5. Total energy evolution in time for the tested algorithms: PIC-NGP, PIC-CIC, hybrid and hybrid with the non-physical heating reduction. Sampling rate: 10 steps. Please note the logarithmic scale for energy.

Table 2. Comparison of the speed of the algorithms.

	Computation time per step [ms]	
Algorithm	2 000 particles	200 000 particles
PIC-NGP	12	36
PIC-CIC	13	70
Periodic Coulomb [a]	5 825	Not tried [b]
Hybrid	16	5 749

[a] Periodic Coulomb is considerably (at least 13, maybe 100 times) slower than ordinary Coulomb.
[b] Our estimate is 58 million ms, i.e. 16 hours.

almost the same. So it is probably useless to use such complicated algorithms in practical plasma models, at least in the case of two-dimensional models.

Acknowledgments
This work is a part of the research plan MSM0021620834 that is financed by the Ministry of Education of the Czech Republic. The authors acknowledge the support of the Grant Agency of Charles University in Prague, grant No. GAUK-51808/2008.

References
[1] Maršálek O 2008 *Advanced Techniques of Computer Modelling in Plasma Physics*, Thesis (Prague: Charles University)
[2] Birdsall C K and Langdon A B 1991 *Plasma physics via computer simulation* (Bristol: IOP Publishing Ltd)

Monte Carlo simulation of a sputtering hollow-cathode discharge for laser applications

S Karatodorov[1,2], D Mihailova[3], J van Dijk[3], J van der Mullen[3], and M Grozeva[1]

[1] Institute of Solid State Physics, Bulgarian Academy of Sciences, Bulgaria
[2] Faculty of Physics, Sofia University, Bulgaria
[3] Department of Applied Physics, Eindhoven University of Technology, The Netherlands

E-mail: s.karatodorov@gmail.com

Abstract. We report on a kinetic model that computes the electron behaviour in a hollow cathode discharge. It is a part of the PLASIMO toolkit and is based on a Monte-Carlo technique. The model is tested by varying the input parameters and by comparing the output with the output obtained by the freeware Boltzmann equation solver BOLSIG+. The results show that the Monte-Carlo model gives reliable information about the behavior of the electrons in the discharge. The Monte-Carlo module is applied to the case of a hollow cathode discharge for laser applications. Analysis of the output data and its adequateness is done. Future developments of the model are discussed.

1. Introduction

In a hollow cathode discharge (HCD) device the closed cathode shape results in more efficient use of the charged particles: the higher ion current to the cathode give rise to higher second electron emission and higher cathode sputtering rate. These characteristics make the HCD appropriate for a variety of applications: atomic spectrometry, UV light generation, plasma processing, ion sources, laser technologies and others. One of the important applications of HCDs is as excitation medium of ion metal vapour lasers using cathode sputtering for metal atom production.

The research of these lasers has for long time engaged modeling techniques side by side with experiments. Two main modeling approaches exist: fluid and kinetic. The fluid approach is often very accurate when sufficient collisions take place in the plasma [1], i.e. when the particles mean free path is smaller than the plasma size.

A number of models simulating HCD have been published so far (see [1] and references therein). One of them developed in recent years is PLASIMO's simulation platform MD2D model [2] - a time dependent two-dimensional multi-fluid model. The results of the MD2D model are found to agree well with experimental and analytical observations on HCD [3]. However not all discharge phenomena can be simulated with a pure fluid model – e.g. "fast" electrons kinetics, ion inertia in cathode fall region, sputtering yield dependence on ion energy. In order to investigate these issues and to test their influence on the simulation results a kinetic model is required. Such model provides description of the discharge by closely following a large number of simulated particles, which is possible at the expense of a long simulation time. The kinetic model can be used in two ways. First, it can be employed as a stand-alone model where the simulated particles move and collide in plasma with defined properties. Such a PLASIMO Monte Carlo (MC) sub-module has been previously developed to simulate drift-diffusive plasma [4]. Second, the kinetic model can be used in hybrid simulations [5]. In this case, some of the characteristics of the studied system are described by a fluid model while others are delegated to the MC module.

As a first attempt in the field our aim is to develop a stand-alone MC PLASIMO sub-model for the electrons in a HCD for laser applications. By closely following the electrons in both cathode fall and bulk regions, the kinetic model will provide as output more accurate data for the electron energy distribution function and the mean electron energy. These values are supposed to be used for input by the fluid MD2D model. In such a way the kinetic model will enable the development of a more realistic model of the HCD.

2. Theoretical and experimental setup

The developed kinetic model is based on the MC statistical technique. This technique is usually applied for gas discharges, in which one or more of the particle species are far from thermodynamic equilibrium [4]. In HCD plasma the electrons are such species.

The model requires as input the discharge geometry, the profile of the electric field, a list of species and a set of reaction among the defined species. As electrons are the followed particles, the model treats only reactions in which they participate: the simulated electrons undergo excitation, ionization and elastic collisions with the buffer gas particles and ionization and elastic collisions with the sputtered metal atoms.

The environment buffer gas is a mixture of He and 5% Ar with pressure of 2.7 kPa [1]. The sputtered Cu metal atoms percentage is varied from 0÷2%.

The number of electrons in the simulations is chosen to be 10^9 and it is based on data for the electron density from the fluid model [3] and the cathode volume.

The kinetic model is engaged in the performing of two distinct experiments. The first aims to verify the kinetic model by comparing it with a known model. The second experiment is the actual simulation of the HCD.

2.1. Drift experiment - Infinite "vessel" with uniform electric field

The kinetic model is compared with the freeware Boltzmann equation solver BOLSIG+ [6]. There are two main reasons for this comparison. First, Boltzmann equation solvers are an alternative of the kinetic models in solving the Boltzmann equation (BE). Second, so far BOLSIG+ is used to deliver the electron energy distribution function (EEDF) for the fluid model, which uses the EEDF to obtain electron rate and transport coefficients.

The conditions for the comparison are: infinite "vessel" and uniform electric filed. The value for the electric field is varied between 0÷2000 Td.

2.2. HCD – Cylinder with non-uniform electric field

The simulated HCD is active medium of a copper vapour laser [3]. The experimental realization of this HCD consists of an anode-cathode-anode configuration of the type shown in figure 1(a). Due to the existing symmetry the model simulates just one fourth of the cathode – figure 1(b). The sizes of the cathode are the following – inner diameter 4 mm and length 30 mm.

The electric field in HCD plasma is not uniform [1,3]. The profile of the applied electric field is constructed out of numerical data from the fluid model. It is decomposed in two projections: axial and radial – figure 2, which are fitted with appropriate functions and used in that form in the MC simulation.

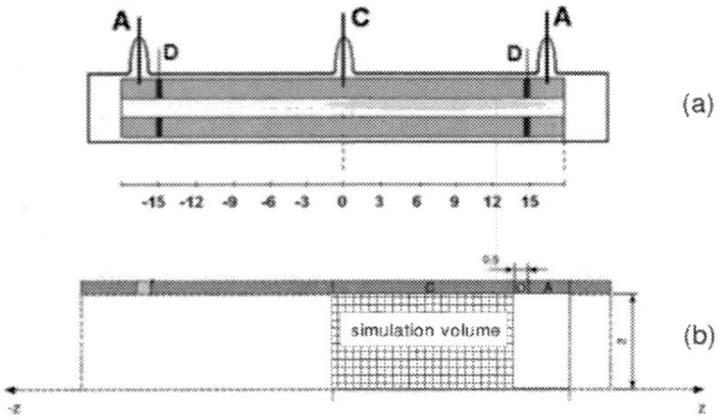

Figure 1. Experimental setup: (a) anode-cathode-anode configuration of HCD; (b) the simulated part of the cathode volume.

Figure 2. Applied profiles of the electric field in the simulation: (a) radial profile; (b) axial profile.

3. Results

3.1. Comparison of MC model and BOLSIG+

In order to verify the kinetic model we compare it with a known model as stated in section 2.1 under constant electric field. The electric field value is chosen according to the typical values for HCD. As known the HCD comprises of almost field-free negative glow (NG) with electric field in the order of 10 Td and cathode fall (CF) region with much higher field – 1500 Td. We present comparison results for the two typical cases in figure 3: NG with 15 Td and CF with 1700 Td. There is reasonable agreement between the results of the two models. It is better for small values of the electric field – figure 3(a). For large values of the electric field – figure 3(b), we probably go out of the region of validity of the two-term approximation, which is used by BOLSIG+ [6]. That is most likely the reason for the discrepancy between the BOLSIG+ and MC model in figure 3(b).

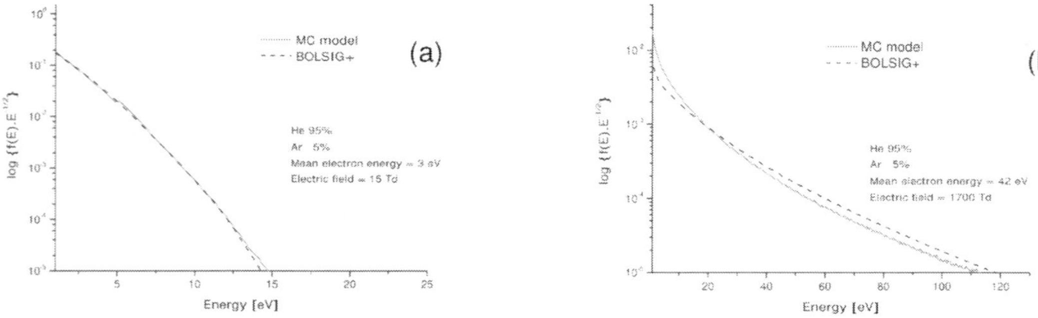

Figure 3. Comparison of MC model and BOLSIG+ EEDFs under constant electric field: (a)15 Td; (b) 1700 Td.

Journal of Physics: Conference Series **516** (2014) 012019 doi:10.1088/1742-6596/516/1/012019

3.2. Simulation of HCD cylinder tube

The simulations are carried out under the set of conditions described in section 2.2. The EEDF, obtained from the simulation under optimal lasing conditions, is shown in figure 4. From this EEDF the mean electron energy (MEE) is computed.

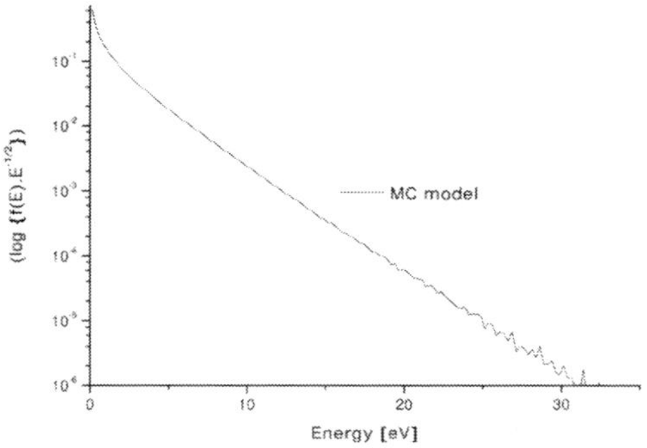

Figure 4. EEDF obtained from the kinetic model under optimal lasing conditions

The EEDF obtained from the kinetic model is a better source for input data for the fluid MD2D model. This is due to the fact that the MC model takes into account the strong non-uniformity of the electric field and the specific cathode geometry of the HCD for lasing. Moreover, the values that the MC model produces are supposed to be more accurate in the region of high electric fields and according high particle energies.

Conclusions

The development of a kinetic Monte-Carlo model for the calculation of electron behaviour in HCD for laser application is presented. The model is a part of the PLASIMO toolkit. The MC model is capable of obtaining the EEDF and from it the MEE for the HCD.

The model was compared to a Boltzmann solver – BOLSIG+. As both methods provide a solution of the BE, the comparison provided the opportunity of validating the kinetic model. There is good agreement for small values of the electric field, as for the case of high electric fields the differences are analyzed.

The future development of the model is mainly oriented towards the creation of a hybrid model for the HCD by joining it with the fluid MD2D model. Moreover, the kinetic model can be used independently for the analysis of the behaviour of every kind of charged species in the discharge – e.g. to obtain sputtering yield, second electron emission coefficients.

Acknowledgments

The work was done in the framework of the Erasmus Programme № 256/2009 and with financial support of NSF of Bulgaria DO 02-274/2008.

References

[1] Mihailova D, Grozeva M, Hagelaar G, van Dijk J, Brok W and van der Mullen J 2008 *J.Phys.D: Appl. Phys.* **41** 245202.

[2] J. van Dijk, K. Peerenboom, M. Jimenez, D. Mihailova, and J. van der Mullen 2009 *J. Phys. D: Appl. Phys.* **42** 194012

[3] Mihailova D, van Dijk J, Grozeva M, Hagelaar G J M and van der Mullen J J A M 2010 *J. Phys. D: Appl. Phys.* **43** 145203.

[4] Brok W J M 2005 Modelling of Transient Phenomena in Gas Discharges, PhD Thesis, Eindhoven University of Technology.

[5] Bogaerts A and Gijbels R 2002 *J. Appl. Phys.* **92** 6408.

[6] Hagelaar G and Pitchford L 2005 *Plasma Sources Sci.Technol.* **14** 722.

Journal of Physics: Conference Series **516** (2014) 012020 doi:10.1088/1742-6596/516/1/012020

Random transition between two temperatures profiles in magnetized plasma

N. Pometescu [1], **B. Weyssow** [2]

[1] Association EURATOM-MECTS, University of Craiova, Department of Physics, 13 A.I.Cuza Str., 200585 Craiova, Romania

[2] Assoc. EURATOM-Etat Belge, Université Libre de Bruxelles, Belgium and EFDA-CSU Garching, Germany

E-mail: npomet@yahoo.com, boris.weyssow@efda.org

Abstract. A set of Langevin equations are used to describe the transport of the guiding centers of charged particles perpendicularly to the main magnetic field and the stochastic transition between the two temperature states is described by a randomly interrupted noise. Using this model, the variation of the running diffusion coefficients in the radial direction is analyzed for two cases of temperature profiles with applications to fusion plasma.

1. Introduction

The Langevin equations was used before to study the temporal variation of the difusion coefficients in one temperature plasma and to analyze the regime of diffusion of particles in fluctating magnetic field, see e.g. [1]. Here we aim to study the spatial variation in perpendicular direction of the equilibrium magnetic field in random transition between two temperature state of the plasma using a randomly interrupted noise [2]. We give here two examples when the electron temperature presents transition between two temperature values: in pre-ELM-phases, see [3], and in pellet fuelling of Ohmic and Lower Hybrid - driven discharges in Tore Supra, see [4]. We can find many examples of random transitions between two given states both in hot plasmas and in cold plasmas where this model can be used.

2. Characteristics of the model

The caracteristics of the model, sumarized here, are given in detail in [5].

The transport of the guiding centers of charged particles is described by using the following Langevin equations:

$$\frac{d\boldsymbol{x}_{\perp}(t)}{dt} = \beta\mathbf{b}[\boldsymbol{x}_{\perp}, z]\, v_z(t) \tag{1}$$

$$\frac{dz(t)}{dt} = v_z(t) \tag{2}$$

$$\frac{dv_z(t)}{dt} = -\nu_z v_z(t) + \alpha_z(t) \tag{3}$$

Journal of Physics: Conference Series **516** (2014) 012020 doi:10.1088/1742-6596/516/1/012020

The stochastic process of jumps between two given temperatures at random times is describing by using a randomly interrupted noise [2]

$$\alpha_z(t) = \frac{1}{b-a}\left[b-\eta(t)\right]\alpha_{za}(t) + \frac{1}{a-b}\left[a-\eta(t)\right]\alpha_{zb}(t) \tag{4}$$

where a and b refer to the temperature T_a (low) and T_b (high), respectively. With the stationary solution for the probability transitions we obtain

$$\langle\eta(t)\rangle = 0; \quad \langle\eta(t)\,\eta(u)\rangle = a^2\left(\frac{1}{\lambda\tau_0}-1\right)\exp\left[-\frac{|t-u|}{\tau_0}\right] \tag{5}$$

The components $\alpha_{za}(t)$ and $\alpha_{zb}(t)$ verify the properties, e.g. [1], $\langle\alpha_{zi}(t)\rangle = 0$, $\langle\eta(t)\,\alpha_{zi}(t)\rangle = 0$ and

$$\langle\alpha_{zi}(t)\,\alpha_{zj}(u)\rangle = 2\delta_{ij}\, v_{Ti}^2\nu_i\delta(t-u) \tag{6}$$

where $i,j = a,b$ and v_{Ti} , ν_i are respective the thermic velocity and collision frequency at temperature T_i. Then $\alpha_z(t)$ is such that $\langle\alpha_z(t)\rangle = 0$ and

$$\langle\alpha_z(t)\,\alpha_z(u)\rangle = 2\left\{(1-\lambda\tau_0)\,\nu_{za}\, v_{T_a}^2 + \lambda\tau_0\nu_{zb}\, v_{T_b}^2\right\}\delta(t-u) \tag{7}$$

where $\lambda\tau_0$ is the weighting factor of collisions of kind b and $(1-\lambda\tau_0)$ for the collisions of kind a.

2.1. Model for the magnetic field

The magnetic field is stochastic

$$\overrightarrow{B} = B_0\left[\overrightarrow{e}_z + \beta b_x(z)\,\overrightarrow{e}_x + \beta b_y(z)\,\overrightarrow{e}_y\right] \tag{8}$$

Stochastic process for the components b_x, b_y is described by the equation of second order correlations:

$$\left\langle\widehat{b}_m(k)\,\widehat{b}_n(k')\right\rangle = \widehat{\mathcal{B}}_{\|}(k)\,\delta(k+k')\,\delta_{mn} \tag{9}$$

where the spectral density of the magnetic field fluctuations is

$$\widehat{\mathcal{B}}_{\|}(k) = \lambda_{\|}(2\pi)^{-1/2}\exp\left(-\frac{1}{2}\lambda_{\|}^2k^2\right) \tag{10}$$

2.2. Second order correlation of parallel velocity

With $f = T_b/T_a$ and $f > 1$ it result that $v_{Tb}^2 = f\ v_{Ta}^2$. The relation between collision frequencies in the hot plasma limit read as, e.g. [6], $\nu_{zb} = f^{\ -3/2}\ \nu_{za}$ in the limit approximation $n_a = n_b$. The stationary correlation (in the limit $t\to\infty$) is

$$\langle\ v_z(t)\ \ v_z(t+\tau)\rangle = \frac{1}{2}\ v_{Ta}^2\exp\left(-\nu_{za}|\tau|\right)C(\tau;\lambda\tau_0, f) \tag{11}$$

where

$$C(\tau;\lambda\tau_0,f) = \frac{1-\lambda\tau_0\left(1-f^{\ -1/2}\right)}{1-\lambda\tau_0\left(1-f^{\ -3/2}\right)}\exp\left[\lambda\tau_0\left(1-f^{\ -3/2}\right)\nu_{za}|\tau|\right] \tag{12}$$

2.3. Perpendicular transport

The perpendicular transport is given by the mean square displacement,

$$\left\langle (\delta x_m(\tau))^2 \right\rangle = \int_0^\tau d\tau_1 \int_0^\tau d\tau_2 \left\langle b_m[z(\tau_1)]\; v_z(\tau_1)\, b_m[z(\tau_2)]\; v_z(\tau_2)\right\rangle \tag{13}$$

We assume that the magnetic field fluctuations and the velocity fluctuations are statistically independent: $\langle b_m(z)\; v_z(t)\rangle_{b,\, v_z} = 0$. The parallel collisional diffusion coefficient $\chi_\parallel$ and the anomalous perpendicular running diffusion coefficient $D_\infty(t)$ in our case read as, see [5],

$$\chi_\parallel = \frac{v_{Ta}^2}{2\nu_{za}} \frac{1-\lambda\tau_0\left(1-f^{-1/2}\right)}{\left[1-\lambda\tau_0\left(1-f^{-3/2}\right)\right]^2}, \qquad D_\infty(t) = \chi_\parallel \beta^2 \frac{\phi(\nu_z t)}{\left[1+2\overline{\chi}_\parallel \mu(\nu_z t)\right]^{\frac{1}{2}}} \tag{14}$$

where $\mu(\theta) = \theta - \phi(\theta)$, $\phi(\theta) = 1 - \exp(-|\theta|)$, $\theta = \nu_z\tau$.

3. Discussions and conclusions

We apply the model in the following two cases:

Case 1: We consider the electron temperature profiles (see figure 1):

$$T_a = T_0\ \exp[-3 * \tanh[7(x-0.8)]] \tag{15}$$
$$T_b = T_0\ \exp[-3 * \tanh[7.8(x-0.82)]] \tag{16}$$

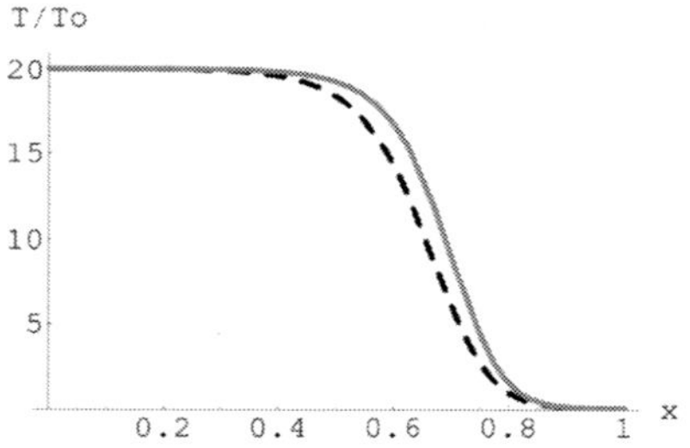

Figure 1. Temperature profiles: T_a given in (15) - (dashed line) and T_b given in (16) - (full line).

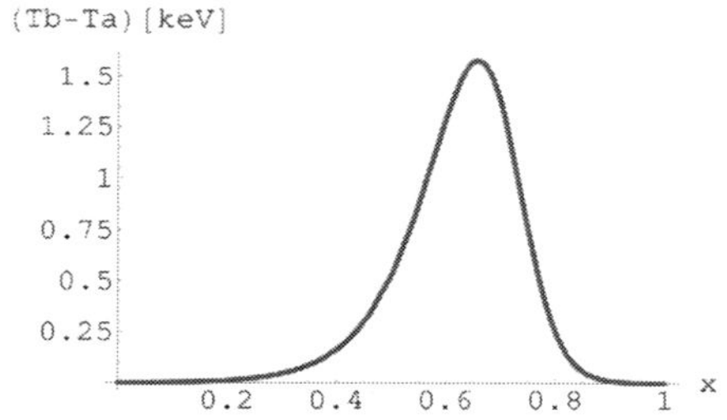

Figure 2. Profile of $T_b - T_a$ for the case 1.

The difference between the two temperatures has an amplitude of about 1.5 keV, - see figure 2- corresponding to the experimental observations, e.g. [3].

Case 2: We consider the following temperature profiles (see figure 3):

$$T_a/T_0 = \exp[-3 * \tanh[3(r/a-0.84)]] \tag{17}$$
$$T_b/T_0 = \exp[-3 * \tanh[3.5(r/a-0.8)]] + 1.5\exp[-40 * (r/a-0.8)^2] + 0.5\exp[-480 * (r/a-0.9)^2] \tag{18}$$

with $T_b - T_a$ plotted in figure 4. We will draw the graph for the dimensionless running diffusion coefficient $D(\theta) = \left(\chi_{\parallel,a}\beta^2\right)^{-1} D^\infty\left(\theta; \overline{\chi}_{\parallel,a}\right)$ for $\overline{\chi}_{\parallel,a} = 0.01$ which corresponds to the strong collisionality regime.

In the case 1, the running diffusion coefficient (see figure 5) is maximum for about $r \approx 0.8a$ corresponding to the maximum for $T_b - T_a$ at $r \approx 0.7a$. In the case 2, the running diffusion

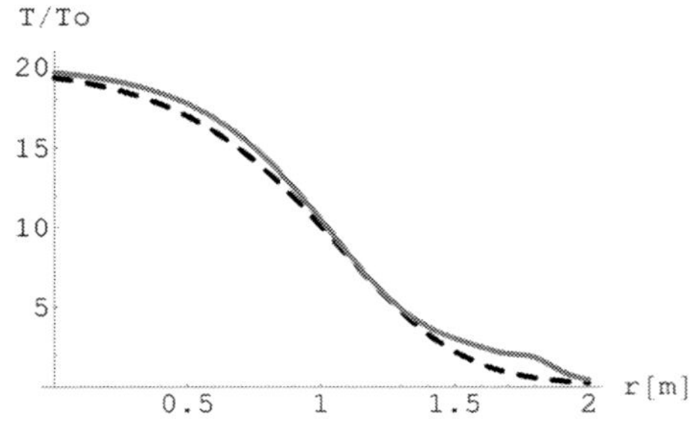

Figure 3. Temperature profiles: T_a given in (17) - (dashed line) and T_b given in (18) - (full line)

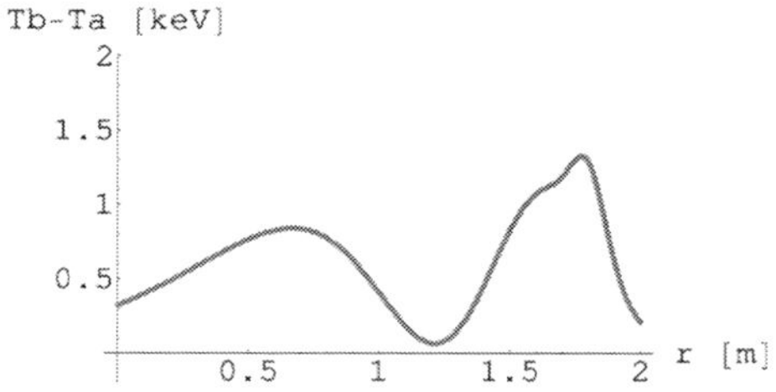

Figure 4. Profile of T_b - T_a in the case 2.

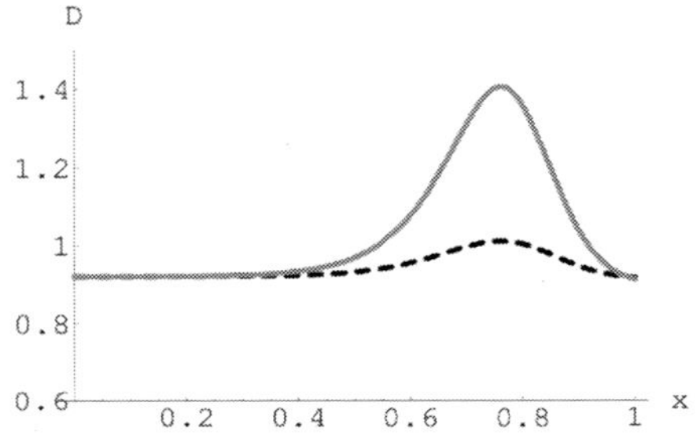

Figure 5. Dimensionless running diffusion coefficient D for $\overline{\chi}_{\parallel,a} = 0.01$ with $\lambda\tau_0 = 0.2$ (dashed line) and $\lambda\tau_0 = 0.8$ (full line).

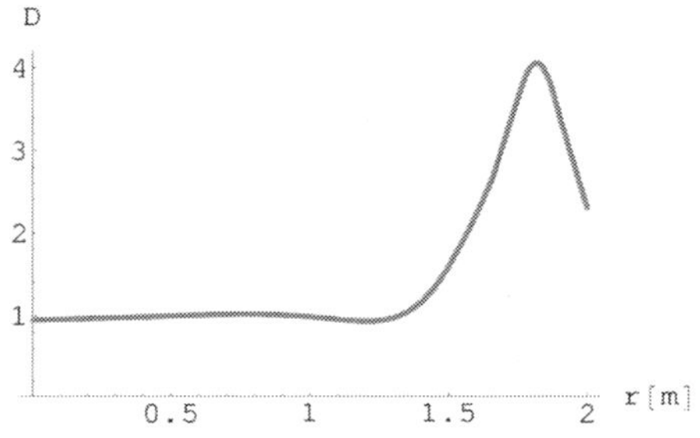

Figure 6. The dimensionless running diffusion coefficient D for $\overline{\chi}_{\parallel,a} = 0.01$ (corresponding to strong collisionality) is plotted for $\lambda\tau_0 = 0.8$.

coefficient (see figure 6) has one maximum value for about $r \approx 0.9a$ and $T_b - T_a$ has two maximum values, for $r \approx 0.3a$ and for $r \approx 0.85a$.

From here we conclude that the running diffusion coefficient is less sensible to the difference between the two temperatures in the central region than the difference between temperatures in the plasma edge region.

Acknowledgments

The present work has been supported by EURATOM-MECTS Romania for NP, and EURATOM-Belgium state and Wallonie-Bruxelles International for BW.

References

[1] Radu Balescu, *Aspects of Anomalous Transport in Plasmas*, IoP (Taylor & Francis, Series in Plasma Physics), London, 2005
[2] Crispin.W Gardiner : Handbook of Stochastic Methods: for Physics, Chemistry and the Natural Sciences (Springer Series in Synergetics)
[3] R A Pitts et all, *The impact of large ELMs on JET*, Journal of Nuclear Materials 390-391 (2009) 755-759
[4] A Geraud, J Bucalossi et all, *Pellet fuelling in Tore Supra long discharges*, Journal of Nuclear Materials 337-339, (2005) 485-489
[5] N Pometescu, B Weyssow, *Modeling the random transition between two temperature profiles in magnetized plasma*, Phys. Scr. **81** (2010)
[6] J Wesson, *Tokamaks*, Oxford: Clarendon, 1987, p 40

Experimental research of sliding surface distributed nanosecond discharge in supersonic air flow

I Belysheva, I Mursenkova and A Chvyreva
M.V. Lomonosov Moscow State University, Faculty of Physics; Leninskie Gory, Moscow, 119992, Russia

E-mail: alife_@mail.ru, murs_i@mail.ru

Abstract. Results are presented from the experimental study of the spatial structure and spectrum of the nanosecond distributed sliding surface discharge (plasma sheet) in quiescent air and in a supersonic and transonic gas flow behind a plane shock wave. Discharge investigations were carried out in air flow with velocities up to 1600 m/s within density range of 0.08-0.40 kg/m3. Different structures of plasma glow were observed for the laminar and turbulent flow regimes in the boundary layer. Analysis of the obtained images clearly shows that character of the glow of the discharge changes with change of the boundary layer structure. The analysis of the experimental data shows that the radiation spectrum of sliding surface discharge changes in gas flow after a shock wave

1. Introduction

A great number of papers devoted to different approaches of non-equilibrium plasma flow control in aerodynamics have appeared recently [1, 2]. Applications of plasma to flow control involve managing flow detachment or attachment to the surface, which results in managing lift and drag force of the airfoil.

A sliding surface discharge (plasma sheet) is the distributed formation of parallel channels, sliding on a dielectric surface. Such discharges are used as plasma electrodes, UV radiation sources in gas-discharge and excimer lasers [3, 4]. The development of the channels of a sliding discharge is characterized by high values of the electric field. The growth rate of the channels is $\sim 10^6$ m/s. Breakdown of a gas layer near a dielectric surface is accompanied by an abrupt increase in the current (as much as several kA) and the formation of bright channels [3-5]. It is this phase of a high-current discharge that determines the energy input into the surface gas layer. Initiation of the plasma sheet in the gasdynamic flow allows providing the pulse energy deposition in a boundary layer region [5, 6]. The electric discharge energy is converted into gas heating, molecule rotation, vibration, ionization, and electron excitation.

Research of discharge plasma and gasdynamic flow interaction should be done for the high speed flow control possibility analysis. The pulse surface discharge can strongly influence the gasdynamic flow structure. Shock waves arise in a flow field as a result of quick energy input [5, 6]. It is important to study plasma-dynamic characteristics of the sliding discharge in air flow for a better understanding of the fundamental problems in the physics of nonequilibrium plasma and the optimal solution of engineering problems.

The analysis of spatial structure of a luminescence and spectral characteristics of radiation of the distributed sliding surface discharge in a supersonic and transonic gas flow behind a shock wave was a research objective.

2. Experimental Set-up

A detailed description of the experimental set-up has been presented in [5, 6]. A special discharge chamber mounted with a shock tube was used for the experiments (figure 1). The rectangular cross-section of the discharge chamber was the same as that of the shock tube (48×24 cm^2). The shock tube allowed plane shock waves, with transonic and supersonic flows behind them, to be generated with Mach numbers up to 5. Plasma sheets were initiated on two opposite walls of the discharge chamber at a distance of 24 mm from each other. Two other walls of the experimental discharge test section were quartz glasses; 170 mm in length (see figure 1).

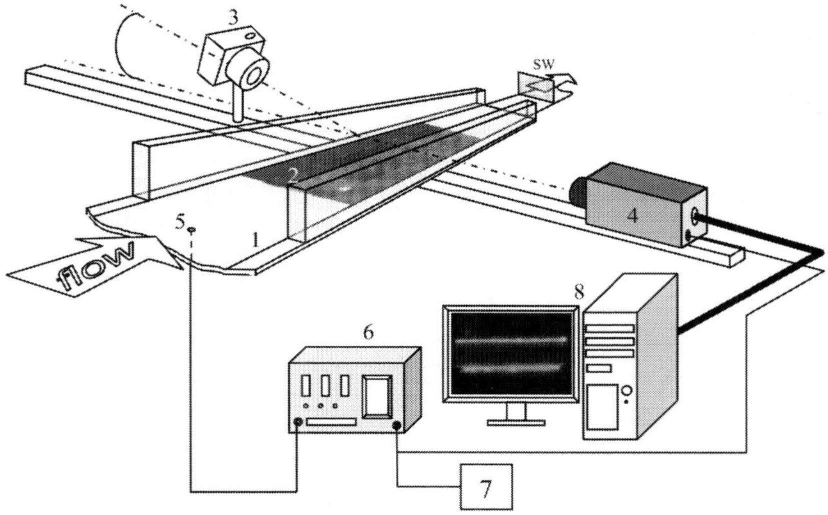

Figure 1. Experimental set-up: 1 – Discharge chamber; 2 – Discharge area (black colour corresponds to electrodes); 3 – Photo camera; 4 – ICCD camera; 5 – Pressure transducer; 6 – Timing electronics; 7 – Spark-gap controller; 8 – Image PC.

In order to organize a plasma sheet, the special electrode configuration was used with discharge gap dimensions of 30×100 mm^2 [5]. A high-voltage pulse applied to the discharge gap, was transmitted from the main capacity to the high-voltage electrode of the discharge chamber by means of a spark-gap controller. Experiments were carried out at 25 kV pulse voltage. A reduced electric field was realized at parameter E/N = 500-1000 Td, where E was the electric field strength and N was the gas number density. The current parameters were measured using a special current shunt. Pulse discharge time was about 200 ns, which was much less than the characteristic gasdynamic time in the shock tube (~10 μs). The discharge current reached a maximum of 1-2 kA during the first 50 ns of the discharge development. The current pulse signals showed that the main energy deposition in the discharge occurred during the time ~ 100 ns.

Plasma sheets in air flow were initiated after the shock wave left the discharge area. The signal from the piezoelectric gauge of pressure started the spark-gap controller through timing electronics.

Plasma sheet glow was registered integrally and with nanosecond resolution by digital cameras. The exposure time during integral recording was equal to discharge glow time. A Nanogate-2 ICCD camera (spectral response 380-800 nm) was used to determine spatial-temporal characteristics of the discharge. Special digital image processing was then used to determine space characteristics of the surface discharge in the flow. Spectrometer AVASpec-2048FT (174-1100 nm) was used for the spectral analysis of discharge glow.

Discharge glow investigations were carried out in motionless air and in air flow with velocities up to 1600 m/s within density range 0.08-0.40 kg/m^3. Flow Reynolds numbers were in the (4.5-12.5) × 10^4 range. The thickness of the plasma layer in motionless air, measured from the emission intensity profiles in the plane of the plasma sheet, decreased from 0.8 to 0.4 mm with density increasing. The plasma layer thickness in air flow was about 0.4 mm [5].

Journal of Physics: Conference Series **516** (2014) 012021 doi:10.1088/1742-6596/516/1/012021

a)
b)
c)

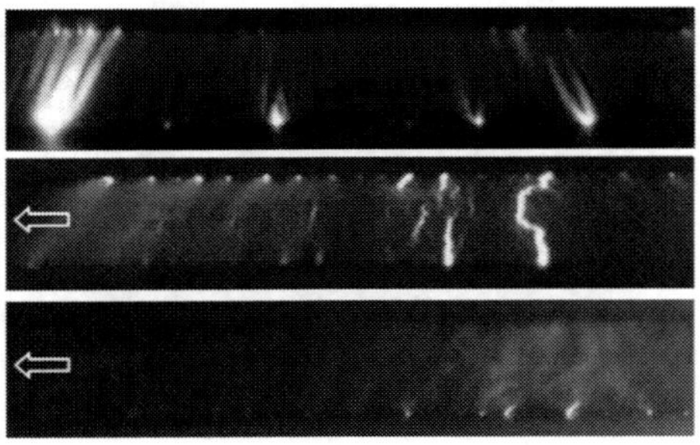

Figure 2. Images of the discharge glow: a) motionless air, density 0.11 kg/m^3; b) transonic flow, Mach number 0.8, density 0.20 kg/m^3; c) supersonic flow, Mach number 1.5, density 0.11 kg/m^3. Arrows show the flow direction.

3. Experimental Results

Space and time distribution analysis of surface discharge glow was carried out in the supersonic and transonic gas flows with flow Mach numbers up to 1.6 after the plane shock waves (Mach numbers up to 4.5).

Glow evolution analyses revealed, that the time of discharge current (200 ns) and the time of discharge glow were close (300 ns). However gasdynamic time in the shock tube was much longer than the time of the discharge radiation. Thus, flow shifted 0.3 mm during registration time of the discharge glow. The structure of the high-speed flow didn't change during integral glow registration.

Figure 2 presents the images of nanosecond surface discharge glow. The registered integral plasma sheet glow shows the discharge structure in motionless air (figure 2 a). One can see that the discharge has a quasi-planar structure and consists of a set of channels. The plasma sheet region is uniformly filled with straight-line diffuse channels, with the presence of brighter channels. Greater diffuse distribution of plasma glow was observed after the shock wave in transverse (relative to the propagation direction of the discharge channels) air flow. Some channels were curved (figure 2 b, c). An analysis of the experimental patterns showed clearly, that discharge glow character changes with the changing of the flow boundary layer structure. The structure of the boundary layer flow could be determined by the obtained images of the plasma discharge glow.

The sliding surface discharge radiation in the visible and ultraviolet spectral region was determined generally by the 2$^+$ system of nitrogen corresponding to N_2 ($C^3\Pi_u \rightarrow B^3\Pi_g$) transition (figure 3). One can see typical atoms lines of hydrogen, oxygen and nitrogen in the spectra. The long-wave part of the spectrum was more intensive in supersonic flow after the shock wave. In this case the discharge was initiated in heated air behind the shock wave front. The vibrational temperature was determined from relative intensities of emission lines of 2$^+$ system of nitrogen with the sequence $\Delta v = -2$. The distribution of energy over vibrational degrees of freedom with a reasonable degree of accuracy appears the Boltzmann distribution. The vibrational temperature was equal to (3400 ± 400) K in motionless air (gas density 0.11 kg/m^3) and (3900 ± 400) K in supersonic gas flow behind the shock wave (Mach number 3,6; gas density 0.11 kg/m3, temperature 1000 K).

The distribution of the plasma discharge radiation intensity corresponds to the instant distribution of gas density due to the strong dependence of radiation intensity on local value of reduced electric field *E/N* and due to the short time of the discharge. Consequently the heterogeneity of density near the surface influences on the character and the geometry of the sliding surface discharge development.

Figure 4 presents the main flow elements behind the shock wave in the shock tube. The thickness of the boundary layer in the shock tube channel increases from zero at shock wave front. The boundary layer becomes turbulent at certain distance from shock wave front. The thickness of the laminar boundary layer was about 0.5 mm in supersonic flow.

One can see that, immediately behind the shock-wave front, the glow has a diffuse homogeneous and smooth character, without visible discontinuities (figure 4, on the left). At a certain distance from the shock-wave front, the character of the glow exhibited a change; i.e., the glow became inhomogeneous, showing chaotic structures and separate curvilinear channels (figure 4, on the right). An analysis of images in the flow yielded convincing evidence that the character of the glow was altered due to a modification of the boundary layer structure, which changed from laminar (behind the shock-wave front) to turbulent. Laminar and turbulent zones were distinctly visible. Homogeneous area of the glow corresponds to laminar boundary layer and heterogeneous area (curved channels) visualizes turbulent one [6]. The turbulent medium turns out to be ionized and contains plasma structures, with their scale corresponding to the turbulence scale.

Journal of Physics: Conference Series **516** (2014) 012021 doi:10.1088/1742-6596/516/1/012021

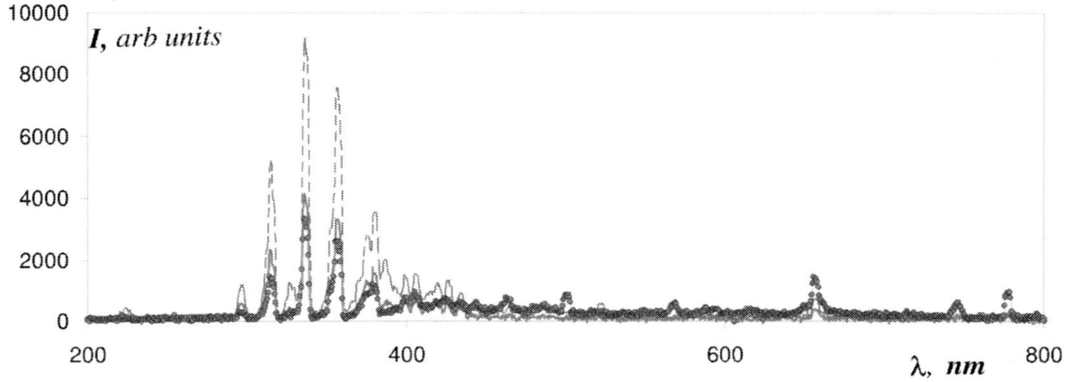

Figure 3. Measured emission spectra of the distributed sliding surface discharge in motionless air at 0.05 kg/m^3 (dashed line) and at 0.11 kg/m^3 (solid line) and in supersonic air flow (symbols) at density 0.11 kg/m^3 and Mach number 1.5.

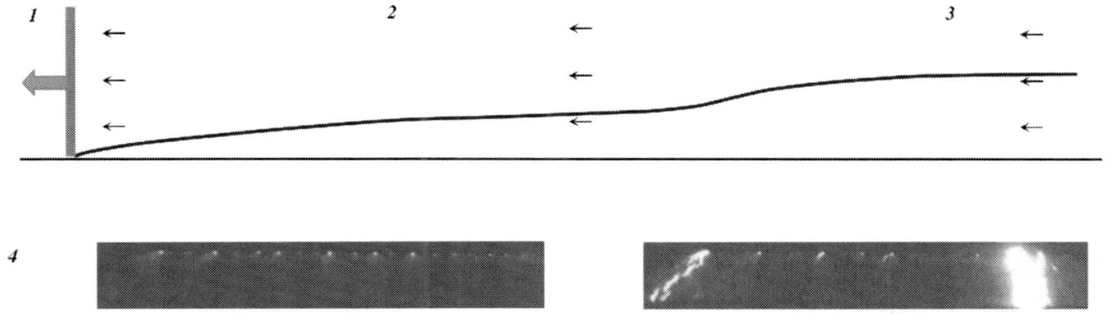

Figure 4. Flow diagram for the shock-tube channel: (1) front of the propagating shock wave; (2) laminar region in the boundary-layer flow; (3) turbulent region in the boundary-layer flow; (4) integral glow of the plasma sheet (the flow Mach number is 1.2 and the gas density is 0.19 kg/m^3)

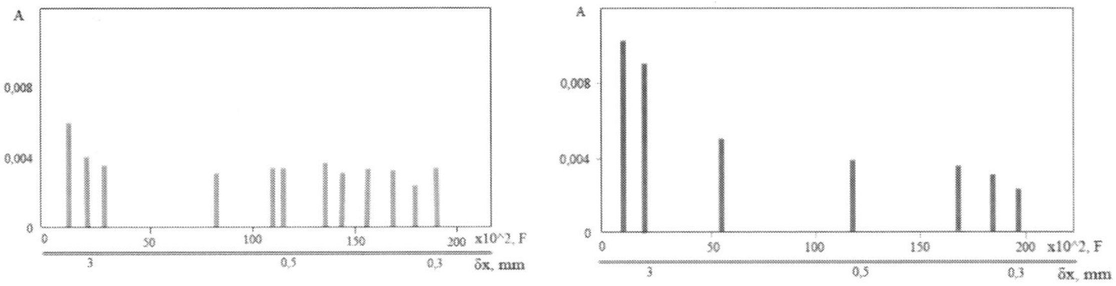

Figure 5. The results of Fourier-analysis: frequencies and amplitudes of discharge glow intensity in supersonic boundary layer in the laminar region (left) and in the turbulent region (right).

On the other hand the discharge initiation essentially affects the flow structure. The pressure and temperature of the surface thin gas layer increases during one microsecond due to fast energy input. Consequently perturbations propagate away from the plasma sheets [5, 6]. However, insofar as the discharge-glow duration is many times shorter than the characteristic times of gas-dynamic processes in the shock wave tube, the structure of the near-surface flow layer remains unchanged during the discharge-glow time.

Due to the strong dependence of radiation intensity on local value of reduced electric field E/N and due to the short time of the discharge compared with gas dynamics times the distribution of the plasma discharge radiation intensity does correspond with the instant distribution of gas' density. This makes it possible to get spatial calibers typical for laminar and turbulent flow areas of boundary layer by relevant mathematical analysis of the spatial distribution of discharge glow. The spectrum of frequencies and amplitudes of discharge glow intensity in supersonic boundary layer varies for different types of flow (see figure 5). In the laminar flow there aren't any specialized structures, while in the turbulent boundary layer there are some characteristic structures with the size of 3-5 mm. Sliding-windows method was used for image processing and investigation of the fractal dimension. The investigated picture is covered by random-radius circles and mass for each circle is counted due to chosen criteria. As a result, we have got the fractal dimension of the image in the flow as tangent of mass-radius dependence. The fractal dimension of the discharge glow images changes from 1.8 in the laminar boundary layer to 1.9 in the turbulent boundary layer.

4. Conclusions

The investigation of the sliding surface discharge of nanosecond duration in the supersonic and transonic flow behind the shock wave was carried out. Our studies have demonstrated that the type of boundary layer affects the character of discharge channels developed across the flow. The thickness of the plasma sheet was about 0.4 mm that is an order of a thickness of a boundary layer. The turbulence in the boundary layer results to the formation of curvilinear channels due to the appearance of inhomogeneities of density and the ionization coefficient. We have got spatial calibers typical for laminar and turbulent flow areas of boundary layer by relevant mathematical analysis of the spatial distribution of discharge glow. Also fractal dimensions were counted for images of laminar and turbulent flows. In the parameter range under study, the radiation spectra of the sliding surface discharge changed in the supersonic flow. It was shown that the long-wave-line part of spectrum is more intensive in supersonic flow after the shock wave. It was found out that the distribution of energy over vibrational degrees of freedom is represented by Boltzmann distribution.

References

[1] Bletzinger P., Ganguly B.N., Van Wie D. and Garscadden A. "Plasmas in High Speed Aerodynamics". J. Phys. D: Appl. Phys., Vol. 38, No. 4, pp. R33-R57, 2005.
[2] E. Moreau. "Airflow control by non-thermal plasma actuators". J. Phys. D: Appl. Phys., Vol. 40, No. 3, pp. 605-636, 2007.
[3] Karlov N. V., Kuz'min G. P. and Prokhorov A. M. "Gas Discharge Lasers with Plasma Electrodes". Izv. Akad. Nauk SSSR, Ser. Fiz., Vol. 48, No. 7, pp. 1430-1436, 1984.
[4] R.E. Beverly III. "Electrical, gasdynamic, and radiative properties of planar surface discharge". J. Appl. Phys., Vol. 60, No.1, pp. 104-124, 1986.
[5] Znamenskaya I.A., Latfullin D.F., Lutsky A.E., Mursenkova I.V. and Sysoev N.N. "Development of gasdynamic perturbations propagating from a distributed sliding surface discharge". Technical Physics, Vol. 52, No. 5, P. 546–554, 2007.
[6] Znamenskaya I. A., Latfullin D. F. and Mursenkova I. V. "The Surface Transverse Distributed Nanosecond Discharge in the Supersonic Boundary Layer". Doklady Physics, 2009, Vol. 54, No. 7, pp. 309–311.
[7] Latfullin D.F., Mursenkova I.V. and Znamenskaya I.A. "Supersonic Boundary Layer Investigation using Surface Nanosecond Discharge". Proceedings of 13th International Symposium on Flow Visualization (ISFV-13) and 12th French Congress on Visualization in Fluid Mechanics (FLUVISU-12), Nice, France, July 1-4, 2008. CD Rom Proceedings, 146-080414.pdf.

Journal of Physics: Conference Series **516** (2014) 012022 doi:10.1088/1742-6596/516/1/012022

Simulation of defects in fusion plasma first wall materials

Troev T[1], Nankov N[1] and Yoshiie T[2]

[1] Institute for Nuclear Research and Nuclear Energy, 72 Tzarigradsko Chaussee Blvd., 1784 Sofia, Bulgaria

[2] Research Reactor Institute, Kyoto University, Kumatori-Cho, Sennan-Gun, Osaka-Fu 590-0494, Japan

E-mail: troev@inrne.bas.bg

Abstract. Numerical calculations of radiation damages in beryllium, alpha-iron and tungsten irradiated by fusion neutrons were performed using molecular dynamics (MD) simulations. The displacement cascades efficiency has been calculated using the Norgett–Robinson–Torrens (NRT) formula, the universal pair-potential of Ziegler–Biersack–Littmark (ZBL) and the EAM inter-atomic potential. The pair potential overestimates the defects production by a factor of 2. The ZBL pair potential results and the EAM are comparable at higher primary knock-on atom (PKA) energies (E > 100 keV). We found that the most common types of defects are single vacancies, di-vacancies, interstitials and small number of interstitial clusters. On the bases of calculated results, the behavior of vacancies, empty nano-voids and nano-voids with hydrogen and helium were discussed.

1. Introduction

The paper is devoted to numerical calculations of cascade effects due of 14 MeV neutron irradiation and the defect creation processes in fusion materials. The purpose of this study is to perform numerical calculations to understand the defect creation processes in beryllium, alpha-iron and tungsten as a result of fusion 14 MeV neutron irradiation.

The fusion reaction (D, T) produces 14 MeV neutrons and energetic 3.5 MeV alpha particles.

$$\text{D-T Fusion Reaction: } D + T \rightarrow {}^{4}He + n \tag{1}$$

Alpha particle (He nucleus):
20 % of reaction energy
Neutron:
80 % of reaction energy

In the last years the study of the evolution of radiation defects in tungsten under fusion neutron irradiation increased, due to the requirements of fusion reactor technology. Beryllium alpha-iron, tungsten and tungsten alloys are considered as one of the major materials for the first wall of the upcoming new generation fusion prototype reactors [1] ITER (International Tokomak Experimental Reactor) and DEMO aimed at demonstrating the technical feasibility of electrical power production by means of (D,T) fusion reaction. Plasma facing materials have to resist a high neutron flux and a high thermal flux. The thermal flux is expected to be 10 MW m-2 in the ITER [2]. The plasma facing materials will protect the fist wall from high particle flux and they will transfer the thermal energy away from the first wall surface. The reason for the selection of Be is the low effect on plasma contamination due to the low Z, low radiative power losses, good oxygen gettering ability, absence of chemical sputtering in comparison with carbon and the possibility of in situ repair of damaged surfaces. On the other hand in the fusion reactor beryllium could play an important role as a neutron multiplier to sustain tritium breeding. This is especially important for the present design of the solid breeder blankets of the DEMO reactor [1] since the resulting low energy neutron flux spectrum necessary for tritium breeding via ^{6}Li(n, n,) ^{4}He + ^{3}H reaction is due to beryllium neutron moderation. Iron is considered as one of the major materials for the first wall of the upcoming new generation fusion prototype reactors. Its function would be the protection of actively cooled wall structures from high heat fluxes in order to sustain the component lifetime and the plasma compatibility. Another reason for the selection of Fe would be its ability to produce only stable or very short lived Mn and Cr isotopes due to nuclear transmutations caused by fusion neutrons. Additionally iron and the stainless steel are widely used low cost materials which would significantly lower the costs of the first wall (FW) blanket. Several nuclear facilities are proposed [2] for experimental studies of irradiation defects on material due to the neutron flux from ITER and DEMO. Another approach relays on numerical simulations of the beryllium response to the intensive high-energy neutron irradiation. The validation of the performed defect simulations for vacancies and interstitials in first wall fusion materials could be done by experimental transmission electron microscopy (TEM) or positron annihilation spectroscopy measurements (PAS). TEM cannot be used for investigation of nano voids smaller than 2 nm, while PAS allows the local investigation of single vacancies di-vacancies and nano-voids smaller than 2 nm.

2. Methodology of the simulation

Different approaches are necessary for the simulation of defects in fusion materials under irradiation with high energy neutron fluence at about 2.5×10^{22} n/cm^2. The method of molecular dynamic (MD) and kinetic Monte-Carlo (kMC) could be applied for calculation of the damage evolution of PKA down to 10 keV. The fast neutrons create recoils, whose kinetic energy is in the range from 1 keV to 1–2 MeV. In order to achieve our goal we have used the multi-scale approach [3]. Most of the produced defects in fusion materials posses PKA energies below 50 keV [4]. It is interesting to note that the created displacement cascades are also at the same PKA energy range. At higher PKA energies the crystal structure plays not remarkable role. We created a computer Monte-Carlo code based on Geant 4 CERN package to simulate ion and neutron interactions in homogeneous target at high energies (E > 0.1 MeV). High-precision Geant4 NDL ver. 3.20 database for neutron interaction cross sections (up to 20 MeV) was used. The PKAs (primary knock-on atoms) were calculated in $30\times30\times30$ nm^3 volume on the neutron beam axis. SRIM calculations using ZBL [5] potential were carried out to determine the defect distribution. Initially we studied the number of displacements per atom (dpa) created by the fusion neutrons. For this purpose we used the NPRIM code [6]. The pair-potential that depends only on the atomic separation of two atoms can be used for defect calculations. At higher energies (>100 keV) we applied the universal pair-potential of ZBL for calculation of defects creation in fusion materials. In our simulations for energies below 50 keV the modified MD code MDCASK [7], was used for calculation of the radiation induced damage. The time required to compute a single cascade depends on the size of simulation domain, the neutron energy, and the inter-atomic potentials. The size of the simulation domain was chosen according to the PKA energy. The number of atoms in the simulation domain was in the range from 4.32×10^5 up to 1.5×10^6. This corresponds to sample sizes from 60 to 90 inter-atomic distances in all directions. The periodic boundary conditions were assumed for all simulations. We simulated the defect creation and time evolution at 300 K. Additionally, reference calculations at a temperature of 4K were carried out in order to determine the defects distribution due to the displacement cascades. The embedded atom method (EAM) potential, developed by Daw and Baskes [8] has provided a good potential format for metal atoms. The crystal energy according to the EAM is the sum of pair-wise potential and an energy required to embed an atom into a local electronic medium in which the interaction between the atoms depends on the position of their neighbours. The embedding energy in EAM for local environment of the potential depends on the local electron density and therefore has been successfully used to solve problems like surface structures and defect states. We easily adapted the generalized EAM potential for our purpose to calculate defect creation in fusion neutron irradiated beryllium, alpha-iron and tungsten, containing hydrogen and helium.

Using Monte-Carlo (MC) simulation we have calculated the atom displacements caused by fusion materials PKA. At the end of the simulation the final position of the atoms involved in damage cascades have been recorded. The statistical methods were applied to calculate the atom displacements and the number of the produced vacancies and interstitials as a function of the PKA energy for the created isotopes. The processes like neutron elastic and inelastic scattering, neutron capture with and without emission of charged particles were taken into consideration in the Fusion materials matrix. The low energy processes like gamma ray and secondary electron emissions as well as all types of electromagnetic interaction of charged particles and photons with matter were included in the present calculations. Nuclear transmutations which play an important role in helium, deuterium and tritium formation were also taken into account. The number N of Frenkel pairs produced per cascade has been estimated by the NRT formula [9].

$$\mathrm{NRT}=\frac{0.8E_{PKA}}{2E_d}, \tag{2}$$

where E_{PKA} is the initial PKA energy and E_d is the displacement energy for the creation of a stable Frenkel pair.

3. Results and discussions

First we have simulated the interaction of neutrons with macroscopic Be target at high energy ($E > 0.1$ MeV). For this purpose the fusion neutron spectrum data was taken from the experiment as an input for the simulation. In the present work the initial neutron energy distribution for the simulation was taken from [5]. The total number of primary events used in the simulation was 2×10^9 which resulted in a neutron fluence of about 1.6×10^{20} n/cm^2 ($E > 0.1$ MeV). Nuclear transmutations were also taken into account [10]. The ^{7}Li production has threshold of 12 MeV and therefore it takes place in the fusion reactor. If we scale the result for the neutron flux of 1.3×10^{15} n/cm^2/s expected for the DEMO first wall (FW) [2] our data are in agreement with the available ones in the literature [11,12] although our results are obtained by diferent procedure. It is well known that the main parameter which carries the information about the radiation resistance of the material is the dpa (displacement per atom). It has been established that this parameter reaches the value of 30 dpa/fpy (dpa per full power year of operation) for iron irradiated by fast neutron fluxes up to 1.3×10^{15} n/cm^2/s [13]. The results for dpa cross section vs. various neutron energy are shown in figure 1.

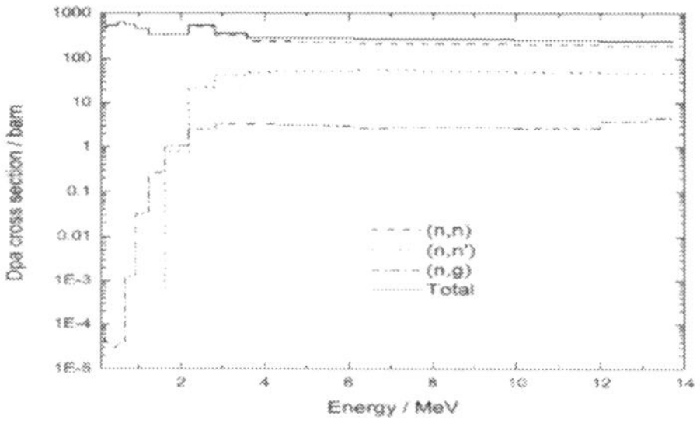

Figure 1. dpa cross section for a typical deuterium-tritium reactor for diferent types of neutron interactions as a function of the neutron energy

Important process is the neutron capture in Be with gamma ray emission. Its magnitude is relatively small but it is important for the tritium production. Our calculations show that the threshold for the (n,n^0) induced dpa is about 2 MeV and for the (n,n^0) is about 1 MeV. The dpa due to elastic scattering was present at the whole energy range (0.1–14 MeV) in the present simulation. The dpa cross sections for (n,n^0) grow rapidly up to about 4 MeV, after that they reach saturation. The next step in our computer simulations was to carry out nano-structural damage calculations using the results from the described high-energy neutron interactions. For this purpose we used the binary collision approximation (BCA) and the SRIM MC code [14] because it relays on the universal pair potential of ZBL [5, 15]. In this approach the crystal structure of Be has been neglected. Beryllium has been considered as a homogeneous continuous medium. For each cascade we consider the displacement collisions, vacancy production, replacement collisions and interstitial atoms, as described below. The number of displacement collisions records how many target atoms were set in motion in the cascade with energies above their displacement energy which is specifed at the input. The Be target vacancies are next items which were recorded. The stopped recoil atom can take position at some distance from its vacancy. When after inducing a displacement cascade it leaves the target, the sum of interstitials will be less than the number of vacancies by the loss of that atom. In our computer simulation both Be and He PKA could induce displacement cascades. In the simulation the average displacement energy Ed has been chosen to be 40 eV. The lattice binding energy was taken to be 3.32 eV equal to the cohesive energy of the Be crystal. Tritium created by neutron irradiation of Be behaves as trapped ones. The implanted tritium never difuses into the bulk of the material and becomes trapped at ITER operating temperatures [16]. According to calculations [17, 18] the total tritium inventory in the beryllium first wall armor due to implantation, difusion, trapping, and breeding after 12000 pulses will be of 100–250 g [16]. Tritium could be well localized in bulk Be at ITER and DEMO temperatures but the tritium concentration could increase locally due to trapping of tritium by single vacancies and vacancy clusters (nano-voids). These types of defects are due to the large amount of displacement damage cascades as a result of intensive neutron irradiation. We have shown that for neutron fluences ($E > 0.1$ MeV) up to 2.5×10^{22} n/cm^2 the damage level reaches the value of 11 dpa/fpy. This value is in agreement with that found in [19, 20]. It was established in our calculations that the total damage of 0.06 dpa for neutron fluence of 1.6×10^{20} n/cm^2 scales well with already reported results for 2.5×10^{22} n/cm^2 [6]. As discussed in [21] the smallest He gas bubbles with sizes in the range of 1–4 nm appear when the annealing temperature was 500 °C. Helium was not found at temperatures below 400 °C. They concluded that the He atoms were trapped in subatomic complexes. According to [21] the defects in neutron irradiated Be should be interstitials and dislocation loops although they can be distributed throughout the structure non-uniformly.

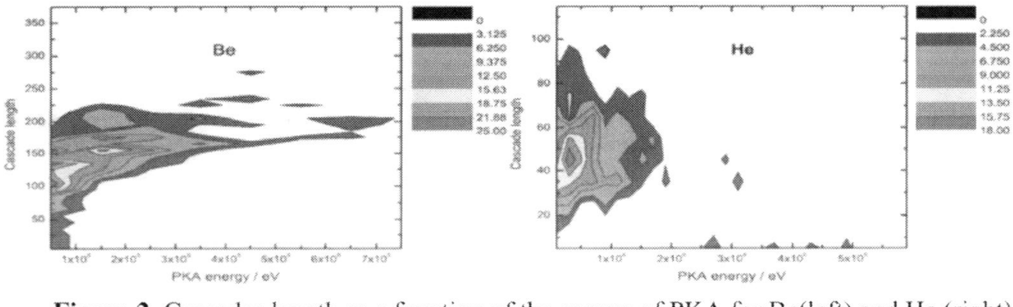

Figure 2. Cascades length as a function of the energy of PKA for Be(left) and He (right)

The neutron irradiation of beryllium over the temperature range up to 500 °C leads to formation of radiation-induced plane defects. The cascades length distribution for both Be and He PKA is presented in figure 2. In the Figure it is seen that He recoils have higher initial kinetic energy but they cause approximately 2–3 times less dpa. We have established that the calculated number of the replacement collisions which lead to recombination of vacancies and self-interstitials was only 1.8 % compared to the total number of displacements. This result indicates that high-energy recoils generate atomic collision cascades in Be in which a fraction of the defects recombine during the cooling phase.

The calculated results of the Be sub-cascades as a function of the PKA energy are shown in figure 3. They indicate a correlation between PKA energies and the number of sub-cascades. It it seen in figure 3, that PKA with higher energies produce large number sub-cascades. It is seen that Be PKA energies in the range 20–120 keV have higher probability to induce sub-cascades. Obviously the most probable number of sub-cascades is around 50. Similar results are discussed in [15, 21, 22]. The authors pointed out that for PKA with energy higher than some critical value in the range of 10–40 keV the formation of sub-cascades is more probable. Our simulation results show that low energy cascades ($E < 100$ keV) are responsible for about 70 % of the defects production.

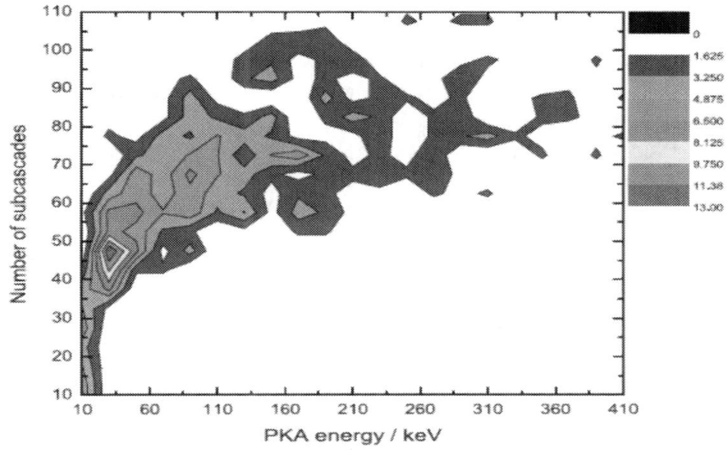

Figure 3. Correlation between the number of Be PKA sub-cascades and the energy.

The distribution of vacancies created by beryllium atom displacements and helium atoms created by neutron capture in Be is shown in figure 4. The crystal structure of Be was not taken into account for the calculation of the Be dpa in the present model. It has been used the pair potential of ZBL [5].

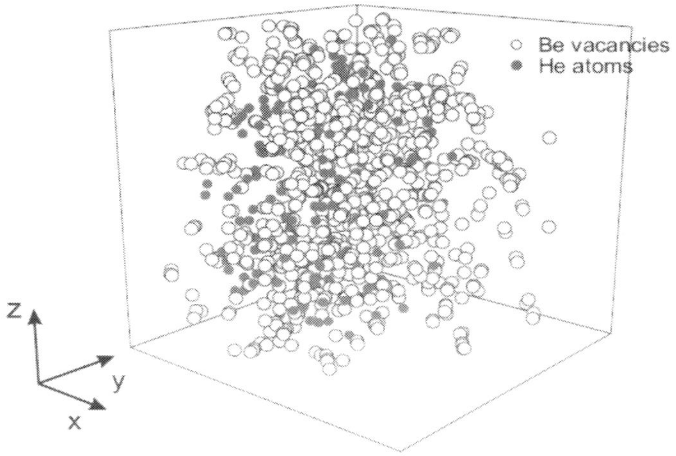

Figure 4. Distribution of vacancies and helium created by beryllium recoil the domain size is $60\times30\times30$ nm^3.

It is seen from figure 4 that large amount of helium atoms are stopped outside of the single vacancies or vacancy clusters after the ballistic phase of the cascades. The He atoms we considered as interstitials which contribute to the interstitial clusters concentration. The solution of this problem has been done by using the methods of gas difusion in solids, the kinetic Monte-Carlo (kMC) approach [23] and the kinetic rate-theory [24]. Experimental data of the He and T kinetics in neutron irradiated Be were presented in [25]. Most of the He atoms precipitated into bubbles and are released during abrupt of pores at 1500 K. The explanation of He nucleation in bubbles and its release at high temperatures was done in [25]. The same authors applied the kinetic-rate theory for calculation of the time evolution of the He concentration in bulk and in bubbles. This is an important macroscopic property which gives information about the He migration in the Be sample 2000 seconds after the displacement cascade.

We apply the multi-scale approach [26] to study the damages in Fe lattice caused by intensive fusion neutron irradiation. First we have simulated the interaction of neutrons with macroscopic iron target at high energy ($E > 0.1$ MeV). For this purpose the fusion neutron spectrum data was taken from the experiment as an input for the simulation. We have recorded the initial energy and position of iron primary knock-on atoms (PKA) and of all isotopes created in the nuclear reactions at the output of the calculation. The obtained results allowed us to calculate the concentration of Fe, Mn and Cr isotopes as well as the concentration of helium and deuterium and used it as an input for the next step in our simulation.

In figure 5(a) is shown the total neutron interaction cross section in ^{56}Fe taken from the ENDF/B database and used in our simulation. The resonance structure is dominant for the neutron interactions at energies lower than 3 MeV. At neutron energies higher then 5 MeV up to 14.1 MeV the neutron interaction cross section is almost constant and has the value of about 5 barn. In figure 5(b) are shown the production cross sections of ^{4}He and ^{3}H used in our numerical modeling. The integral cross section for ^{4}He production is about 200 times lower than the total neutron interaction cross section in ^{56}Fe. Therefore one could expect that the most of the created defects are due to Fe recoils as a result of neutron elastic and inelastic collisions.

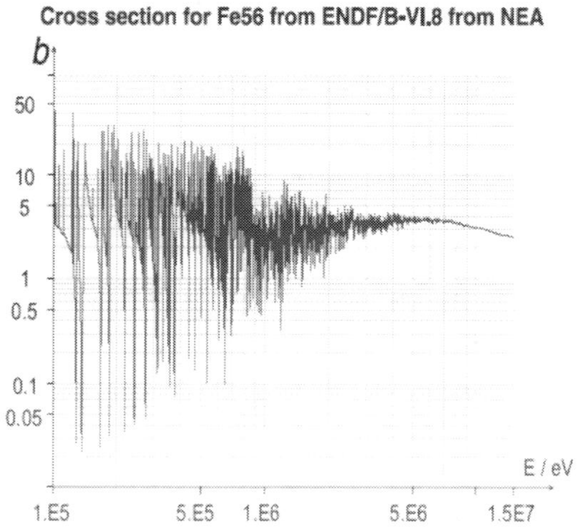

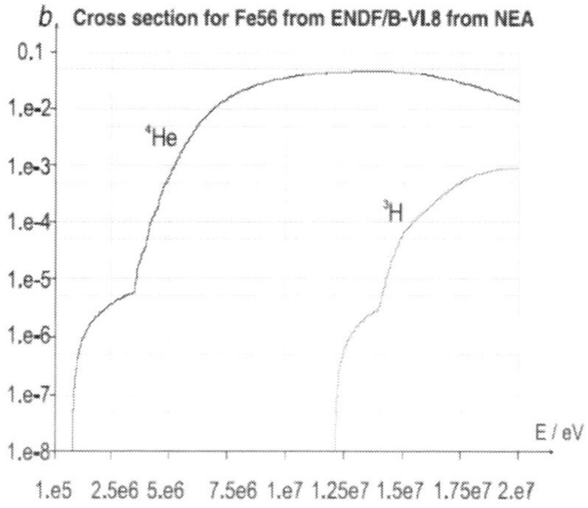

Figure 5(a). Total neutron interaction cross section. for neutrons at different energies.

Figure 5(b). Cross sections of 4He and 3H

The next step in our computer simulations was to carry out structural damage calculations using the results from the described high-energy neutron interactions. Every PKA from the high energy neutron interaction with Fe atoms was tracked down to its stop in the medium. For each cascade we consider the displacement collisions, vacancy production, replacement collisions and interstitial atoms, as described below. The number of displacement collisions records how many target atoms were set in motion in the cascade with energies above their displacement energy which is specified at the input.

Our SRIM-based calculations show that the damages, produced by fast neutron fluence up to 8×10^{19} n/cm^2, correspond to 0.151 dpa. The total damages of 37.2 dpa/fpy were calculated using the NPRIM code [5]. ^{4}He and ^{53}Cr concentration was calculated to be about 0.4 appm. According to calculations using the NPRIM code [5] it reaches 375 appm/fpy. Those values are in a good agreement with the results, obtained in [13]. The ^{2}H and ^{55}Mn concentrations were determined to be about 7 % of the He concentration. In order to investigate the influence of the energy distribution of the ^{56}Fe PKAs, created as a result of fusion neutron induced nuclear reactions, we have calculated the total number of defects in displacement cascades as a function of the PKAs energy. The result is shown in figure 6. Most probable cascades have between 50 and 150 displacements. We have found that they are well concentrated in the PKA energy range up to 10 keV. The maximal damage is caused by PKAs with energy around 30 keV. PKAs with higher energies could produce more displacements but the number of such cascades is small probably due to the fact that the most efficient energy transfer in 56Fe occurs at lower energies. If the PKA has enough energy it can produce secondary recoils which could initiate additional cascades.

In our simulation we obtained values for the number of such sub-cascades as a function of the PKA energy. The results indicate that 56Fe PKA energies up to 50 keV have higher probability to induce sub-cascades. We determined that the most probable number of sub-cascades is around 150. Similar results are discussed in [12,13]. We established that low energy cascades (E < 100 keV) are responsible for about 90 % of the defects production. In our simulation we investigated the vacancy and self-interstitial recombination due replacement collisions. We found that the number of recombination processes was about 2.5 % of the total number of displacements. In the last step in our multi-scale model calculation we determined the depth profile of the vacancy distribution inside the ^{56}Fe sample. The result is shown in figure 7. It indicates that although the damages are created in a volume of $30\times30\times30$ nm^3 the damage extends up to 400 nm.

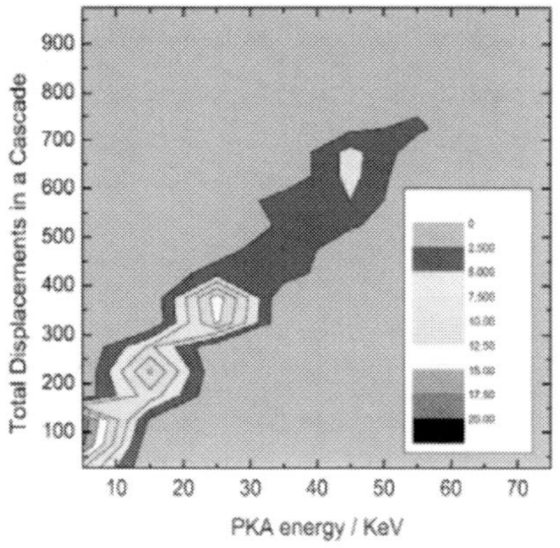

Figure 6. Correlation between PKA energy and the total displacements in a cascade in Fe after neutron irradiation fusion

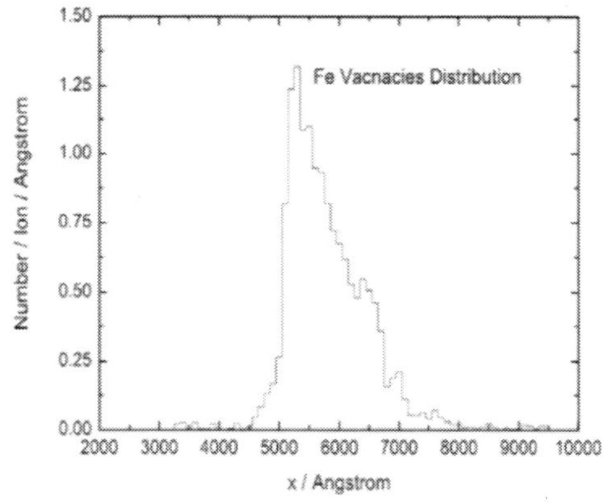

Figure 7. Vacancy distribution in ^{56}Fe

The distribution has a sharp maximum in the area where PKAs are created as a result of the neutron interactions. As we already shown this fact can be explained with the large probability of the damage creation due low energy PKAs. The small maximums in the distribution indicate the presence of sub-cascades.

Tungsten has low hydrogen isotopes retention. Helium is induced in the tungsten sample by the 14 MeV neutrons through the (n, α) nuclear reaction. Irradiation by 14 MeV fusion neutrons produces displacement damage in the form of Frenkel pairs i.e. self-interstitial atoms and vacancies. The high energy Primary knock-on atoms (PKA) in tungsten created as a result of elastic and inelastic neutron scattering may leave their lattice sites and transfer their energy to other lattice atoms thus forming displacement cascades. In the present work we use MD simulations combined with EAM potential to study defects formation in the presence of displacement cascades in tungsten. In addition we calculate the time evolution of the defects in tungsten after neutron irradiation from 5 ps to 30 ps at a temperature of 300 K. The PKA energies, range from 10 keV to 40 keV.

The EAM potential for tungsten found in [27] has been applied in our numerical studies. Calculations were performed for PKA with initial kinetic energies of 10, 20, 35 and 40 keV. The initial positions of PKA were chosen in such a way that the crystal lattice has the smallest initial energy. The initial directions of the tungsten PKA atoms were selected to be different from the main crystallographic axis in the BCC lattice, in order to avoid channelling. The tungsten atoms and PKA positions in the lattice were recorded at regular intervals of 5 ps. The whole simulations were set up to calculate the lattice dynamic up to 30 ps. The output of the MD results were analyzed, using the atom positions, calculated at 4K as a reference. Vacancies and interstitials positions were obtained by the Wigner-Seitz cell method [5]. We calculated the number of displacement atoms (dpa) in tungsten for Full Performance Year in unit dpa/FPY as a function of fast neutrons ($E > 0.1$ MeV) irradiation at a neutron fluence of 2.5×10^{22} n/cm^2. A value of 2.85 dpa/FPY has been determined. When compared to beryllium and iron which are 11 and 30 dpa/FPY, respectively, we find that tungsten has the greatest radiation resistance against the fusion neutron irradiation among the candidates for first wall materials. This could be due to the large value of the crystal lattice bonding energy of tungsten. In order to investigate more precisely the radiation resistance of tungsten we calculated the density of stable Frenkel pairs created by PKA with different initial energy and momentum orientation. The number of created Frenkel pairs due to the displacement cascades is first determined by the NRT formula [9]. In our model for Ed we used the value of 90 eV found in [10]. At the next step we compare our results for tungsten obtained by SRIM [12] and MD calculations using ZBL and EAM potentials [27]. The cascade efficiency is defined as the ratio between the number of Frenkel pairs, calculated by MD or SRIM models and the theoretical number of Frenkel pairs in the displacement cascade determined by the NRT formula.

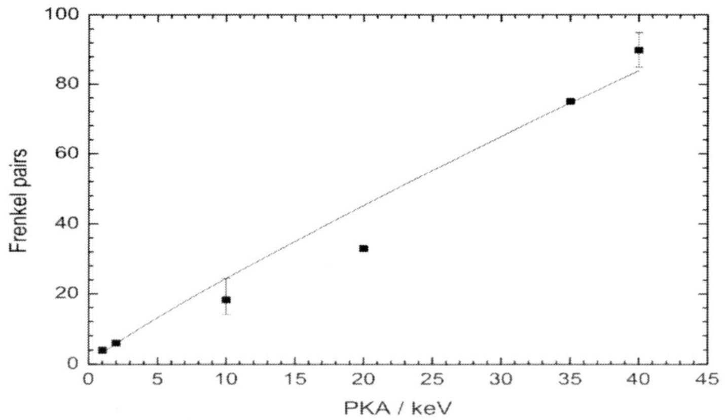

Figure 8. Frenkel pairs as a function of PKA energy in tungsten at 300 K

The cascade efficiency obtained by SRIM based on binary collision approximation shows large deviation for lower PKA energies ($E < 50$ keV) relative to the one of the MD with EAM potential results. The binary collision approximation overestimates the damages by factor of 2 at energies below 30 keV. Note that both models predict lower concentration of damages compared to the NRT formula [9]. Similar results have been obtained by Caturla et al [28], although they apply different method. We calculated by our MDCASK the number of stable Frenkel pairs as a function of the PKA energy 30 ps after the displacement cascade at temperature of 300K. The results are shown in figure 8.

A remarkable similarity is found between the cascade dimensions and PKA energy for (100), (010) and (001) lattice directions for vacancies. These results are similar to ones reported earlier for copper [29]. The differences exists in the cascades dimensions for tungsten which are generally 30 % lower. The time evolution of the cascade dimensions for PKA energies from 20 keV to 40 keV is shown in figure 9, for vacancies and interstitials. In the case of vacancy calculation the interstitials and interstitial clusters are removed before the simulation. We start the simulations with the 10 keV cascade in tungsten as shown in figure 9. It is seen that there is a variation in the dimensions for vacancies at about 10 % for time of 25 ps after the irradiation. The variation of the cascade dimensions in the case of interstitial is more prominent especially at lower energies. This is probably due to the high vacancy mobility and recombination, during the first stages of the displacement cascades in tungsten. Similar behaviour has been reported in [34, 35] although the interatomic potential and the simulation procedure have been different. On the other hand one can expect formation of single vacancies and di-vacancies, but not larger vacancy-clusters (nano-voids). The rise of the interstitial dimensions at 25 ps could be explained with the formation of sub-cascades which are difficult to be separated. The formation of sub-cascades leads to larger damage volumes compared to the case of a single cascade. We established that the formation of sub-cascades starts at PKA energy of 20 keV. Sub-cascades are dominant at energy of 40 keV. Our calculations using the Wigner-Seitz method show that in the early stages of the cascade the dominant are vacancies, interstitials and interstitial-clusters containing more than 3 atoms. The interstitial and interstitial clusters formation is about 2 times higher than those of vacancies at the same time scale.

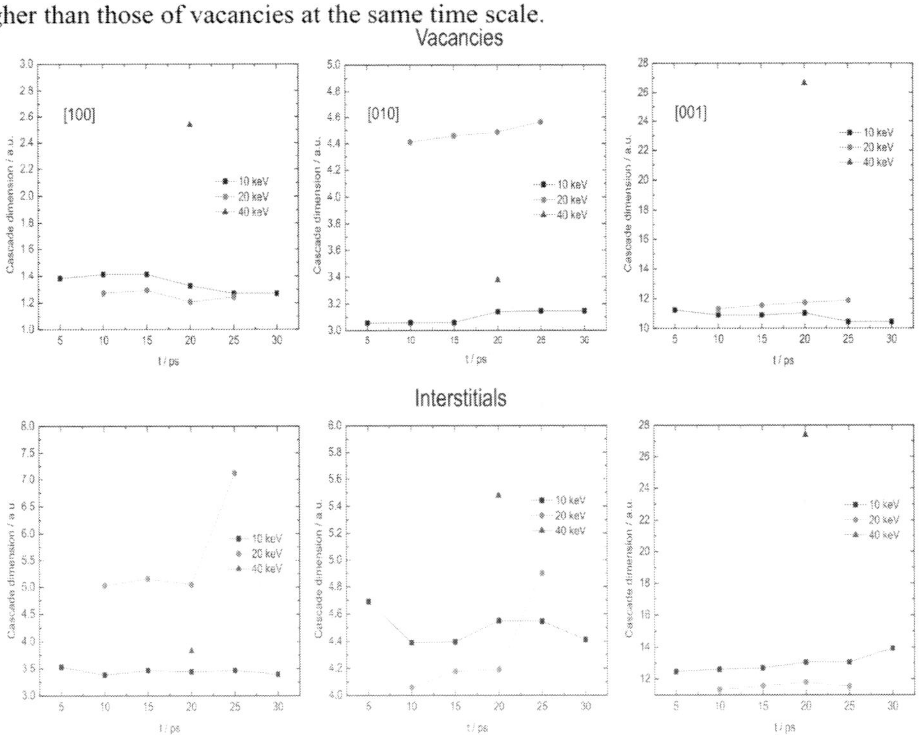

Figure 9. Time dependence between the cascades dimensions evolution time for vacancies and interstitials for PKA energies from 20 to 40 keV in tungsten at 300 K.

4. Conclusions

Numerical simulations of creation of point defects in beryllium, alpha-iron and tungsten irradiated by intensive fast (E > 0.1 MeV) neutron fuence up to 8×10^{20} n/cm^2 has been performed. Neutron interactions with first wall fusion materials like elastic and inelastic scattering, gamma ray emission and nuclear transmutations have been considered. The calculated damage of 0.06 dpa for Be was compared to the available data and a good agreement was found. He content was determined to be 168 appm. ^{6}Li content which is important for the tritium production was estimated to be 5 % of the He content. The probable cascade length for Be is between 100 and 150 dpa. The simulation results show that 70 % of the damages occur at Be PKA energies below 100 keV. The number of sub-cascades increases with the PKA energy and reach saturation at about 110 keV. The sub-cascade number is 50 for PKA energies close to 50 keV. The results indicate that the damages caused by He atoms are about 3 times less than damages caused by Be PKA. The final Be and He atoms distribution confirm that the most likely defects are single vacancies, divacancies and vacancy clusters. String-like small nano-voids with 5 and more vacancies are present but they are rare.

Point defects in ^{56}Fe, irradiated by intensive fast (E > 0.1 MeV) neutron fluence up to 8×10^{19} n/cm^2 has been performed. The calculated damage of 0.151 dpa and 37 dpa/fpy was compared to the available data in the literature and a good agreement was found. 4He content was determined to be 0.4 appm. ^{2}H content was estimated to be 7 % of the ^{4}He content. The correlation between the PKAs energy and the damages was established. It was shown that the most vacancies are created by PKA with energy below 50 keV with maximum around 30 keV.

The damages produced by displacement cascades in tungsten, irradiated with intensive neutron flux have been calculated to be 2.85 dpa/FPY for a neutron fluence of 2.5×10^{22} n/cm^2. The displacement cascades efficiency calculated by using the theoretical NRT formula is suitable only for rough estimation of the damages. The pair potential overestimates the defects production by a factor of 2. At higher PKA energy (E > 100 keV) the ZBL pair potential results and the EAM are comparable. The results for 3-dimensional cascades structure indicate that cascades are extended in the direction of PKA momentum. The cascade dimensions at lower energies at the early stages after the irradiation indicate that the interstitial volume is about 2 times larger than the vacancies volume. The created defects are single vacancies, di-vacancies, interstitials and interstitial clusters containing more than 3 atoms.

Acknowledgments

This work was supported by European Fusion Development Agreement (EFDA) colaboration and the European Commission under contract with EURATOM/INRNE Fusion Association. The authors gratefully acknowledge Dr. K. Sato, RRI-KU Japan and Drs. P. Staykov, E. Popov, S. Peneva and K. Berovski – EURATOM/INRNE FU-Association for their participation in discussions and their contribution to the technical preparation of the manuscript.

References

[1] Mishima Y et al. 2007 J. Nucl. Mater. **367-370** 1382-1386.
[2] Bolt H et al, 2004 J. Nucl. Mater. **307** 66.
[3] Becquart . S 2005 Nuclear Instruments and Methods in Physics Research B **228** 111-121
[4] Troev T, Nankov N, Petrov L and Popov E 2008 Research Letters in Physics doi:10.1155/2008/746892
[5] Ziegler J, Biersack J P, Littmark U 1985 *The Stopping and Range of Ions* Vol. 1, Pergamon, New York
[6] S. Shimakawa et al., Reactor dosimetry in the 21st century Proceedings of the 11th International Symposium on Reactor Dosimetry Brussels, Belgium, 18- 23 August 2002,
[7] MDCASK, https://asc.llnl.gov/computing_resources/ purple/archive/benchmarks/mdcask/mdcask.readme.html
[8] Daw S and Baskes M I 1984 Phys. Rev. B **29** 6443
[9] Norgett M J, Robinson M T, Torrens I M, 1975 Nucl. Eng. Design **33** 50.
[10] Gelles D et al. 1994 J. Nucl. Mater. **212-215** 29.
[11] Khomotov A et al. 2002 J. Nucl. Mater. **307-311**630-637
[12] *Materials Assessment Report* G 74 MA 10 01-07-11 W 0.2, Ch 2.1, pp 29
[13] Vladimirov P, Möslang A 2004 J. Nucl. Mat. **329-333** 233–237.

[14] SRIM-2006, www.srim.org
[15] Robinson M T 1992 Nucl. Instr. Meth. Phys. Res. B **67** 396
[16] Syslov D N et al, 2002 J. Nucl. Mater. **307-311** 664.
[17] Chakin V P, Ostrovsky Z Ye 2002 J. Nucl. Mater **307-311** 657-663.
[18] Troev T et al., 2006 J. Nucl. Mater **359** 93-101.
[19] Troev T et al., 2006 Nucl. Instr. Meth. Phys. Res. B **248** 297-304.
[20] Yoshie, T et al.
[21] Stoller R E 2000 J. Nucl. Mater. **276** 22
[22] Bacon D J et al., 2000 J. Nucl. Mater. **276** 1.
[23] Wirth B D, Bringa E M 2004 Physica Scripta **T108** 80-84.
[24] Caturla M J et al, 2000 J. Nucl. Mater. **276** 13.
[25] Rabaglino E et al. 2002 J. Nucl. Mater. **307-311** 1424-1429.
[26] Troev T, Nankov N, Petrov L and Popov E 2008 Research Letters in Physics 746892 (doi:10.1155/2008/746892)
[27] Zhou X W et al. 2001 Acta mater. **49** 4005-4015
[28] Caturla M J et al. 2001 J. Nucl. Mat. **296** 90-100
[29] Hou M 1981 Nucl. Instr. Meth. Phys. Res. B **182/183** 153

Journal of Physics: Conference Series **516** (2014) 012023 doi:10.1088/1742-6596/516/1/012023

Non-destructive research methods applied on materials for the new generation of nuclear reactors

I Bartošová, V Slugeň, J Veterníková, S Sojak, M Petriska, A Bouhaddane

Slovak University of Technology, Faculty of Electrical Engineering and Information Technology, Ilkovičová 3, Bratislava 812 19, Slovakia

E-mail: iveta.bartosova6@gmail.com

Abstract. The paper is aimed on non-destructive experimental techniques applied on materials for the new generation of nuclear reactors (GEN IV). With the development of these reactors, also materials have to be developed in order to guarantee high standard properties needed for construction. These properties are high temperature resistance, radiation resistance and resistance to other negative effects. Nevertheless the changes in their mechanical properties should be only minimal. Materials, that fulfil these requirements, are analysed in this work. The ferritic-martensitic (FM) steels and ODS steels are studied in details. Microstructural defects, which can occur in structural materials and can be also accumulated during irradiation due to neutron flux or alpha, beta and gamma radiation, were analysed using different spectroscopic methods as positron annihilation spectroscopy and Barkhausen noise, which were applied for measurements of three different FM steels (T91, P91 and E97) as well as one ODS steel (ODS Eurofer).

1. Introduction

GEN IV is an assemble of six theoretical concepts of fission reactors. Comparative advantages include reduced capital cost, enhanced nuclear safety, minimal generation of nuclear waste and reduction of the risk of proliferation. In comparison with present nuclear power plant technology the claimed benefits of the GEN IV reactors are [1]:

- Reduced lifetime of nuclear waste from 10^5 to 10^2.
- 100–300 times more energy yield from the same amount of nuclear fuel.
- The ability to use MOX and thorium fuel in a closed fuel cycle.

High temperatures, irradiation stress and consequential change in the mechanical properties of the components of these reactors, urge scientists over the world to examine and develop construction materials that would resist these severe conditions with minimal changes in their mechanical properties. Our research is aimed on FM and ODS steels. These materials are actually foreseen also for the use in the area of thermonuclear fusion in ITER construction [2].

2. Irradiation damage

Fusion neutrons, which originate from the Deuterium-Tritium (D–T) fusion reaction, interact with the walls, blanket and other structures that surround the plasma. Neutron damage weakens the construction materials and induces secondary radioactivity, which will require conscientious salvage procedures or deposition of the used materials as waste. The radioactivity in fusion power plants will only concern construction materials, because the product of the fusion reaction is only harmless helium. When a neutron collides with an atom of the primary wall or blanket, it can shoot out the atom from its ordinary position in the crystal lattice. The displaced atom can occupy the space between other atoms in their ordinary positions as interstitial and in his former position in the lattice a vacancy appears. This atom may have enough energy to distort the positions of other atoms before he finds his place in the lattice. Therefore the distortion usually occurs in a form of a cascade. The damage is quantified by dpa (displacements per atom) during the durability of the construction. On the primary wall, where the neutron flux is most intensive we expect that the damage state will be in order of hundred dpa [2]. With this kind of high dpa damage the material strength decreases and some parts of the wall will have to be repeatedly replaced during the functioning. The neutron can interact with the atom in the crystal lattice and leave a transmuted atom or atoms. The new atom is generally radioactive and is the primary source of radioactivity. The neutron can also shoot out a proton or alpha particle from the stricken atom. These reactions can be inscribed as (n,p) and (n,α) reactions. The reactions (n,α) are essential to neutrons with energies of 14 MeV that occur from the D–T fusion reaction. Protons and alpha particles from these reactions capture electrons and produce hydrogen and helium atoms in the crystal lattice. Individual atoms then form into clusters and produce bubbles of gas that decrease the strength of the construction. The process of irradiation damage is sufficiently studied in fission reactors and their understanding provides a solid base for studying and evaluating similar problems in the future fusion power plants.

3. Reduced activation materials

One of the most important advantages of fusion in comparison with fission is the absence of highly radioactive waste in the fusion fuel cycle. The construction will be irradiated due to neutrons and be therefore radioactive, but with a cautious selection of construction materials we can provide low activation so the radioactivity of a fusion power plant may decrease on a very low level in a period of less than 100 years. The blanket, primary wall and the system of diverters (in case of magnetic confinement) will have a definite durability and will have to be maintained with the help of robots. Important source of activation are also small impurities in the construction materials, so the testing of quality and purity of materials will be essential in the effort to achieve low activation. One of the construction materials that may be considered as low activated (LA) are known as FM steels, which are frequently used where we can encounter high temperatures for example in fission reactors. Iron and chromium, the main components of FM steels, have relatively good activation properties. The main source of activation comes from minor impurities, for example niobium. Therefore LA materials commute niobium for vanadium or titan or we can commute molybdenum for wolfram [2].

4. FM steels

FM steels consist of a feritic matrix and particles of martensit. The 9–12% Cr FM steels are being recently considered as main candidate materials for the first wall components of future fusion reactors. These materials can be used in components that have to resist high temperatures up to 550 °C. After exceeding 18% of chromium the material inclines to fracturing. Recent developments are aimed on the chemical composition to achieve low activation, fast decay properties after irradiation (RAFM steels) and also on the reduction of the shift of the ductile-brittle transition temperature after irradiation.

4.1. T91

T91 presents valuable advantages related to their excellent elevated temperature strength, creep behavior and improved resistance to thermal fatigue, and good transfer and low expansion coefficient compared to austenitic steels. This material has been identified as the ultimate candidate for DEMO in the frame of the international fusion program. Despite very limited data within MYRRHA (Multi-purpose hybrid research reactor for high-tech applications) operating conditions on neither the effect of irradiation, liquid metal and their synergy, T91 is the prime candidate material to be in contact with the LBE (Lead-Bismuth-Eutectic) and to contain the fuel in MYRRHA. This material seems to be the best compromise between fabricability and resistance to severe irradiation conditions in the operating temperature range from 200 to 450 °C [3].

Table 1. Chemical composition of material T91 [3]

element	C	Si	Mn	P	S	Cr	Ni	Nb	N	V	Mo
intensity (%)	0.11	0.4	0.4	-	-	8.3	0.13	0.08	0.02	0.2	0.95

4.2. P91

P91, is widely used in the construction of power plants and other sectors involving temperatures higher than 500 °C. The heat resistant martensitic steel P91 is being utilized in conventional power plants for piping systems with operating temperatures of 560 – 600°C and high pressures in the range of 270 bars. The design of welded structures requires resistance to high temperatures and pressures for up to 30 years. Therefore, the selection of the correct welding consumables is of great importance to prevent premature equipment failure. Although the creep strength is the prime consideration for elevated temperature applications, notch toughness is also important. P91 steel weld fusion zone toughness depends on factors such as welding process, chemical composition, and flux composition. Niobium and vanadium are the main alloying elements that significantly influence the toughness as well as creep strength.

Table 2. Chemical composition of material P91 [4]

element	C	Si	Mn	P	S	Cr	Ni	Nb	N	V	Mo	Al
intensity (%)	0.1	0.45	0.5	0.019	0.002	9.12	0.05	0.06	0.04	0.21	0.96	0.004

4.3. Eurofer97(E97)

E97 has been developed as a possible first wall and breeder blanket structural material for fusion applications. E97 is a representative of RAFM steels, manufactured on an industrial scale. Further reduction of impurities and therefore achieving lower activation is possible. The properties of E97 meet the requirements of the test blanket module of ITER. The mechanical and compatibility properties are adequate, also in the neutron flux exposure environment expected in ITER. The potential for applying E97 in the blanket for a demonstration fusion power plant is there, but experimental verification that simulates the near plasma environment remains essential. E97 has a tempered martensitic microstructure which permits operation at relatively high temperatures (500 °C), offers good dimensional stability under high neutron doses and exhibits higher swelling resistance in comparison to austenitic steels. Future developments are aimed on obtaining a fine microstructure with small prior austenite grains and a low number of inclusions which should provide an optimum combination of toughness, strength and high temperature properties .

Table 3. Chemical composition of material E97 [6]

element	Fe	W	Mn	P	S	Ta	C	V	N_2	Cr
intensity (%)	88.805	1.1	0.4	<0.005	<0.005	0.12	0.11	0.2	0.03	9.0

5. ODS Eurofer

Oxide dispersion strengthened (ODS) steels are considered as leading candidates of structural materials for advanced nuclear systems such as GEN IV fission power plants and DEMO fusion reactor because of their excellent creep strength [7], corrosion [8] and radiation resistance [9]. To apply ODS steel in these advanced nuclear systems with huge and complex structures, suitable bonding and welding techniques need to be developed. These techniques must provide such a process that the microstructures with very fine grain and homogeneous distribution of nano-scaled oxide particles are not remarkably changed by the joining processing [10].Whereas the temperature window of FM steels is limited to 550 – 650 °C by degradation of mechanical strength, ODS steels have the potential to increase this limit considerably up to 1000 °C. The main problem in the development of ODS steels is the inefficient production of material by the means of powder metallurgy and thermal isostatic pressing. ODS Eurofer is strengthened by particles of Y_2O_3. E97 mainly differs from ODS Eurofer in properties that relate with the formation of defects. One of the first applications of ODS Eurofer might be in the dual coolant lead lithium (DCLL) BB concept. ODS Eurofer layers will be plated onto the first wall made of E97 steel. The feasibility of joining ODS Eurofer and E97 by diffusion welding was recently demonstrated [11].

6. Positron annihilation lifetime spectroscopy (PALS)

Positron (e^+) is the antiparticle of electron (e^-). The principal of PALS is the ability of a positron emitted from the source due to β^+ decay thermalize and diffuse into the material, until it gets trapped by a defect and encounters an electron with which it annihilates due to electromagnetic interactions. The annihilation results in two annihilation photons γ with energies of 511 keV. When the energetic positrons ($E_{\rm kin} \approx 0.1– 1$ MeV) enter the condensed medium, they rapidly lose their energy. At highest positron energies, the positron excites core electrons when it collides with the host atoms, so the main mechanism of energy losses is ionization. At lower energies the electron-hole excitations is more present. When the positron energy has declined to a fraction of eV, the scattering off phonons dominates. Nevertheless positrons reach thermal equilibrium with the medium, due to phonon emission and absorption. The initial kinetic energy of positrons during thermalization drops below 0.1 eV. Despite the rapid loss of energy the positron thermalization time $t_{\rm th}$ is only a few picoseconds [12]. During diffusion, the positron eventually annihilates with an environmental electron. In homogeneous defect-free media all positrons annihilate with the same rate $\lambda_{\rm b}$ which is a characteristic of the given material. In a medium with defects the positron may become trapped in open-volume defects (monovacancies, larger vacancy clusters, dislocations etc.) because they represent a potential well for positrons. The transition from the delocalized state to the localized one is called positron trapping. Thermal detrapping is practically impossible from such deep wells and positrons remain trapped until annihilation. There may occur phonon-assisted detrapping if a positron trap is shallow enough ($E_{\rm b}$ < 0.1 eV). The lifetime of the trapped positrons (LT2) is longer than the lifetime of positrons which annihilate within the material (LT1) due to the lowered density of electrons near the defect. Positron can form a quasi-stable state with a host electron called positronium (Ps) which is an analogue of the hydrogen atom [12]. The time interval between the injection of positrons into a sample and the instant of annihilation is known as the positron lifetime spectrum and has a statistical distribution. In homogeneous defect-free materials, the lifetime spectrum is a simple decaying exponential. When the material contains defects, the lifetime spectrum becomes a superimposition of decaying exponentials. The principal of PALS is based on injecting positrons into a sample and measuring their lifetime spectrum. Our positron source is a β^+ radioactive isotope ^{22}Na (figure 1).

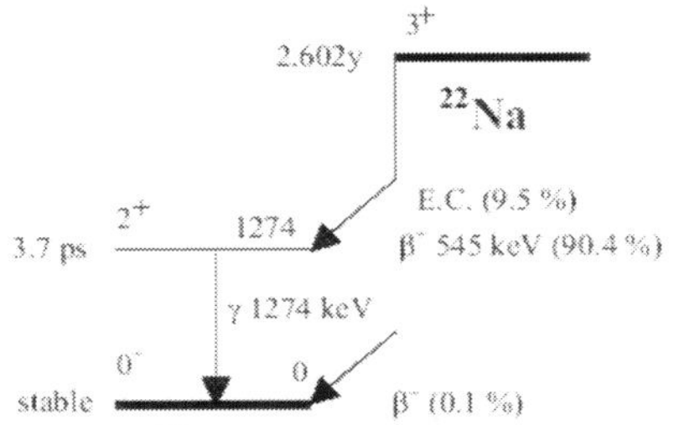

Figure 1. ^{22}Na decay scheme

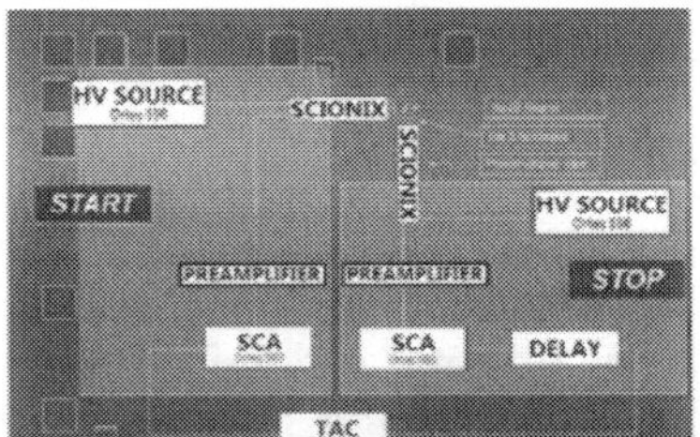

Figure 2. Basic scheme of PALS equipment setup in PAS laboratory at STU Bratislava [13]

The advantage of this isotope is the emission of a photon with the energy of 1.274 MeV in coincidence with the positron which gives the signal that indicates the birth of the positron (START signal). The signal of annihilation is either one of the two annihilation photons with energies of 0.511 MeV (STOP signal). Typical measuring equipment contains two detectors (scintilator + photomultiplier), two energy analyzers (SCA), time to amplitude converter (TAC) and a multichannel analyzer (MCA) that records particular events. Simple scheme of this setup is shown in figure 2.

7. Barkhausen noise

Ferromagnetic materials consist of domains. These domain are magnetized along a certain crystallographic direction. Domains are separated from one another by domain walls. AC magnetic fields cause domain walls to move back and forth. In order for a domain wall to move, the adjacent domain of the wall has to increase in size while the domain on the opposite side of the wall shrinks. As a result the overall magnetization of the sample changes. If we place a coil of conducting wire near the sample while the domain wall moves, the resulting change in magnetization will induce an electrical pulse in the coil. Barkhausen effect proves that the magnetization process, which is represented by the hysteresis curve, is not continuous, but is made up of small, abrupt steps caused when the magnetic domains move under an applied magnetic field. When the electrical pulses produced by all domain movements are summed, a noise-like signal called Barkhausen noise is generated. Due to magnetoelastic interaction, in materials with positive magnetic anisotropy, compressive stresses will decrease the intensity of Barkhausen noise while tensile stresses increase it. This fact can be exploited so that by measuring the intensity of Barkhausen noise the amount of residual stress can be determined. The measurement also describes the direction of principal stresses.The micro-structure of the sample is another important material characteristic affecting Barkhausen noise. The noise intensity continuously decreases in microstructures characterized by increasing hardness. Barkhausen noise measurements provide information on the microstructural condition of the material.

8. Measurements

The source of positrons contains particles that annihilate positrons, therefore we have to calibrate the source before every use to estimate the amount of positrons annihilating in the source. The calibration is performed by measuring the spectrum of an almost defect-free reference material, for which we know the characteristic lifetimes of the positrons, in our case pure silicium. After the measurement of silicium we measure the spectrum of the investigated material and consecutively compare the lifetimes of positrons with theoretical values of known defect types.

Meanings of components in table 4: LT1 – lifetime of positrons in bulk, LT2 – lifetime of positrons in defects, I1– Intensity of LT1, I2 – Intensity of LT2, MLT – mean lifetime of positrons, FV – Fit´s variance.

From figure 3 and table 4 we can see, that every material was described by two LT's, which means that all materials contain defects in their microstructure. From theoretical values we can deduce that materials T91 and P91 contain divacancies and E97 monovacancies with dislocations.

Figure 5 represents the intensities of defects in particular samples. The intensities are as follows: in the material E97 74%, in P91 43% and in T91 52%. The next non-destructive method we used to study materials was the method of Barkhausen noise. The output noise, produced by the rotating domain walls, can be described by an envelope. The bigger the envelope the less amount of residual stress the material exhibits, because the domains move more freely due to lack of defects. The measurement confirms the accuracy of the previous results. The material P91 exhibits minimal residual stress and thus contains least defects. After the material P91 follow materials T91, E97 and ODS Eurofer. Due to the lack of time, material ODS Eurofer was not measured by PALS so we cannot confirm the accuracy of results for this material.

Journal of Physics: Conference Series **516** (2014) 012023 doi:10.1088/1742-6596/516/1/012023

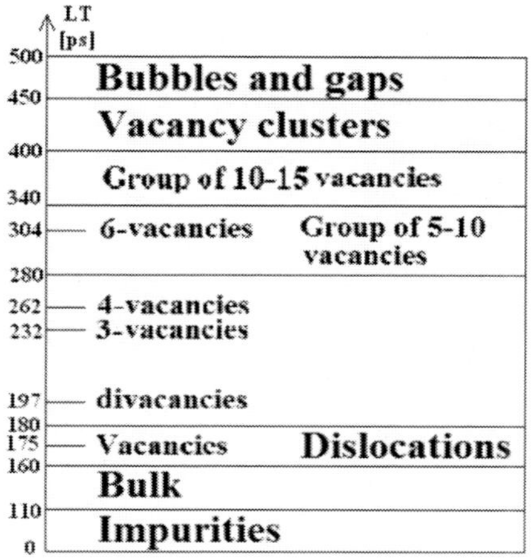

Figure 3. Theoretical lifetimes of positrons in defects [14]

Table 4. Measurements of materials T91, P91 and E97

Sample	T91	P91	E97
LT1 (ps)	110 ± 7	117 ± 7	83 ± 15
I1 (%)	48 ± 5	57,2 ± 1	24,4 ± 5
LT2 (ps)	199 ± 9	207 ± 2	178 ± 8
I2 (%)	51,5 ± 5	42,5 ± 1	73,8 ± 5
FV	1,07	1,02	1,01
MLT (ps)	160	159	161

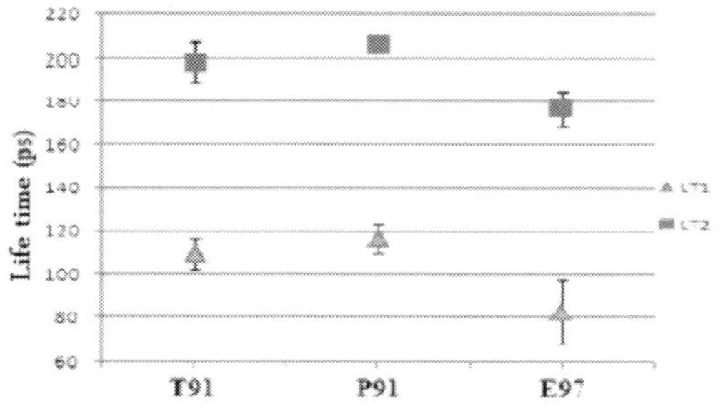

Figure 4. Lifetimes of positrons.

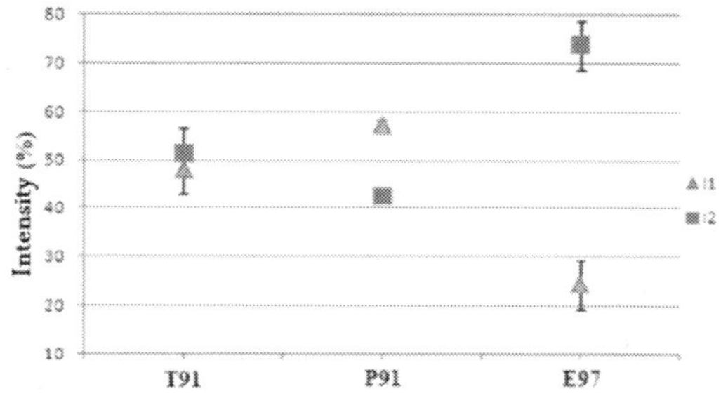

Figure 5. Intensities of positrons.

9. Summary

10.Materials P91, T91 and E97 were studied by PALS and the method of Barkhausen noise. The material P91 showed most resistant to the formation of defects. Unfortunately, this material can resist temperatures only in the range of 560- 600 °C, whereas ODS steels can resist temperatures exceeding 1000°C. All the materials are prime candidates for components for the GEN IV, ITER, DEMO and other applications. Nevertheless, we have to keep improving and developing construction materials, so we can provide efficient contribution to the safe design of advanced nuclear facilities.

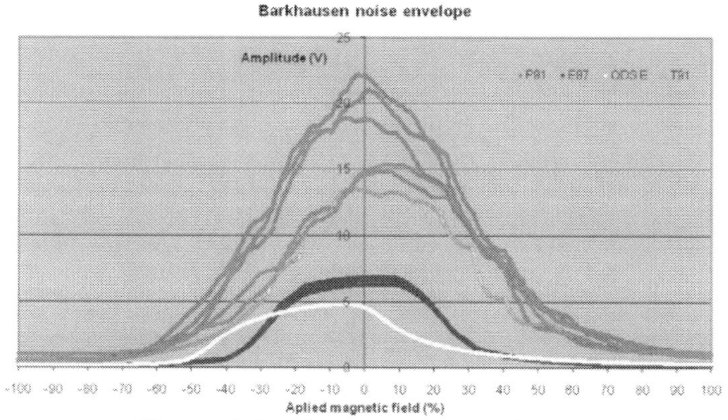

Figure 6. Barkhausen noise envelope.

References

[1] US DOE Nuclear Energy Research Advisory Committee, *A Technology Roadmap for Generation IV Nuclear Energy Systems* 2002 GIF-002-00. http://nuclear.energy.gov/genIV/documents/gen_iv_roadmap.pdf.

[2] McCracken G, Stott P 2005 *Fusion - The Energy of the Universe* **183**

[3] Sapundjiev D, Mazouzi A Al, Van Dyck S 2006 *Journal of Nuclear Materials* **356** 229-236

[4] P J Ennis et al. 2002 *In OMMI* **1** 1

[5] Petersen C 1997 *Proceedings of IEA W.G. Meeting on RAF/M Steels*, JAERI, Tokai, Ibaraki, Japan 134.

[6] Lindau R, Möslang A, Rieth M, Klimiankou M, Materna-Morris E, Alamo A, Tavassoli A-A F, Cayron C, Lancha A-M, Fernandez P, Baluc N, Schäublin R, Diegele E, Filacchioni G, Rensman J W, Schaaf B v d, Lucon E, Dietz W 2005 *Fusion Engineering and Designs* **75-79** 989-996

[7] Ukai S, Okuda T, Fujiwara M, Kobayashi T, Mizuta S and Nakashima H 2002 *J. Nucl. Sci. Technol.* **39** 872–879.

[8] Cho H S, Ohkubo H, Iwata N, Kimura A, Ukai S and Fujiwara M 2006 *Fusion Eng. Design* **81** 1071–1076.

[9] Yamashita S, Akasaka N, Ukai S and Ohnuki S 2007 *J. Nucl. Mater.* **367–370** 202–207.

[10] Lindau R, Möslang A, Rieth M, Klimiankou M, Materna-Morris E and Alamo A *et al.* 2005 *Fusion Eng. Design* **989–996** 75–79.

[11] Lindau R, Möslang A, Rieth M, Klimiankou M, Materna-Morris E, Alamo A, Tavassoli A A F, Cayron C, Lancha A M, Fernandez P, Baluc N, Sch¨aublin R, Diegele E, Filacchioni G, Rensman J W, Schaaf B v d, Lucon E, Dietz W 2005 *Fusion Eng. Des.* **75–79** 989–996.

[12] Procházka I 2001 *Materials Structure* **8** 55-60

[13] Kršjak V, Slugeň V 2008 *Positron annihilation study of advanced nuclear reactor materials,* Bratislava

[14] Natesan K, et al. 2006 *Argonne National Laboratory Report,* Chicago, ANL/EXT-06/45

Journal of Physics: Conference Series **516** (2014) 012024 doi:10.1088/1742-6596/516/1/012024

Investigation of materials for fusion power reactors

A. Bouhaddane, V. Slugeň, S. Sojak, J. Veterníková, M. Petriska and I. Bartošová

Department of Nuclear Physics and technology: Faculty of Electrical Engineering and Information Technology, Slovak University of Technology, Ilkovičova 3, 812 19 Bratislava 1, Slovakia

E-mail: amine.bouhaddane@centrum.sk

Abstract. The possibility of application of nuclear-physical methods to observe radiation damage to structural materials of nuclear facilities is nowadays a very actual topic. The radiation damage to materials of advanced nuclear facilities, caused by extreme radiation stress, is a process, which significantly limits their operational life as well as their safety. In the centre of our interest is the study of the radiation degradation and activation of the metals and alloys for the new nuclear facilities (Generation IV fission reactors, fusion reactors ITER and DEMO). The observation of the microstructure changes in the reactor steels is based on experimental investigation using the method of positron annihilation spectroscopy (PAS). The experimental part of the work contains measurements focused on model reactor alloys and ODS steels. There were 12 model reactor steels and 3 ODS steels. We were investigating the influence of chemical composition on the production of defects in crystal lattice. With application of the LT 9 program, the spectra of specimen have been evaluated and the most convenient samples have been determined.

1. Introduction

Rising energy needs of human society require the creation of modern electric sources. These are for example the Generation IV fission reactors and thermonuclear reactors. Running these modern facilities is a difficult issue in terms of requirements for materials. Increased radiation damage to materials significantly limits operational life as well as the safety of reactors. It is necessary to search and explore more resistant substances that find their application in these facilities. This paper is focused on the fusion reactors and materials to use for their construction. The detection of the effect of chemical composition of construction materials on their structure and hence the resistance to radiation stress plays an important role. This work will refer the possibility of analysis using positron annihilation spectroscopy and program LifeTime 9. Proper interpretation of the results obtained on the basis of broad knowledge related to the topic is essential.

2. Radiation damage

2.1. Interaction of radiation with matter

During operational lifetime of nuclear facilities there is radiation damage. The radiations of concern here include charged particles such as electrons (beta particles), protons, alphas and fission fragment ions, and the neutral radiations including photons (gamma and X rays) and neutrons.

Table 1 compares some key characteristics of the radiation, including charge, mass and range in air. For a kinetic energy of 1 MeV, the electron acts as a relativistic particle (0.94*c*). For the same energy, the heavier particles are slower, stopped easier and deposit their entire energy over a shorter distance [1].

Table 1. Comparison of ionising radiation [1]

Characteristic	Radiation (E_k = 1MeV)				
	Alpha (α)	Proton (p)	Beta (β) Electron (e)	Photon (γ or X ray)	Neutron (n)
Symbol	$^{4}_{2}\alpha$ or He^{2+}	$^{1}_{1}p$ or H^{1+}	$^{0}_{-1}e$ or β	$^{0}_{0}\gamma$	$^{1}_{0}n$
Charge	2	1	-1	0	0
Ionization	Direct	Direct	Direct	Indirect	Indirect
Mass (u)	4.001506	1.007276	0.0005486	-	1.008665
Velocity (m/s)	6.94E+05	1.38E+07	2.82E+08	3.00E+00	1.38E+07
Speed of light [% of c]	2.30	4.60	94.10	100	4.60
Range in air [m]	$5.6x10^{-3}$	$1.81x10^{-2}$	$3.19x10^{-2}$	820 *	392.5 *

* range based on a 99.9% reduction

2.2. Radiation effects in materials

Due to influence of irradiation on structural lattice and its interaction there are these processes [1]:

- ***Impurity production***, that is, transmutation of nuclei into other nuclei which themselves may be radioactive; this mechanism is caused by neutrons through fission and activation (capture). Impurities can also be deposited from the creation of hydrogen or helium when a proton or an alpha particle, respectively, becomes neutralized in the material of passage.
- ***Atom displacement*** from their normal position in the structure of the material; displacement atoms may leave lattice vacancies and lodge in interstitial locations or cause interchange of dissimilar atoms in the lattice structure.
- ***Ionization***, that is, the removal of electrons from atoms in the material and the formation of ion pairs in the path of the charged particles
- ***Large energy release*** in a small volume, which can result in thermal heating of the material. This may be especially important in those cases where the material is a radiation shield [1].

2.3. Types of defects in materials

Few, if any, crystals are perfect in that all unit cells consist of the ideal arrangement of atoms or molecules and all cells line up in a three dimensional space with no distortion. Some cells may have one or more atoms less whereas others may have one or more atoms than the ideal unit cell. The imperfections in crystal are called crystal defects. Defects may be classified into four categories depending on their dimension [2]:

- **0D, Point defects**: atoms missing or in irregular places in the lattice vacancies, interstitials, impurities),
- **1D, Linear defects**: groups of atoms in irregular positions (e.g. screw and edge dislocations),
- **2D, Planar defects**: the interfaces between homogeneous regions of the material (grain boundaries, external surfaces),
- **3D, Volume defects**: extended defects (voids, Stacking Fault Tetrahedra, pores, cracks).

3. Structural materials of fusion reactors

Nuclear fusion is a process by which multiple atomic nuclei join together to form a single heavier nucleus. It normally takes place on the surface of sun, when Hydrogen nuclei collide and create heavier Helium atoms while releasing tremendous amounts of energy in the process. To achieve such a process you need to get Hydrogen atoms close to each other to overcome repulsive Coulomb force. It is possible by accumulating energy. A fusion reaction requires nonetheless temperatures of ~$150x10^6$ K to take place - ten times higher that the H-H reaction occurring at the Sun's core. At extreme temperatures, electrons are separated from nuclei and a gas becomes plasma - a hot, electrically charged gas. In a star as in a fusion device, plasmas provide the environment in which light elements can fuse and yield energy. Currently there are few facilities which experiments with fusion reaction (JET) and new facility ITER (International Thermonuclear Experimental Reactor) which is in building process. The fusion reaction will be achieved in a TOKAMAK (Toroidal chamber with magnetic coils) device that uses magnetic fields to contain and control the hot plasma [3].

In fusion reactors, the plasma facing (first wall, divertor) and breeding blanket components will be exposed to plasma particles and electromagnetic radiation and will suffer from irradiation by an intense flux of 14.1 MeV neutrons. The high-energy fusion neutrons will produce atomic displacement cascades and nuclear transmutation reactions within the irradiated materials. From the point of view of materials science, atomic displacement cascades induce the formation of point defects (i.e. vacancies, interstitial atoms, vacancy and interstitial clusters) and segregation of alloying elements, while nuclear transmutation reactions produce helium and/or hydrogen gas atoms. The final microstructure of the irradiated materials results from a balance between environmental conditions, especially radiation damage and temperature, and stress/strain histories [4].

Extreme conditions of a fusion reaction lead into high demands on structural materials. What we expect from these materials is high resistance to radiation damage, so they can be in operation for a long period and a low level of activation. It means that it will be possible to handle them in a shorter period after removing from a power plant. For this purpose, new materials were designed.

3.1. RAFM steels

The development of ferritic/martensitic steels for applications in fusion devices emanates from the limitations of the austenitic stainless steels and the promising high dose experience with ferritic/martensitic steel fuel cladding in liquid metal cooled fast reactors. The austenitic steels suffer from helium embrittlement at elevated temperatures and swell to a degree not acceptable for fusion reactor components. The ferritic/martensitic steels exhibit superior performance to austenitic steels in both swelling and helium embrittlement resistance. At a temperature of about 575 K the swelling rate of ferritic/martensitic steels is about 1 vol% after 100 dpa, while it is about 1 vol% after 10 dpa for typical austenitic steels. Ferritic/martensitic steels also exhibit a better surface heat capability than austenitic steels (5.4 kWm^{-1} at 675 K, i.e. about three times that of austenitic steels) [5], favourable cost, availability and service experience, and their good compatibility with aqueous, gaseous and liquid metal coolants permits a range of design options. At the same time ferritic/martensitic steels with alloying elements such as chromium, some tungsten, vanadium and tantalum activate little compared to conventional austenitic stainless steels with nickel and molybdenum. In addition, manganese-stabilized austenitic stainless steels are not attractive due to high decay heat (safety) concerns. In Europe, Japan and Russia it has been demonstrated that it is now feasible to produce RAFM steels on an industrial scale with sufficiently low impurity levels [6–11]. The promise remains to produce low activation varieties that allow recycling within a century [12, 13]. Main RAFM steels that are being investigated include the Chinese low activation martensitic (CLAM) steel, the European EUROFER 97 alloy, the Japanese F82H and JLF-1 alloys, and the Russian RUSFER-EK-181 alloy. Their composition lies in the following range: Fe-(7.5–12)Cr-(1.1–2)W-(0.15–0.25)V, in weight percent.

The major issues remaining with the RAFM steels are:

- Limited strength at high temperatures. RAFM steels show a drop in tensile strength at about 825K and a strong reduction in creep strength at T = 875 K. In addition, softening occurs during cyclic loading, which may lead to maximum allowable loads much smaller than the limits predicted by the current design rules.
- Irradiation-induced hardening and embrittlement effects at temperatures T <675 K.
- The production of high amounts of helium and hydrogen under fusion neutron irradiation conditions.

- Fabrication techniques, in particular joining, as RAFM steels are more difficult to weld than austenitic stainless steels.
- Limited database at high irradiation doses [4].

3.2. ODS steels

Oxide dispersion strengthened (ODS) ferritic alloys (e.g. MA957, MA956) developed by the International Nickel Company (INCO) are also considered for application in future nuclear facilities. The oxide dispersion, which is in case of MA957 & MA956 yttrium, is intending to provide high temperature strength. The alloys are produced by "mechanical alloying", which is a powder metallurgy process which distributes the alloy elements homogeneously on a very fine scale [14]. Chemical composition of yttrium strengthened ODS alloys with addition of titanium and molybdenum in order to improve strength, ductility and oxidation resistance are shown in Table 3. They belong to main candidate materials for first wall (plasma facing and structural material) and breeding blanket components.

4. Measurement

Practical part is aimed on investigation of influence of chosen elements and impurities on mechanical behaviour and sensitivity to radiation damage of model reactor steels used in present reactors (VVER-1000) and of ODS steels. All primary material preparation of reactor steels was coordinated by the scientific technological company OOO "Nucleon-M" in Moscow, Russia in 2004.

4.1. Chemical composition of samples

The nominal base compositions of the 12 model steels are derived from typical Russian and Western RPV materials for:

1. WWER-1000 (15Cr2NiMoV steel grade) and
2. PWR (ASTM A533-B steel grade).

This choice is explained by striving to reveal the possible distinction of different compositions in their sensitivity to the deleterious element components. The studying material includes mainly various Ni and Mn combinations and nearly the same range of Si content (see table 2).

Table 2. Chemical composition of model reactor steels (% of mass) [16].

Sample	C	Si	Mn	Cr	Ni	Mo	V	Cu	S	P
Ap631	0.11	0.28	0.43	2.22	<0.02	0.71	0.10	0.09	0.008	0.010
Bp632	0.11	0.26	0.38	2.19	0.99	0.70	0.10	0.10	0.008	0.010
Cp633	0.12	0.24	0.38	2.13	2.00	0.69	0.10	0.10	0.008	0.010
Dp634	0.11	0.23	0.83	2.13	2.00	0.68	0.10	0.09	0.008	0.009
Ep641	0.12	0.33	0.77	2.16	1.02	0.70	0.10	0.10	0.008	0.009
Fp642	0.12	0.33	1.37	2.15	1.02	0.70	0.10	0.10	0.008	0.010
Gp643	0.11	0.32	1.36	2.06	1.99	0.69	0.10	0.10	0.008	0.009
Hp644	0.12	0.51	1.31	2.07	2.00	0.69	0.10	0.10	0.008	0.010
Kp301	0.17	0.35	0.78	0.10	0.58	0.64	-	0.07	0.005	0.009
Lp302	0.18	0.35	0.77	0.08	0.96	0.63	-	0.05	0.005	0.010
Mp303	0.16	0.37	0.74	0.09	1.90	0.61	-	0.05	0.005	0.010
Np304	0.16	0.33	1.27	0.07	1.97	0.63	-	0.06	0.005	0.010

Table 3. Chemical composition of ODS steel MA957 (mass %) [15].

Sample	C	Si	Mn	Ni	Cr	Mo	Ti	Al	Y_2O_3*
MA957	0.03	0.04	0.09	0.13	13.7	0.30	0.98	0.03	0.28
MA956	0.03	0.05	0.06	0.11	21.7	<0.05	0.33	5.77	0.38

* *Estimated from yttrium content assuming it is contained in the form of Y_2O_3*

These limits reflect the common range of Ni and Mn variations in WWER and PWR RPV materials. Potentially, silicon plays an active role in the radiation damage of ferritic steels but its content in RPV materials varies within rather a narrow range and that is why this element was concluded to be of secondary importance. For this reason the most part of the prepared materials should contain the nominal 0.35 % of Si excluding one model steel of WWER type. In the latter case the Si content is increased up to about 0.6 % simultaneously with the maximum 2 % of Ni and 0.5 % of Mn, respectively. The impurity elements (as Cu, P, S) content should also reach the required level close to the upper concentration limit typical for WWER-1000 RPV materials [16].

4.2. Investigation method and evaluation program
Positron annihilation lifetime spectroscopy (PALS) was chosen for this measurement, because of its sensitivity on vacancy defects. Obtained annihilation spectra were processed by program LifeTime 9 [17]. From LT9 we gain output data like positron lifetime, intensity of positrons or mean lifetime. Positron lifetimes were divided into: τ_1 (positrons annihilating in bulk according to standard trapping model), τ_2 (positrons annihilating in defects) and τ_3 (in-flight annihilation). Similar partition is valid for intensities of positrons (I_1, I_2 and I_3). For the ODS steels investigation, 2-component analysis was chosen. We neglected τ_3 and I_3 (which refers to in-flight annihilation), because their values were very low and brought an error into evaluation.

5. Results

5.1. Model reactor steels
Figure 1 and 2 show results for model reactors steels. The analysis is based mainly on values τ_2 and I_2. In samples Ap631-Dp634, content of Ni was rising and so τ_2 rose too (while amount of the rest elements was not changing significantly). From lower I_2 values results presence of larger clusters of defects.

Samples Ep641-Hp644 have apparently lower values of τ_2. It is probably caused by the highest amount of Ni and Mn, which are known for decorating defects from the inside and so they seem to be smaller.

The highest I_2 values in this measurement reveal us, that defects in these samples are distributed homogenously.

Remaining samples Kp301- Np304 are slightly different in comparison to previous 2 groups, because of zero V content and slightly increased C content. Characteristic for them is significant drop of Cr content (≤ 0.1). Evaluation was the most complicated due to multiple changes in composition (share of many elements was changing at the same time). Reduction of Cr caused increased τ_2 values. By Np304 we expected lower value of τ_2 than by Mp303, but it did not happen, probably due to obliteration of marking into the sample.

5.2. ODS steels
Figure 3 represents results for ODS steels. The evaluation is based on τ_1, τ_2, I_1 and I_2 values. In case of τ_1 we cannot talk about annihilation only in bulk, because values are moving around 202 ps. The increase of τ_1 is caused by strengthening oxides (Y_2O_3) and partly by 2-component analysis. According to I_2, contribution of large defects was very low. Variances of intensities were less than 1%.

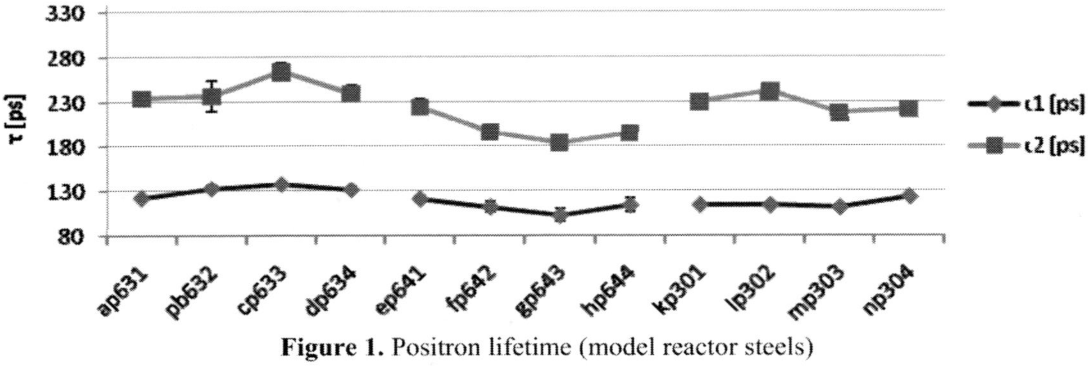

Figure 1. Positron lifetime (model reactor steels)

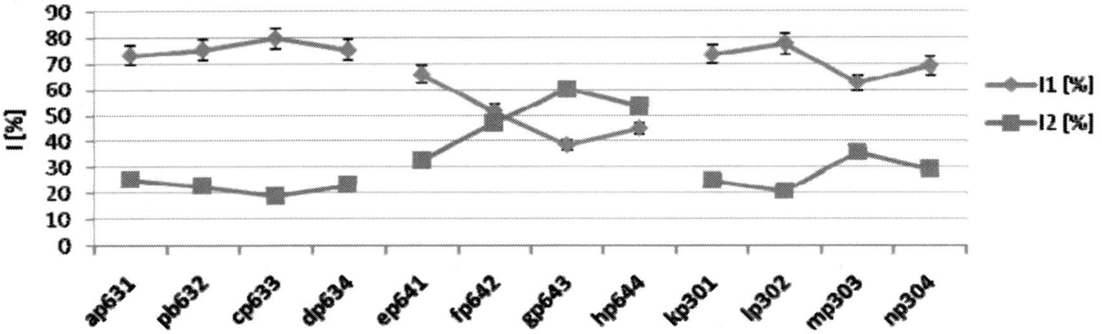

Figure 2. Intensity of positrons (model reactor steels)

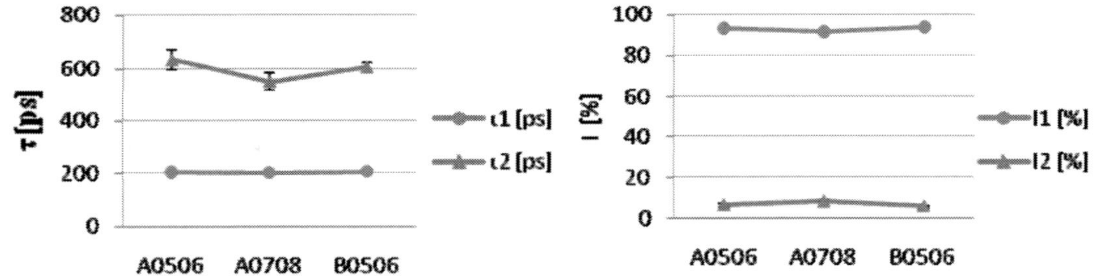

Figure 3. Positron lifetime (left) and intensity (right) of positrons (ODS steels

6. Conclusion

This paper was aimed on investigation of radiation damage to structural materials of nuclear facilities with increased concern over materials for fusion power reactors. The experimental part included 2 measurements – model reactor steels and ODS steels. Results indicate the influence of alloying elements on defects occurrence. Higher content of Cr decreases amount of defects so the steel has better mechanical and thermal properties. By samples with higher amount of Ni and Mn, defects seem to be smaller, but it is probably caused by the ability of Ni and Mn to decorate interior of defects.

Second group of model reactor steels was evaluated as the best according to PALS (based on τ_2 comparison). By ODS steels, almost no defects were observed. Positron annihilation spectroscopy is a powerful tool of investigation vacancy type defects. It is necessary to do more measurements to confirm the influence of alloying elements on defects appearance and hence on properties of steels.

References

[1] Holbert K E 2006 *Radiation Effects on Materials: Displacement Damage* lectures from *Nuclear Concepts for the 21st Century* course, Arizona State University, http://www.eas.asu.edu/~holbert/eee460/RadiationEffectsDamage.pdf

[2] Koutský J, Kočik J 1994 *Radiation Damage of Structural Materials* Prague: Academia

[3] http://www.iter.org/sci/Pages/WhatisFusion.aspx

[4] Baluc N et al. 2007 Status of R&D activities on materials for fusion power reactors *Nucl. Fusion* **47** 696-717

[5] Wong C P C et al. 2000 Helium-cooled refractory alloys first wall and blanket evaluation *Fusion Eng. Des.***49–50** 709–17

[6] Rieth M et al. 2003 EUROFER 97—tensile, Charpy, creep and structural tests *Scientific Report FZKA* 6911

[7] Lindau R et al. 2005 Present development status of EUROFER and ODS-EUROFER for application in blanket concepts *Fusion Eng. Des.* **75–79** 989–96

[8] Kimura A et al. 2006 Ferritic steel-blanket systems integration R&D—compatibility assessment *Fusion Eng. Des.* **81** 909–16

[9] Materna-Morris E *et al* 2004 Mechanical and structural characterization of EUROFER 97-2 *Scientific Report FZKA* 7117 p 188

[10] Moeslang A 2005 Towards reduced activation structural materials data for fusion DEMO reactors *Nucl. Fusion* **45** 649–55

[11] Ioltukhovskiy A G et al. 2005 The heat resistant RAFMS RUSFER-EK-181 for fusion and fast breeder power reactors applications *12th Int. Conf. on Fusion Reactor Materials (Santa Barbara)* Contribution 08-45

[12] Ehrlich K et al. 1996 The development of structural materials for reduced long-term activation *17th Int. Symp. on Effects of Radiation on Materials ASTM STP 1270* ed D.S. Gelles *et al* pp 1109–22

[13] Zucchetti M et al. 2005 The feasibility of recycling and clearance of active materials from a fusion power plant *12th Int. Conf. on Fusion Reactor Materials (Santa Barbara)* Contribution 03B.3

[14] Hamilton M L, Gelles D S, Lobsinger R J, Paxton M M and Brown W F 2000 Fabrication Technology for ODS Alloy MA957 PNL-13165 10

[15] Klueh R L, Shingledecker J P, Swindeman R W and Hoelzer D T 2005 Oxide dispersion-strengthened steels: A comparison of some commercial and experimental alloys, *J. Nucl. Mater* **341** 103-114

[16] European comission , *Mechanical Testing of As-Cast Model Steels with Parametric Variation of Ni, Mn and Si Content* 4-44

[17] Kansy J 2007 LT for Windows, Version 9.0, Silesian University, Poland

Influence of the oblique magnetic field on the secondary electron emission from the plasma facing materials in fusion reactor

I E Berezina and I V Tsvetkov

National Research Nuclear University MEPhI Moscow 115409, Kashirskoe sh. 31, Russia

E-mail: ira.berezina@mail.ru

Abstract. The effect of the magnetic field inclination on the energy and angular distribution of electrons impinging on the first wall is modeled. It demonstrates the necessity to take into account the dependences of average energy and incident angle of electrons bombarding the wall on the magnetic field inclination in calculation of coefficient and the yield of secondary electron-electron emission.

1. Introduction

The studying of plasma – surface interaction processes is a very important task for realization of the international project ITER. To determine hydrogen isotopes implantation and reflection, plasma – facing materials sputtering and secondary emission, as well as to analyze the experimental data in the plasma sheath it is necessary to determine correctly the inflow and outflow of charged particles at the surface [1].

The oblique magnetic field has a substantial influence on the electron motion in the sheath. Taking into account the oblique magnetic field is important for the calculation of the sheath potential drop. It determines the energy of incident ions and therefore influences the surface sputtering. The electric field potential distribution depends on the reflection and secondary electron – electron emission, since these processes change the electron fluxes to and from the wall. The reflection and secondary electron – electron emission are dependent on parameters of electrons incidence onto the surface, in particular, the energy and angular distributions of impinging electrons [2]. That's why our purpose is to estimate dependences of parameters of electrons incidence onto the surface on the magnetic field inclination and then reveal the magnetic field influence on the coefficient and the yield of secondary electron emission.

2. Simulation model

The transport of electrons in the sheath layer is calculated by solving the equation of motion for electrons in the electric and magnetic fields near the wall. In the modelling it is assumed that electrons are injected from the boundary of the pre-sheath layer in the XY plane with the velocities **V** at angles β with respect to the normal. Emitted electrons have uniform angular distribution and the Maxwell velocity distribution. The plasma ion and electron temperatures are equal. The magnetic field is uniform with a certain inclination angle α with respect to the surface normal $\mathbf{H} = \mathbf{H}(H_x,0,H_z)$. The electric field is perpendicular to the wall $\mathbf{E} = \mathbf{E}(E_x,0,0)$.

The following form of the equation of electron motion is used:

$$\begin{cases} \frac{d^2}{dt^2}x = -\eta \cdot E - \frac{\eta}{c} \cdot \left(\frac{d}{dt}y\right) \cdot H_z \\ \frac{d^2}{dt^2}y = \frac{\eta}{c} \cdot \left(\frac{d}{dt}x \cdot H_z - \frac{d}{dt}z \cdot H_x\right) \\ \frac{d^2}{dt^2}z = \frac{\eta}{c} \cdot \left(\frac{d}{dt}y \cdot H_x\right) \end{cases} \tag{1}$$

where $\eta = q/m = 5.3\times10^{17}$ SGSE/g, $c = 3\times10^{10}$ cm/s, $H_x = -H\cos(\alpha)$, $H_z = H\sin(\alpha)$.

To describe the potential distribution in the plasma sheath we use Poisson's differential equation:

$$\left(\frac{d}{dx}\phi\right)^2 = -8\cdot\pi\cdot e\cdot n_o \cdot \left[2\cdot\phi_o\cdot\left(\sqrt{\frac{\phi(x)}{\phi_o}} - 1\right) - \frac{k\cdot T_e}{e}\cdot\left[\exp\left[e\cdot\frac{(\phi(x) - \phi_o)}{k\cdot T_e}\right] - 1\right]\right] \tag{2}$$

with boundary condition on the plasma side, which is derived from Bohm criterion:

$$|\phi_0| \geq \frac{k\cdot T_e}{2\cdot e} \tag{3}$$

and boundary condition on the wall, which is derived using the fact that the ion flux must be balanced with the electron flux (the expression for the floating potential):

$$\phi_w = \frac{k\cdot T_e}{2e}\cdot\left[\ln\left[2\pi\cdot m_e\cdot\frac{(T_e + T_i)}{m_i\cdot T_e}\right] - 1\right] \tag{4}$$

For solving the Poisson's equation we use the dimensionless form of equation (2):

$$\frac{d}{d\xi}\lambda = \sqrt{8\cdot\left[\sqrt{-\lambda} + e^{(\lambda+1)\div 2} - 2\right]} \tag{5}$$

where $\lambda = \varphi/\varphi_0$ represents the normalized potential and $\xi = x/r_d$ gives the distance from the surface in units of the Debye length. The equation was solved by Runge–Kutta method of third power.

Then we approximate the equation solution by extent function of the forth order:

$$A(\xi) = a_0 + a_1\cdot\xi + a_2\cdot\xi^2 + a_3\cdot\xi^3 + a_4\cdot\xi^4 \tag{6}$$

$a_0 = -5.98$
$a_1 = 2.08$
$a_2 = -0.374$
$a_3 = 0.032$
$a_4 = -1.044\cdot10^{-3}$

The approximation of this kind is quite useful for modeling the charged particles behaviour in the sheath. If you know plasma parameters, you can obtain the potential distribution in the sheath without solving Poisson's equation. The solutions of the dimensionless equation (5) by numerical method and by approximation function (6) are shown in figure 1. These solutions agree satisfactorily along the whole plasma sheath length equal to 10 Debye lengths.

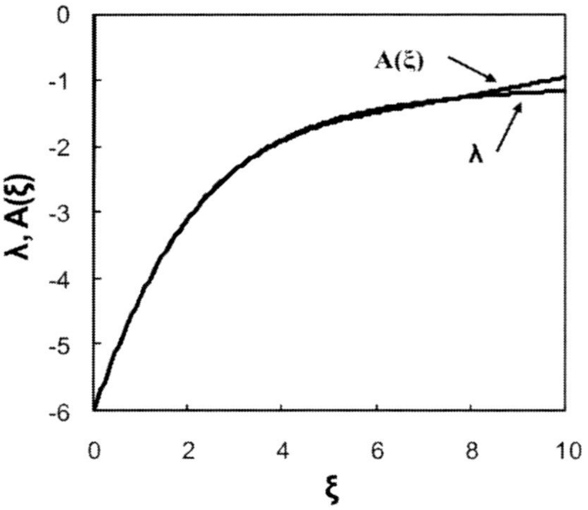

Figure 1. Potential distributions obtained by numerical solution λ of equation (5) and by approximation function A(ξ) (6).

Then we return to the dimension values of potential and the distance from the surface and obtain the dependences of potential and electric field in the sheath on the distance from the surface and on the plasma parameters:

$$\phi(x) := -8.64\cdot10^{-3}\cdot T + 4.28\cdot10^{-6}\sqrt{n\cdot T}\cdot x - 1.1\cdot10^{-9}\cdot n\cdot x^{2} + 1.34\cdot10^{-13}\cdot\frac{n^{\frac{3}{2}}}{\sqrt{T}}\cdot x^{3} - 6.26\cdot10^{-18}\cdot\frac{n^{2}}{T}\cdot x^{4} \quad (7)$$

$$E(x) = -4.27\cdot10^{-6}\cdot\sqrt{n\cdot T} + 2.19\cdot10^{-9}\cdot n\cdot x - 4.02\cdot10^{-13}\cdot\frac{n^{\frac{3}{2}}}{\sqrt{T}}\cdot x^{2} + 4.56\cdot10^{-18}\cdot\frac{n^{2}}{T}\cdot x^{3} \quad (8)$$

The values of distance, plasma density and temperature are given in cm, cm^{-3}, eV. The potential and electric field distribution in the sheath can be obtained from these approximation functions without solving Poisson's equation.

3. Influence of magnetic field inclination on the energy distribution of electrons reaching the wall

To determine a minimal initial energy of electrons impinging on the wall the electrons emitted along the surface normal were considered. These electrons reach the wall faster than other electrons emitted in XY plane with different velocities. In the absence of the magnetic field or when the magnetic field inclination equals zero the minimal initial energy of impinging on the wall electrons corresponds to the potential drop in the sheath layer. For given plasma parameters $T = 30$ eV, $n = 10^{12}$ cm^{-3} the potential drop equals 64.5 eV. However, with the magnetic field inclination the minimal initial energy increases. In this case the electron motion in the electric and the magnetic field is considered. Electron drifts along the wall and moves to the wall along the magnetic lines with velocity equal to its velocity projection on the magnetic field direction $V_{\|} = V\cdot\cos(\alpha)$. The condition of reaching the wall is excess of kinetic electron energy above the potential drop in the sheath layer. Using this condition we derive the formula of dependence of minimal initial energy of electrons impinging on the wall on the magnetic field inclination angle α:

$$W_{min}(eV) = e(\varphi_w - \varphi_0)/\cos(\alpha)^2. \quad (9)$$

The values calculated from this formula (9) and the results of numerical experiment for angles α less than 60° are in good agreement. For following calculations we consider the electrons from "Maxwell tail" with energies $W_0 > W_{min}$.

Energy distribution of electrons reaching the wall for different cases of the magnetic field inclination are shown in figure 2. Electrons are injected in the XY plane. The distributions are normalized to the number of emitted by plasma electrons. The magnetic field strength is H = 5 T. Electrons have the Maxwell velocity distribution on the boundary of the sheath layer. The electrons with energies from 64.5 eV to 370 eV were considered. These values correspond to minimal initial energy of electrons reaching the wall for α = 0° and energy equal to $E_{av}+3\sigma$, where σ represents the root-mean-square deviation and E_{av} gives the average electron energy in Maxwell distribution for T = 30 eV. One can see that the quantity of electrons bombarding the wall reduces. The fraction of the electrons reaching the wall is calculated as an area under the curve. So, for the magnetic field angle equals 30° the fraction of the electrons reaching the wall is equal to 0.03, and for 60° it is equal to 10^{-4}.

The minimal initial energy of electrons impinging on the wall increases with the magnetic field inclination. The values of average energy of electrons impinging on the wall for different angles α were calculated. Figure 3 presents the substantial increasing of the average energy of the electrons bombarding the first wall with the magnetic field inclination.

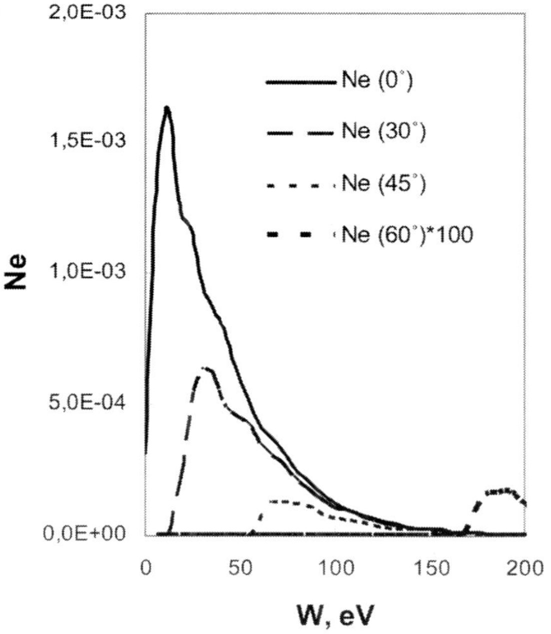

Figure 2. Energy distribution of electrons reaching the wall for different cases of the magnetic field inclination α. W – energy of electrons bombarding the wall, Ne – the fraction of the electrons with the given energy of the number of emitted by plasma electrons

Journal of Physics: Conference Series **516** (2014) 012025 doi:10.1088/1742-6596/516/1/012025

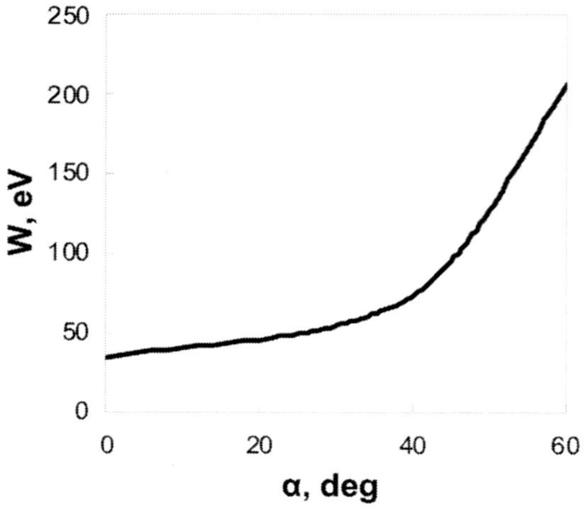

Figure 3. Dependence of the average energy W of the electrons bombarding the first wall on the magnetic field inclination angle α

The average angles of electron incidence for different magnetic field angles are calculated. It is shown that the average angle of electrons incidence does not depend substantially on the magnetic field inclination.

4. The influence of the magnetic field inclination on the coefficient and the yield of secondary electron emission

For each magnetic field inclination angle α the coefficients of secondary electron-electron emission δ_e, reflection η_e and the total coefficient of secondary electron emission $\gamma_e = \delta_e + \eta_e$ were calculated from the empirical formulae [3–5]:

$$\delta(Wp, \theta) := \delta emax \cdot 2.72^{2} \cdot \frac{Wp}{Wmax} \cdot \frac{e^{-2\cdot\sqrt{\frac{Wp}{Wmax}}}}{\cos(\theta)^{1.4}} \tag{10}$$

$$\eta(Wp,\theta) := \left(\frac{Wp^{m(z)}\cdot C(z)}{0.891}\right)^{\cos(\theta)} \cdot 0.891 \tag{11}$$

$$\gamma(Wp,\theta) := \delta(Wp,\theta) + \eta(Wp,\theta) \tag{12}$$

These coefficients are dependent on the electron energy and angle of incidence, that as was shown are dependent on the magnetic field inclination angle. Tungsten was considered as a surface material for calculations. Figure 4 shows the dependences of these coefficients on the magnetic field inclination in the case of electron injection in XY plane. With the magnetic field inclination the total secondary electron emission coefficient increases. The coefficient of secondary electron emission gives the main contribution to the growth of the total coefficient. The reflection coefficient does not substantially depend on the magnetic field inclination angle. However, with the magnetic field inclination increasing the number of reaching the wall electrons decreases due to the suppression of the electron flux to the surface by oblique magnetic field. That is why the yield of secondary electron emission decreases with the magnetic field inclination increasing despite the increasing of the secondary emission coefficient (figure 4). For $\alpha = 0^{\circ}$ the fraction of the secondary electrons from the number of emitted by plasma electrons is 6.7×10^{-2}, for $\alpha = 60^{\circ}$ the fraction of secondary electrons is 2×10^{-4}.

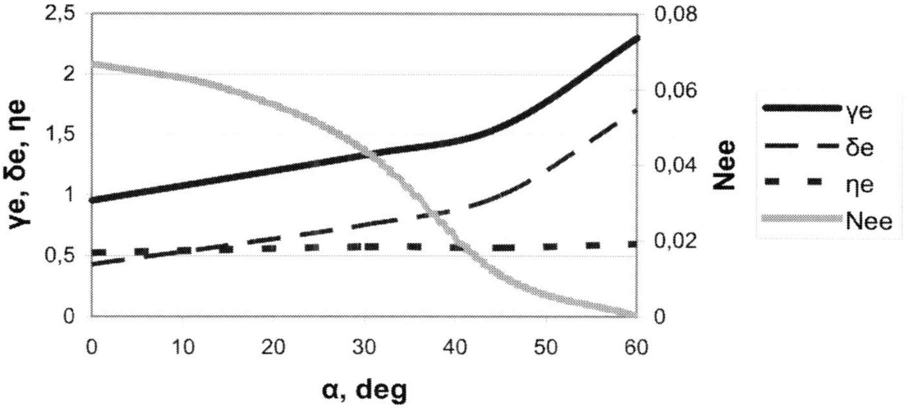

Figure 4. Dependences of the coefficients of secondary electron-electron emission δ_e, reflection η_e and the total coefficient of secondary electron emission γ_e on the magnetic field inclination angle α

5. Conclusions

Thus we study the influence of the magnetic field inclination angle on parameters of electrons incidence onto the surface in the case of flat electron injection. It was shown that the average energy of the electrons substantially increases with the magnetic field inclination. Therefore the secondary electron emission coefficient increases. At the same time due to the suppression of the electron flux to the surface by oblique magnetic field the secondary electron yield decreases with the magnetic field inclination increasing despite the increasing of the secondary emission coefficient. The potential distribution is dependent on the densities of electron inflow and outflow at the wall. That's why it is important to take into account the magnetic field in calculation of a self-consistent potential distribution in the sheath. The self-consistent potential determines the energy of ions bombarding the wall and therefore it is substantial for the calculation of the wall sputtering. Our further researches will be aimed at this problem.

References

[1] Stangeby P 1986 The Plasma Sheath *Physics of Plasma-Wall Interactions in Controlled Fusion* (Plenum Press New York) p 41
[2] Chodura R 1986 Plasma Flow in the Sheath and Presheath of a Scrape-Off Layer *Physics of Plasma-Wall Interactions in Controlled Fusion* (Plenum Press New York) p 99
[3] Kollath R 1956 *Handbuch der Physik* **21** (Springer Verlag Berlin)
[4] Thomas E W 1991 *Atomic and plasma-material interaction data for fusion* **1** p 79
[5] Ertl K and Behrisch R 1986 Electron emission from solid surface *Physics of Plasma-Wall Interactions in Controlled Fusion* (Plenum Press New York London) p 515

Journal of Physics: Conference Series **516** (2014) 012026 doi:10.1088/1742-6596/516/1/012026

Hydrogen retention in volumetric CFC structures

A. Airapetov[1], L. Begrambekov[1], S. Brémond[2], D. Douai[2], A. Kuzmin[2], Ya. Sadovsky[1], P. Shigin[1], S.Vergasov[1]

[1]Moscow Engineering and Physics Institute, 31 Kashirskoe shosse, Moscow, 115409, RF
[2]CEA/IRFM, Association EURATOM-CEA, Centre de Cadarache, 13108 Saint Paul Lez Durance, France

E-mail : arskus@mail.ru

Abstract. The paper experimentally investigates and analyses the features and mechanisms of both of oxygen removal by deuterium glow discharge from CFC, pyrolytic graphite and stainless steel subjected to irradiation in oxygen contaminated plasma. It is shown that oxygen implanted in PG perpendicular to basal plates is removed after sputtering the layer slightly thicker than oxygen stopping zone (≈2nm). Fast deuterium ions penetrating into CFC during GDC transfer the trapped oxygen atoms into the bulk. Thus, much thicker surface layer has to be removed (500-1000 nm) for oxygen release.

1. Introduction

Oxygen impurity presented in plasma of fusion devices leads to f number of negative effects. Among them there are parasitic radiative power, delution of fuel particles, accelerated erosion rate of plasma facing materials, enhanced trapping and recycling of fuel particles.

Various types of discharges providing irradiation of plasma facing materials with weakly ionized plasma (glow discharge, rf-discharge, Taylor discharge) are routinely used for removal of oxygen and other impurities from the first walls of today tokamaks. Nevertheless the mechanisms of oxygen release from the ion irradiated metal and graphite are not underestood satisfuctorly. Its detail investigation seems to allow optimization of conditioning process.

The paper is devoted to the results of experimental investigation of oxygen (O) removal by deuterium (D) glow discharge (D_2-GDC) from carbo fiber composite (CFC), pyrolytic graphite (PG) and stainless steel (SS). Parameters, peculiarities and mechanisms of O release by D_2-GDC from CFC and PG irradiated in O contaminated plasma are presented.

The experiments were performed in the electron initiated discharges. Parameters of irradiating ion flux were rather close to those of glow discharge. Thus, we use the term GDC for surface cleaning process realized in this study.

2. Experimental

The experiments with CFC samples were performed in the thermal desorption stand – (TDS-1, figure1) presented earlier [1]. The main feature of the TDS-1 is a possibility to irradiate the samples in the gas discharge ignited in the plasma chamber of the stand, and to make a thermal desorptional analysis of the samples in the same plasma chamber.

The sample heating rate during TDS measurements was equal to 5 K/s. For an absolute calibration of the mass-spectrometer, a gas output from the implanted sample was modeled by the gas flow into the plasma chamber through a calibrated leakage.

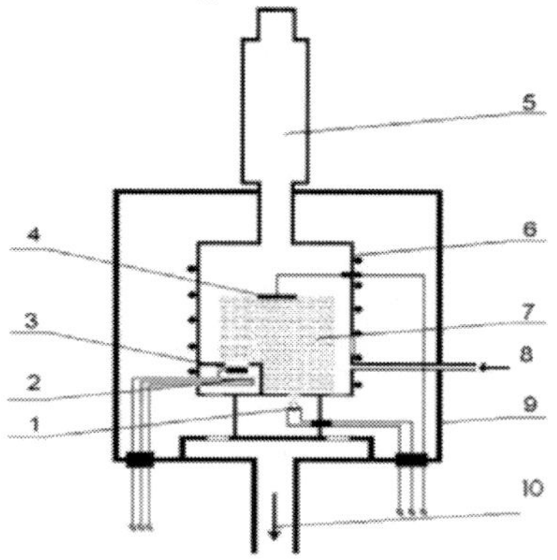

Figure 1. Thermal Desorption Stand (TDS-1). 1 – Heated cathode, 2 – Sample heater, 3 – Sample, 4 –anode , 5 – Monopole mass-spectrometer, 6 – Water-cooled plasma chamber, 7 – Plasma, 8 – Puff in, 9 –Vacuum vessel, 10 – Pipe line.

3. Sample treatment and experimental procedure

The samples of CFC N11 and pyrolytic graphite were used for experiments. After machining they were washed in ultrasonic bath and annealed. The CFC samples were annealed at 1800 K for 5 minutes and SS samples at 1450 K for 5 minutes.

Each experimental cycle included sample annealing, irradiation in (D_2 + 2%O_2)–plasma and irradiation in D_2–plasma (GDC). (D_2 + 2%O_2)-plasma was selected for the experiments, because such composition of plasma in our experimental conditions provided maximum O trapping practically not depending on accidental variatioof O concentration [2]. Irradiation in (D_2 + 2%O_2)–plasma was performed under the following conditions. Ion energy E_i = 100 eV/at (for SS samples), E_i = 100 and 400 eV/at (for CFC samples), ion flux $j_i = 1\times10^{20}$ at/m^{2}s, fluence $\Phi_i = 5\times10^{23}$ at/m^2, sample temperature T = 450 K. For GDC it was E_{GDC} = 100 and 400 eV/at, $j_{GDC} = 2\times10^{19}$ at/m^{2}s, T = 450 K. The experimental series with CFC samples are indicated as 100/100, 100/400, 400/100 and 400/400 combinations.

For determination GDC efficiency, the cycles were made with different GDC duration, and after each cycle an amount of oxygen remained in the sample was measured by TDS.

During TDS examination oxygen was released from CFC samples mainly in the form of CO (99%) and minor portion as CO_2 (1%). That is why the paper considers only CO release.

4. Experimental results and discussions

Oxygen was practically totally removed from PG samples, when surface layer slightly thicker than O stopping range (inset at figure 2) was sputtered (figure 2). Contrary to that, removal of 50 nm thick layers caused only small decrease of O retention in CFC samples (figure 2). Further sputtering led to decrease of O content in the samples with practically constant rate.

In all cases almost total release of O was observed after sputtering the layers many times exceeding O and D ion stopping ranges. The thicker layer (≥1000 nm) was to be sputtered and the shortest time (900–100 min) was needed for O release with 400 eV GDC (combinations 100/400 and 400/400), and vice versa for the combinations 400/100 and 100/100 the thickness 500–600 nm was supposed to be sputtered during 150–200 min. Sample temperature did not influence remarkable the rate of GDC (figures 2 and 3).

It could be assumed that penetration of O into the bulk occurred due to diffusion of implanted O atoms, or due to filling of internal pores with O atoms during implantation in (D_2 + 2%O_2)–plasma.

But in such cases O release rate during GDC in contradiction with experimental results should not depend on energy of impinging D ions, should not be constant but should be independent of the sample temperature.

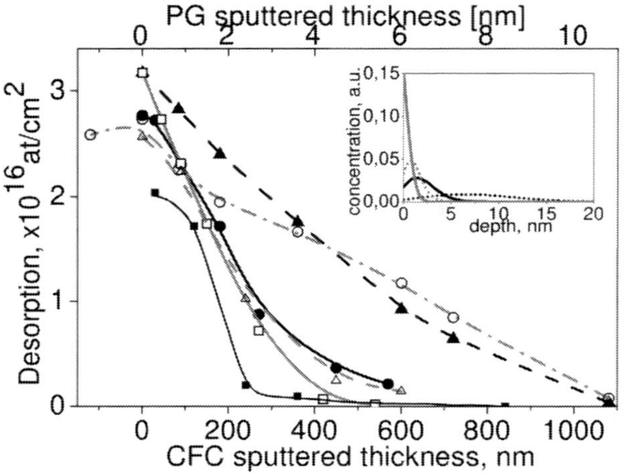

Figure 2. CO desorption from CFC and PG via thickness of sputtered away layers during GDC. Experimental series:■ – PG 400/400 at 200°C, CFC: ● – 100/100 at 120°C, ▲ – 400/400 at 200°C, ○ – 100/400 at 200°C, □ - 400/100 at 200°C, △ – 100/100 at 200°C. The inset shows depth distributions of deuterium and oxygen ions implanting graphite with 100 and 400 eV energies. (Calculation is made with SCATTER code [2], black line – D^+ 100eV, black dots D^+ 400eV, gray line O^+ 100 eV, gray dots O^+ 400 eV)

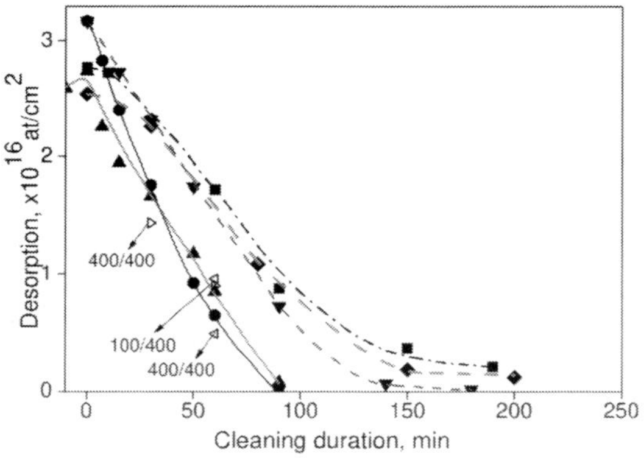

Figure 3. CO desorption from CFC via GDC duration: ■ –100/100 at 120°C, ● – 400/400 at 200°C, ▲ – 100/400 at 200°C, ▼– 400/100 at 200°C, ○ – 100/100 at 200°C, ◁ – GDC j=2x1015 at/cm2, ▷ – implantation j=2x1015 at/cm2

We believe that O was transferred into the bulk due to collision cascades originated by fast D ions during GDC. For ion energies used in our experiments the mean energy of primary recoils (T) could be determined as $T = \frac{1}{2}E_D[4M_DM_C/(M_D+M_C)^2]$, where E_D and M_D are energy and mass of deuterons, and M is mass of target atoms. The value of T is 25 eV, when E_D = 100 eV, and T ≈ 100 eV, when E_D = 400eV.

Each D atom impinging CFC surface with energy 100 eV can displace 4–5 atoms in average, and at least 11–16 atoms can be knocked out and transferred by 400 eV atom. It means that the rate of O atom transfer into deep layers was much higher, than the rate of their release from the surface layers sputtered by D ions with energies 100–400 eV. As a result O was accumulated mainly in the region of ion penetration, the surface layers appeared to be depleted of O, its emission due to surface sputtering was slowed down, and thicker layer needed to be sputtered for total oxygen removal.

The constant rate of O release indicates that O depth distribution in the near surface layers was constant during main part of GDC. It can be provided, if O concentration in permanently transferred "accumulating" layer stayed without changes during main part of GDC. More probably, O concentration in this layer was kept at saturation level, but its thickness thinned down along with the time. As soon as the layer had been exhausted GDC was finished.

The smaller rate of O removal at the first stage of GDC shows that O concentration in the near surface layers was smaller than those at the later stages. The reason of that could be a redistribution of O atoms in D stopping zone after finishing of $D_2 + 2\%O_2$ plasma irradiation and starting of GDC.

The ratio of sputtering yields for D ions with energies 400 eV and 100 eV equals $Y_{400eV}/Y_{100eV} \approx$ 1.5–1.6. This value is smaller than the ratio of the numbers of O atoms displaced by ions of these energies $N_{400}/N_{100} \approx 4$. Thus O concentration in the sputtered surface layers appeared to be higher under 100 eV GDC, and accordingly, higher O amount was released with each sputtered carbon layer. That is why, thinner surface layer should be sputtered away by 100 eV GDC for O removal, than by 400 eV GDC.

Additional reason of lower O concentration in the sputtered layers during 400 eV GDC is radiation induced surface stresses. 400 eV ion irradiation results in formation of higher compress stresses in stopping zone and accordingly in higher tensile stresses in the underneath layers. It may cause easier O penetration (as well as D penetration) from stopping zone into the bulk layers, decrease of O concentration in near surface layers and finally slowing down the rate of O removal.

Different O distribution and concentration in dependence on energy of implanting ions have been established in the near surface layers during ($D_2 + 2\%O_2$)–plasma irradiation. That is why at the first stage of GDC the O retention in 100/400 combination decreased at the same way as in 100/100 combination (figure 2). In a while O distribution in CFC was "reconstructed" under 400 eV D ion irradiation, and further O removal in 100/400 and 400/400 combinations were similar

Fast O removal from PG samples can be explained by strong dependence of particle displacement on graphite structure orientation. One can believe that displacement is negligibly small in directions perpendicular to graphite hexagonal plates and reaches maximum in direction parallel to them. This fact accounts for high O displacement along with the axis of CFC fibers, when the fibers are oriented perpendicular to irradiating surface.

5. Summary

The paper presents results of experimental investigation of oxygen removal by deuterium D-GDC from CFC N11 and pyrolytic graphite PG99.

Oxygen was removed from PG implanted perpendicular to its basal plates after sputtering the layer slightly thicker than D stopping zone (≈ 2 nm). It was necessary to sputter much thicker layer and to spend more time for O removal from CFC. It is ≈1000 nm and t ≈ 90 min, if E_{GDC} = 400 eV/at, and ≈ 500 nm and t ≈ 200 min, if E_{GDC} = 100 eV/at. The results were explained by active O transfer into the bulk of CFC due to participation in collision caskades generated by fast D ions. Increase of surface temperature from 393 to 473 K dtd not influence the rate of GDC. It is concluded that GDC duration could be decreased down to ≈ 50 min, if ≈ 200 eV D ions is used.

The minimum of oxygen retention during D-GDC was achieved, when predominated part of O saturated barrier layer had been removed. After sputtering the barrier layer away O penetrated up to

Journal of Physics: Conference Series **516** (2014) 012026 doi:10.1088/1742-6596/516/1/012026

and trapped in significantly deeper region. Oxygen trapping at this stage, more probably, occurred through potential mechanism from the water molecules absorbed at the surface.

100 eV D-GDC is preferable for SS conditioning. Oxygen retention in minimum is expected to be less, if temperature is higher and O concentration in residual gas is lower.

References

[1] L. Begrambekov, O. Buzhinsky, A. Gordeev, E. Miljaeva, P. Leikin, P. Shigin. Physica Scripta, N108 (2004), p.72-75.

[2] Airapetov, L. Begrambekov, C. Brosset, J.P. Gunn, C. Grisolia, A. Kuzmin, T. Loarer, M. Lipa, P. Monier-Garbet, P. Shigin, E. Tsitrone, A. Zakharov. Deuterium trapping in carbon fiber composites exposed to D plasma. Journal of Nuclear Materials 390–391 (2009) 589–592

Laser guiding through an axially non-uniform collisional plasma channel.

Arvinder Singh and Navpreet Singh

Department of Physics, Dr. B.R.Ambedkar National Institute of Technology, Jalandhar

E-mail: arvinder6@lycos.com

Abstract. This paper presents an investigation of laser guiding through an axially non-uniform collisional plasma channel formed by ionizing laser prepulse. Due to self-defocusing of the ionizing prepulse, an axial non-uniform plasma channel is formed. When delayed second laser beam propagates through such preformed plasma channel, on account of non-uniform intensity distribution of laser beam, non-uniform heating of electrons takes place. Non-uniform heating diffuses the electrons away from the axis which further enhances the plasma channel. Unbalanced diffraction and refraction phenomenon through such an axial non-uniform collisional plasma channel results into periodic beam width variation with the distance of propagation. Second order ordinary differential equations for the beam width parameter of prepulse and the guided beam have been set up by using the moment theory approach. Laser guiding upto several Rayleigh lengths has been observed.

1. Introduction

The propagation of laser beam in plasma is relavant to a wide range of applications, such as laser fusion schemes, X-ray lasers, ultrahigh-gradient electron accelerators, fast ignitor concept and laser induced lightning.[1]-[17] Diffraction broadening of laser radiation is one of the principle phenomenon which limits the significant interaction region of laser beam upto confocal parameter(i.e. two times the Rayleigh length, where Rayleigh length, $R_d = \frac{\pi r_0^2}{\lambda}$, where r_0 is the spot size and λ is the laser wavelength) and hence limits the use of laser interaction in these applications. Therefore the guiding of the laser radiation becomes essential for long distance propagation in plasma. Several methods like capillary discharge, self guiding, two pulse technique etc. have been proposed to extend the propagation distance of pulses beyond the diffraction limit as reviewed by Esarey et al [18], [19]

For optical guiding of laser pulses in plasmas, the plasma channel has to be prepared. For perfect guiding, a plasma density with minimum on the axis and parabolic increase toward the edge is required. Two pulse technique is used for guiding the laser pulse having high intensity of the order of $10^{12} - 10^{20} W/cm^2$. The physics of guiding the laser pulse is as follows, a plasma channel is created by focusing a laser prepulse on picosecond time scale via tunnel ionization. Since the prepulse has a gaussian intensity radial profile, therefore the plasma density that results spontaneously, has a peak on the axis and falls of rapidly with radial distance away from the axis, before the diffusion sets in. As a result, refractive index is minimum on the axis and increases towards the edge and hence the medium behaves as defocusing medium. Therefore plasma channel evolved from such a density profile on a nanosecond time scale, after the laser prepulse is gone, should also have an axial variation of plasma density.

The plasma created by the prepulse diffuses radially away from the axis and therefore the plasma density becomes minimum on the axis and maximum at the edge of the plasma channel. Therefore, the refractive index becomes maximum on the axis and decreases toward edges and as a result the medium behaves like a focusing medium.

In the present paper we have considered an axially nonuniform collisional plasma channel. When a delayed intense second laser pulse is passed through such a preformed plasma channel, then it tends to diverge due to diffraction and converge due to refraction. However, when both the convergence and divergence parameters are equal, then the delayed laser pulse propagates without convergence and divergence, but in an axially nonuniform plasma channel this condition cannot be satisfied throughout the channel and therefore the beam radius changes as it propagates. As the gaussian laser beam propagates through plasma, it causes differential ohmic heating of electrons. Under the effect of temperature gradient, electrons diffuse outwards causing decrease in the density of the electrons on the axis and as a consequence the convergence of the beam takes place.

Laser guidance up to several Rayleigh lengths of guided laser pulse in a plasma channel formed by a laser prepulse has been successfully demonstrated[20],[21]. The self- defocusing of the laser prepulse and laser guiding of the second laser pulse in the evolved axially nonuniform plasma channel has been explained by using paraxial ray approximation[22]. The paraxial theory[23],[24] had been the most popular theory for studying the nonlinear propagation of electromagnetic waves in nonlinear media. Despite its mathematical simplicity, the problem regarding this theory is that it takes into account only the regions close to the beam axis in the self focusing mechanism. Moment theory, which is global approach, does not suffer from such limitations and hence give more accurate results. The importance of non-paraxiality in self focusing phenomenon has already been highlighlited[25]. Optical guiding of a laser beam in an axially nonuniform plasma channel has been studied by moment theory approach[26]. However, at the considered intensity regime, collisional nonlinearity plays a significant role in determining the nonlinear propagation of laser beam through plasma channel. Therefore, we have included the collisional nonlinearity in the present work to study the more realistic situation. Ref[18], review the two pulse technique used by Durfee & Milchberg, in which there is no axial variation of plasma density, as the code does not include the self-defocusing of the prepulse. However, in the present study, we have taken into account the self-defocusing mechanism, which plays important role as far as the laser guidance is concerned. In section 2, we have developed a moment theory of self-defocusing of ionizing prepulse. In section 3, we have investigated the guidance of the second pulse in the preformed plasma channel by moment method approach. Equations of beam width parameters are solved numerically using Runge Kutta method. The results are discussed in section 4.

2. SELF-DEFOCUSING OF PREPULSE

Consider the propagation of a gaussian laser beam of frequency 'ω' through a gas along the z-axis. The laser ionizes the gas via tunnel ionization in a time shorter than the pulse duration of the laser. In the experiment performed by Durfee and Milchberg [20], a laser prepulse($10.6\mu m, \sim 100ps$) forms a channel for the second delayed pulse and the laser intensity used was in the range $10^{13} - 10^{14} W/cm^2$. Sharp plasma density fallout was observed in Durfee and Milchberg's experiment, which was modelled by Liu and Tripathi [22] as

$$\omega_p^2 = \omega_{p0}^2 \exp\left(\frac{-E_a'}{|E|}\right) \tag{1}$$

where $\omega_p = [\frac{4\pi n_e e^2}{m}]^{1/2}$ is the electron plasma frequency, ω_{p0}^2 is a constant depending on the neutral particle density and $E'_a = \frac{2}{3}E_a\left(\frac{E_i}{E_h}\right)^{1/2}$. where E_a is the atomic unit of electric field, $|E|$ is the amplitude of laser field, E_i & E_h are ionization potential of an atom and of a hydrogen atom.

The slowly varying amplitude A of the electric vector $E = \hat{x}A(r,z)e^{\iota(\omega t - k_0 z)}$ of an electromagnetic wave propagating in homogenous medium satisfies the following eqn.[26],[27],[28]

$$\frac{\partial A}{\partial z} + \frac{\iota}{2k_0}\nabla_\perp^2 A + \iota P(A.A^*)A = 0 \tag{2}$$

Where $P(A.A^*)$ is given by

$$P(A.A^*) = \frac{k_0}{2\epsilon_0}(\epsilon - \epsilon_0) = \frac{k_0}{2\epsilon_0}\phi(E.E^*) \tag{3}$$

and ϵ is given by.

$$\epsilon = \epsilon_0 + \phi(E.E^*) \tag{4}$$

where ϵ_0 & $\phi(E.E^*)$ are the linear and nonlinear parts of the dielectric constant.

$$\epsilon_0 = \epsilon|_{r=0} = 1 - \frac{\omega_{po}^2}{\omega^2}\exp(-E'_a f_1/A_{00}) \tag{5}$$

$$\phi(E.E^*) = \frac{\omega_{po}^2}{\omega^2}\left[exp.\left(-\frac{E'_a f_1}{A_{00}}\right) - exp\left(-\frac{E'_a f_1}{A_{00}}e^{\frac{r^2}{2r_1^2 f_1^2}}\right)\right] \tag{6}$$

Intensity distribution of the prepulse is assumed to be gaussian and is given by.

$$|A|^2 = (A_{00}^2/f_1^2)\exp(-r^2/r_1^2 f_1^2) \tag{7}$$

where A_{00} is the maximum amplitude of electric field at $r = 0$ i.e. on the axis of the laser beam, f_1 is dimensionless beam width parameter and r_1 is the initial beam width of the prepulse at $z = 0$, $r_1 f_1$ represents the beam width for $z > 0$.

By using the method of moments [26],[27],[28] , we get normalized Eq. for the beam width parameter f_1 of the prepulse.

$$\frac{d^2 f_1}{d\xi^2} = \frac{1}{f_1^3} + \left(\frac{r_1\omega_{p0}}{c}\right)^2 \frac{E'_a}{A_{00}}\frac{1}{r_1^4 f_1^4} I_1 - \frac{1}{f_1}\left(\frac{df_1}{d\xi}\right)^2 \tag{8}$$

where

$$I_1 = \int_0^\infty r e^{-\frac{r^2}{2r_1^2 f_1^2}} \exp\left(-\frac{E'_a f_1}{A_{00}} e^{\frac{r^2}{2r_1^2 f_1^2}}\right) r^2 dr \quad \& \quad \xi = \frac{z}{R_d}.$$

Initial conditions for numerically solving Eq.(8) are $f_1 = 1$ and $\frac{df_1}{d\xi} = 0$ at $\xi = 0$

3. GUIDED PROPAGATION OF SECOND DELAYED PULSE

The plasma moves radially away from the axis after the passage of prepulse. The second gaussian pulse is guided through the channel with a delayed time less than the recombination time. On the arrival of the delayed second laser pulse the density profile of the plasma has been given by Liu and Tripathi[22]

$$\omega_p^2/\omega^2 = \alpha_1(z) + \alpha_2(z)r^2 \tag{9}$$

where α_1, α_2 are monotonically decreasing functions of z;

$$\alpha_1 \leq \omega_{p0}^2/\omega^2 e^{-(E_a'/A_{00})f_1} \quad , \quad \alpha_2 \leq \alpha_1/r_0^2 f_1^2$$

Permittivity of the collisional plasma [24] is given as

$$\epsilon = \epsilon_0 + \phi(E_0.E_0^*) = 1 - \frac{\omega_p^2}{\omega^2}\left(1 + \frac{\alpha E_0.E_0^*}{2}\right)^{-\frac{5}{2}} \tag{10}$$

Intensity distribution of the delayed second pulse is given as

$$E_0.E_0^* = \frac{E_{00}^2}{f_2^2} e^{\frac{-r^2}{r_2^2 f_2^2}} \tag{11}$$

where r_2 represents beam width at z = 0 and $r_2 f_2$ represents the beam width for $z > 0$ of the guided pulse, f_2 is termed as dimensionless beam width parameter of the guided pulse with $f_2 = 1$ at z = 0.

In Eq.(10) α is given as $\alpha = \frac{e^2 M}{6KTm^2\omega^2}$, T being equilibrium temperature of collisional plasma, K is the boltzmann constant, ω is the angular frequency of the incident laser beam, M & m refer to masses of ion and electron respectively.

$$\epsilon_0 = 1 - \alpha_1\left(1 + \frac{\alpha E_{00}^2}{2f_2^2}\right)^{-5/2} \tag{12}$$

$$\phi(E_0.E_0^*) = \alpha_1\left(1 + \frac{\alpha E_{00}^2}{2f_2^2}\right)^{-5/2} - (\alpha_1 + \alpha_2 r^2)\left(1 + \frac{\alpha E_{00}^2}{2f_2^2} e^{\frac{-r^2}{r_2^2 f_2^2}}\right)^{-5/2} \tag{13}$$

By following the method of moments [26],[27],[28] as in section 2 for the prepulse, we get the following Eq for the beam width parameter f_2 of the second delayed pulse.

$$\frac{d^2 f_2}{d\xi^2} = \frac{1}{f_2^3} - \frac{5\omega^2 \alpha E^2 00}{2c^2 r_2^2 f_2^7}[\alpha_1 I_2 + \alpha_2 I_3] - \frac{2\omega^2 \alpha_2}{c^2 f_2^3} I_4 - \frac{1}{f_1}\left(\frac{df_1}{d\xi}\right)^2 \tag{14}$$

where

$$I_2 = \int_0^\infty \left(1 + \frac{\alpha E_{00}^2}{2f_2^2} e^{\frac{-r^2}{r_2^2 f_2^2}}\right)^{-7/2} .e^{\frac{-2r^2}{r_2^2 f_2^2}} r^3 dr$$

$$I_3 = \int_0^\infty \left(1 + \frac{\alpha E_{00}^2}{2f_2^2} e^{\frac{-r^2}{r_2^2 f_2^2}}\right)^{-7/2} .e^{\frac{-2r^2}{r_2^2 f_2^2}} r^5 dr$$

$$I_4 = \int_0^\infty \left(1 + \frac{\alpha E_{00}^2}{2f_2^2} e^{\frac{-r^2}{r_2^2 f_2^2}}\right)^{-5/2} .e^{\frac{-r^2}{r_2^2 f_2^2}} r^3 dr$$

we have numerically solved Eq.(14) by taking initial conditions, $f_2 = 1$, $\frac{df_2}{d\xi} = 0$ at $\xi = 0$ and for a model $\alpha_2(z)$ (Liu and Tripathi, [22]), $\alpha_2 = \frac{2}{R_d^2(1+\xi^2)^{1/2}}$.

Journal of Physics: Conference Series **516** (2014) 012027 doi:10.1088/1742-6596/516/1/012027

4. DISCUSSION

The differential Eqs.(8) and (14) for the beam width parameter f_1 of the laser prepulse and f_2 of the guided laser pulse respectively have been solved numerically for the following set of parameters.

$$\frac{\omega_{p0}}{c} = 0.426 \times 10^6 m^{-1}, \quad \omega = 1.778 \times 10^{15} S^{-1}, \quad \lambda = 1.06\mu m, \quad r_1 = r_2 = 13\mu m$$

and for different values of prepulse and guided beam intensities.

The results are depicted in the form of graphs in Figure 1 to 3. Eq.(8) is second order nonlinear ordinary differential equation, which governs the variation of beam width parameter f_1 of the laser prepulse with a dimensionless distance of propagation(ξ) as shown in Fig. 1. There are three terms on the right hand side of Eq.(8), the first term is due to diffraction divergence and the second term is due to nonlinear refraction, which is responsible for the self-defocusing of the prepulse. Initial contribution due to the third term at $\xi = 0$ is zero, which evolves with the distance of propagation and hence counteract the diffraction and self-defocusing of the prepulse.

In Fig. 1, Dotted curve corresponds to the case when only diffraction divergence term is present and the other two terms are ignored. Semidotted curve corresponds to the case when only the first two terms, diffraction divergence and nonlinear refraction are considered. Solid curve corresponds to the case when all the three terms are taken into account. From Figure 1, it is observed that in all the cases , the beam width parameter f_1 increases monotonically and leads to defocusing of the ionizing pulse.Due to defocusing of prepulse, intensity of the beam decreases and as a result ionization/density of the plasma decreases and we get axially nonuniform plasma channel.

Eq.(14) is also a second order nonlinear ordinary differential equation, which governs variation of the beam width parameter f_2 of the guided pulse with a dimensionless distance(ξ) of propagation. The first term on the right hand side of Eq.(14) is due to diffraction divergence and the second term is due to nonlinear refraction, which counteracts the diffraction and is also responsible for the guidance of the laser pulse. There is a third term$\frac{1}{f_2}(\frac{df_2}{d\xi})^2$, the initial contribution of this term is zero as $\frac{df_2}{d\xi} = 0$ at $\xi = 0$. However, as the beam propagates, this term supports the second nonlinear refractive term and hence contributes significantly to the guidance of the laser pulse.

Figure 2, represents the variation of the beam width parameter f_2 of the second guided pulse with the dimensionless distance of propagation ξ for two values of the guided pulse intensities, $2.3 \times 10^{13}W/cm^2, 1 \times 10^{14}W/cm^2$ and for a fixed value of prepulse intensity, $4.0 \times 10^{17}W/cm^2$. It is observed that for a second guided pulse intensity $2.3 \times 10^{13}W/cm^2$, f_2 varies in the range $0.1816 < f_2 < 1.414$ as ξ goes from 0 to 36.60. It is further observed that f_2 varies in the range $0.0844 < f_2 < 1.414$ as ξ goes from 0 to 51.20 for a guided pulse intensity $1 \times 10^{14}W/cm^2$. So, as we increase the intensity of the guided pulse from $2.3 \times 10^{13}W/cm^2$ to $1 \times 10^{14}W/cm^2$, laser guiding increases from 36.60 Rayleigh length to 51.20 Rayleigh length. This is due to the dominance of the nonlinear self-focusing term over the diffraction term with the increase in intensity. Due to increase in the intensity of the guided pulse, collisional nonlinearity becomes more prominent, and hence increases the self focusing of the guided pulse , which further leads to increase in the laser guiding.

Figure 3, presents the variation of the beam width parameter f_2 of the second guided pulse with the dimensionless distance of propagation (ξ) for two values of ionizing laser prepulse intensity, $1.0 \times 10^{17}W/cm^2, 4.0 \times 10^{17}W/cm^2$ and for a fixed value of second delayed laser pulse intensity , $1 \times 10^{14}W/cm^2$. It is observed from the figure 3 that for a prepulse intensity, $1.0 \times 10^{17}W/cm^2$, f_2 varies in the range $0.1079 < f_2 < 1.414$ as ξ goes from 0 to 29.40. It is further observed that for a prepulse intensity $4.0 \times 10^{17}W/cm^2$, f_2 varies in the range $0.0844 < f_2 < 1.414$ as ξ goes from 0 to 51.20. This is due to the fact that the second nonlinear

Journal of Physics: Conference Series **516** (2014) 012027 doi:10.1088/1742-6596/516/1/012027

refractive term in Eq. (15) is very sensitive to the prepulse intensity. So, as we increase the prepulse intensity, nonlinear refractive term dominates over the diffractive term and results in increase in the laser guidance. Therefore from the analysis it is predicted that as we increase the intensity of the prepulse from $1.0 \times 10^{17} W/cm^2$ to $4.0 \times 10^{17} W/cm^2$, laser guidance increases from 29.40 Rayleigh length to 51.20 Rayleigh length.

From figures 2 & 3, it is evident that the guided laser pulse propagate through plasma region created by prepulse, without any absorption up to several rayleigh lengths. Results of the present investigation are useful to the physics of laser-induced fusion and other heating experiments, where the guiding of the laser beam in the plasma is very important.

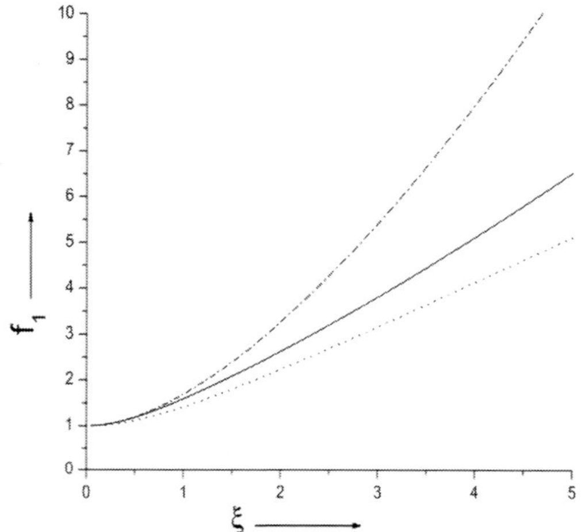

Fig. 1

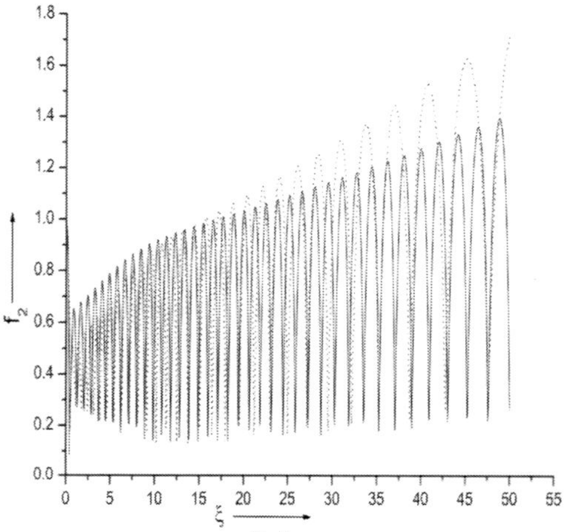

Fig.2

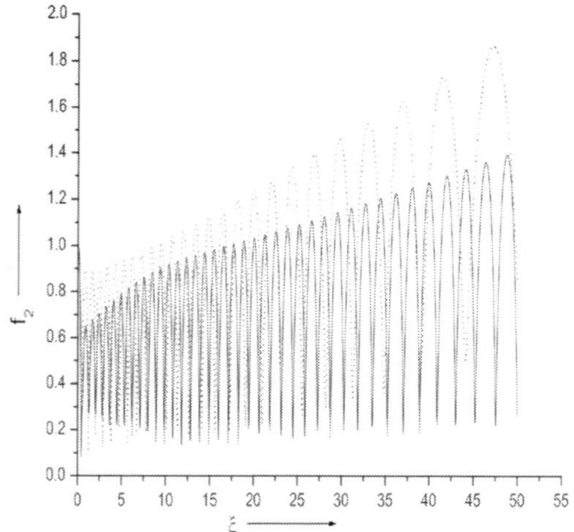

Fig.3

References

[1] Tajima T and Dawson J M 1979 *Phys. Rev. Lett.* 43 267.
[2] Umstadter D, Kim J K and Dodd E 1996 *Phys. Rev. Lett.* 76 2073.
[3] Tabak M, Hammer J, Glinsky M E, Kruer W L, Wilks S C, Woodworth J, Campbell E M and Perry M D, 1994 *Phys. Plasmas.* 1 1626.
[4] Mackinnon A J, Borghesi M, Gaillard R, Malka G, Willi O, Offenberger A A, Pukhov A and Meyer-Ter-Vehn J 1999 *Phys. Plasmas.* 6 2185.
[5] Sprangle P, Esarey E, Ting A and Joyce G 1988 *Appl. Phys. Lett.* 53 2146.
[6] Li X F, L'Huillier A, Feray M, Lompre L A and Mainfray G 1989 *Phys. Rev. A* 39 5751.
[7] Kasparian J, Rodriguez M, Mejean G, Yu J, Salmon E, Wille H, Bourayou R, Frey S, Andre Y B, Mysyrowicz A, Sauerbrey R, Wolf J P and Woste L, 2003 *Sci.* 61 301.
[8] Pepin H, Comtois D, Vidal F, Chien C Y, Desparois A, Johnston T W, Kieffer J C, Fontaine B La, Martin F, Rizk F A M, Potvin C, Couture P, Mercure H P, Bondiou-Clergerie A and Lalande P 2001 *Phys.Plasmas* 8 2532.
[9] Rodriguez M, Sauerbrey R, Wille H, Wostle L, Fujii T, Andre Y B, Mysyrowicz A, Klingbeil L, Rethmeier K, Kalkner W, Kasparian J, Salmon E, Yu J and Wolf J P 2002 *Opt. Lett.* 27 772.
[10] Chen Z L, Unick C, Vafaei-Najafabadi N, Tusi Y Y, Fedosejevs R, Naseri N, Masson-Laborde P E and Rozmus W 2008 *Laser Part. Beams* 26 147.
[11] Neff S, Knobloch R, Hoffmann D H H, Tauschwitz A and Yu S S 2008 *Laser Part. Beams* 24 71.
[12] Yu W, Yu M Y, Xu H, Tian Y W, Chen J and Wong A Y 2007 *Laser Part. Beams* 25 631.
[13] Zhao X M, Yeh C Y, Diels J C and Wang C Y 1993 *Ultrafast phenomenon VIII. Springer-verlag, Berlin.* p.267
[14] Braun A, Korn G, Liu X, Du D, Squier J and Mourou G 1995 *Opt. Lett.* 20 010073.
[15] Yang H, Zhang J, Yu W, Li Y J and Wei Z Y 2001 *Phys. Rev E* 65 016406.
[16] Yang H, Zhang J, Li Y J, Chen Z L, Teng H, Wei Z Y and Sheng Z M 2002 *Phys. Rev E* 66 016406.
[17] Fontaine B La, Vidal F, Jiang Z, Chien C Y, Comtois D, Desparois A, Johnston T W, Kieffer J C, Pepin H and Mercure H P, 1999 *Phys. Plasmas* 6 1615.
[18] Esarey E, Sprangle P, Krall J and Ting A 1997 *IEEE Journal of Quantum Electronics* 33 1879.
[19] Krushelnick K, Ting A, Moore C I, Burris H R, Esaray E, Sprangle P and Baine M 1997 *Phys. Rev. Lett.* 78 4047
[20] Durfee C G and Milchberg H M 1993 *Phys. Rev. Lett.* 71 2409
[21] Durfee C G, Lynch J and Milchberg H M 1995 *Phys. Rev. E* 51 2368.
[22] Liu C S and Tripathi v K, 1994 *Phys. Plasmas* 1 3100.
[23] Akhmanov S A, Sukhorukov A P and Khokhlov R V 1968 *Sov. Phy. Usp.* 10 609.
[24] Sodha M S, Ghatak A K and Tripathi V K, 1976 *Progress in Optics.Volume XIII (edited by Wolf, E.), North Holland, Amsterdam.* 13 171.
[25] Fibich G 1996 *Phys. Rev. Lett.* 76 4356.
[26] Singh A and Singh N 2010 *Laser Particle Beams.* 28 263.
[27] Lam J F and Lippman B 1975 *Opt. Commun.* 15 419.
[28] Sodha M S, Sinha S K and Sharma R P 1979 *J. Phys. D: Appl. Phys* 12 1079.

Modelling and simulation of gyrotrons for ITER

M Damyanova[1], S Kern[2], S Illy[2], M Thumm[2], S Sabchevski[1], I Zhelyazkov[3] and E Vasileva[1]

[1]Institute of Electronics of the Bulgarian Academy of Sciences, Association EURATOM-INRNE, 72 Tzarigradsko Chaussee Blvd., BG-1784 Sofia, Bulgaria

[2]Karlsruhe Institute of Technology, Association EURATOM-KIT, Institute for Pulsed Power and Microwave Technology, Hermann-von-Helmholtz-Platz 1, D-76344 Eggenstein-Leopoldshafen, Karlsruhe, Germany

[3]Faculty of Physics, Sofia University, Association EURATOM-INRNE, 5 James Bourchier Blvd., BG-1164 Sofia, Bulgaria

[1]E-mail: milena@ie.bas.bg

Abstract. Powerful gyrotrons of the megawatt class will be used for electron cyclotron resonance heating (ECRH) and current drive (ECCD) of magnetically confined plasma in the thermonuclear reactor ITER. For computer-aided design (CAD), analysis and optimization of their performance numerical experiments based on adequate physical models are used. In this paper, we outline and illustrate the current status of both the available software tools for numerical simulation of such gyrotrons, as well as the novel computer codes of the problem oriented GYREOSS software package which is under development now.

1. Introduction

Gyrotrons are the most powerful sources of coherent radiation in the sub-terahertz frequency range of the electromagnetic spectrum and are being considered as essential and indispensable components of the electron cyclotron wave system for electron cyclotron resonance heating (ECRH) and current drive (ECCD) of fusion plasmas in various thermonuclear reactors (tokamaks, stellarators etc.) including the ITER, as well as for stability control and diagnostics. The success of the first generation of megawatt class tubes operating in long-pulse and CW regime has motivated a European endeavor for the development of multi-megawatt class gyrotrons [1], most notably the 2 MW, 170 GHz gyrotron with a coaxial cavity. It is quite clear that such an increase of the output power would result in a reduction of the number of required units, power supplies, and transmission lines and, eventually, would decrease the installation cost of the whole system and at the same time would increase its performance. In this short paper we outline briefly the physical models and the computer codes (both available and under development) used for computer-aided design (CAD) and optimization of such powerful radiation sources.

[1] To whom any correspondence should be addressed.

2. Physical models for simulation of gyrotrons

The overall operational performance of the gyrotron depends on the synergy between several subsystems (e.g., electron-optical system (EOS), electro-dynamical system (resonant cavity), quasi-optical output coupler and so on). Below, however, we describe the simulation tools (models and software) pertinent only to the EOS and the resonator. Moreover, the limited space forces us to outline the generic form of the used physical models rather than their detailed computational form.

The longitudinal profile $f_n(z,t)$ of the field intensity in the cavity is given by the following equation

$$\frac{\partial^2}{\partial z^2} f_n(z,t) - \frac{\omega_n^2 - \omega_{\perp n}^2}{c^2} f_n(z,t) - \mathrm{i}\frac{2\omega_n}{c^2}\frac{\partial}{\partial t} f_n(z,t) = R_n, \tag{1}$$

where ω_n is the frequency of the n th mode, $k_{\perp n}^2 = \omega_{\perp n}^2 / c^2$ is the squared transverse wave number, z is the axial coordinate and t is time. The excitation factor $R_n[t, I_\mathrm{b}, U_{\mathrm{b},}\varphi(\alpha, \mathbf{v})]$ depends on the beam parameters (such as the current I_b, the voltage U_b as well as the distribution φ of electron velocities $\mathbf{v}$ and the pitch factor α) and is calculated taking into account the energy balance of the interaction between the beam electrons and the RF electromagnetic field. The solution to equation (1) must satisfy the following radiation boundary conditions at the entrance ($z = z_\mathrm{in}$) and at the exit ($z = z_\mathrm{out}$) of the resonator

$$\frac{\partial f_n(z_\mathrm{in})}{\partial z} = +\mathrm{i}k_{\parallel n} f_n(z_\mathrm{in}), \quad \frac{\partial f_n(z_\mathrm{out})}{\partial z} = -\mathrm{i}k_{\parallel n} f_n(z_\mathrm{out}), \tag{2}$$

where $k_{\parallel n} = [(\omega_n^2 - \omega_{\perp n}^2)/c^2]^{1/2}$ is the axial wave number. The motion of the electrons with the rest mass m_0, velocity $\mathbf{v}(\mathbf{r})$ and momentum $\mathbf{p}(\mathbf{r})$ is governed by the equation

$$\frac{\mathrm{d}\mathbf{p}(\mathbf{r})}{\mathrm{d}t} = \mathbf{F}[f_n(z), \mathbf{E}_\mathrm{HF}(\mathbf{r}), \mathbf{B}_\mathrm{HF}(\mathbf{r}), \mathbf{B}_0(\mathbf{r})], \quad \mathbf{p}(\mathbf{r}) = \gamma m_0 \mathbf{v}(\mathbf{r}). \tag{3}$$

Here $\mathbf{F}(\mathbf{r})$ is the Lorentz force acting on the electrons in the high-frequency electric and magnetic fields $\mathbf{E}_\mathrm{HF}(\mathbf{r})$, $\mathbf{B}_\mathrm{HF}(\mathbf{r})$ and an external magnetic field $\mathbf{B}_0(\mathbf{r})$, and γ is the relativistic Lorentz factor. Both components of the high-frequency electromagnetic field ($\mathbf{E}_\mathrm{HF}(\mathbf{r})$ and $\mathbf{B}_\mathrm{HF}(\mathbf{r})$) are superposition of the corresponding modal fields in the cavity. In fact, an averaged equation for the "slowly varying" quantities is used in the codes instead of equation (3). The more notable simplifications of this self-consistent model are the static (time-independent) model when $\partial f_n / \partial t = 0$ and the so-called cold cavity model ($R_n = 0$), when the influence of the electron beam on the field in the cavity is neglected. The efficiency of the interaction η is calculated from $\eta = \langle (\gamma_0 - \gamma_\mathrm{i})/(\gamma_0 - 1) \rangle$, where γ_0 and γ_i are respectively the initial and the final values of the relativistic factor and $\langle ... \rangle$ denotes an averaging over the whole ensemble of particles representing the electron beam.

The generic physical model, which is implemented in practically all computer codes for simulation (ray tracing) of the electron-optical system (EOS) of gyrotrons, consists of a self-consistent set of equations:

$$\nabla^2 \phi(\mathbf{r}) = -\frac{\rho(\mathbf{r})}{\varepsilon_0}, \quad \forall \mathbf{r} \in \Omega, \quad \mathbf{E}(\mathbf{r}) = -\nabla \phi(\mathbf{r}), \tag{4}$$

$$\phi|_{\Gamma_i} = \phi_i, \quad \frac{\partial \phi}{\partial n}|_{\Gamma_j} = 0, \quad \Gamma = \sum\nolimits_k \Gamma_k = \partial\Omega, \tag{5}$$

$$\frac{d\mathbf{p}(\mathbf{r})}{dt} = \mathbf{F}[\mathbf{E}(\mathbf{r}), \mathbf{B}(\mathbf{r})], \quad \mathbf{p}(\mathbf{r}) = \gamma m_0 \mathbf{v}(\mathbf{r}), \tag{6}$$

$$\mathbf{p}_0(\mathbf{r}) = \mathbf{p}(\mathbf{r})|_{\Omega_{\mathrm{Se}_i}} = f_i(\mathbf{v}_0, \mathbf{r}_0). \tag{7}$$

Here (4) is the Poisson equation with boundary conditions (5) for the electrostatic potential distribution $\phi(\mathbf{r})$ and electric field $\mathbf{E}(\mathbf{r})$ in the domain Ω with a closed boundary Γ composed of the segments Γ_k, $\rho(\mathbf{r})$ is the space charge density and ε_0 is the permittivity of vacuum. Equation (6) is the relativistic equation of motion for electrons with the initial conditions (7) specified on N_S different regions Ω_{Se_i} ($\Omega_{Se} = \bigcup \Omega_{Se_i}$, $\Omega_{Se_i} \bigcap \Omega_{Se_j} = 0$, $\mathrm{diam}(\Omega_{Se_i}) << 1$, $i = 1, 2, ..., N_S$) on which the emitting surface Ω_{Se} of the cathode is decomposed. The space charge density $\rho(\mathbf{r})$ is calculated and allocated to the nodes of the computational grid taking into account the conservation of the beam current. The magnetic field $\mathbf{B}(\mathbf{r})$ is a superposition of the external magnetic field and the magnetic field of the beam current.

3. Computer codes for simulation and CAD of gyrotrons

The physical models outlined above are realized in two groups of software packages for simulation of: (i) the beam–wave interaction in the cavity, and (ii) the formation of the helical electron beam in the EOS, respectively. The first one is represented by the problem-oriented CAVITY software package developed at KIT-IHM by S Kern and co-workers [2]. Its structure and content are illustrated in figure 1. This hierarchy of codes reflects the usual steps in the process of CAD, which habitually starts with an analysis of the geometry and the mode spectrum of the cavity, followed by simplified considerations using for example static, cold cavity and/or single mode physical models and finally utilizes the more adequate (but also the most sophisticated) self-consistent, time-dependent multimode physical model. The latter is implemented in the SELFT code. All components of the CAVITY package are united by a shell (GUI) that allows one to input and edit the data for simulations, to execute a sequence of codes, as well as to visualize the results.

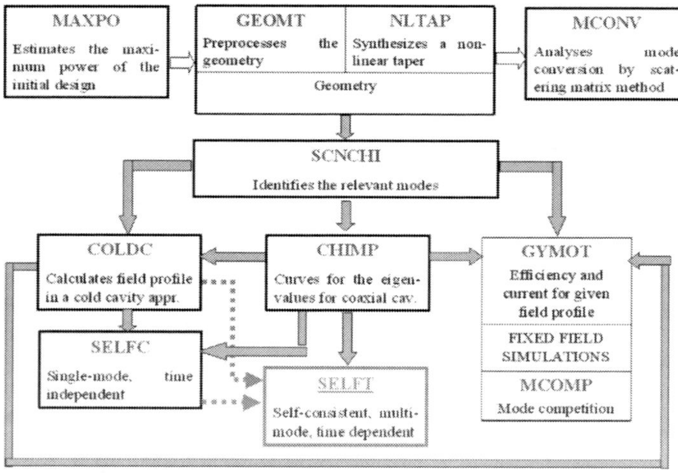

Figure 1. Structure and content of the CAVITY package (KIT-IHM) [2].

Similar functionality have the cavity codes belonging to the GYROSIM package [3] (developed in FIR-FU, Japan), which is also available to our team. In its current version, however, it is restricted only to the conventional resonators, while CAVITY treats coaxial structures (including such with a corrugation insert) as well.

Figure 2 shows a collage of typical screenshots that illustrate both the graphical user interface (GUI) and the visualization capabilities of the CAVITY package.

The physical model, which describes the formation of a helical electron beam in the EOS (equations (4)–(7)) formulated in two spatial dimensions but taking into account the three components of the electrons' momentum (i.e., in a 5-dimensional phase space and thus being a 2.5D physical model) is realized in several packages for trajectory analysis (ray tracing) available to our team, most notably DAPHNE [4] (developed at CRPP-EPFL), ESRAY [5] (developed at KIT-IHM by S Illy) and GUN-MIG/CUSP [3]. Here we outline only the most advanced of them, namely ESRAY. It is characterized by an efficient object-oriented (OO) programming implementation in the C++ language. Both the modular structure of the package (see figure 4) and the OO paradigm of the code make it extensible and allow an easy modification. In ESRAY, the boundary value problem for the electrostatic potential in the physical domain is discretized using a structured boundary-fitted adaptive grid (produced by the GRIDGEN module), which represents the boundary with sufficient accuracy and ensures an appropriate interpolation of all quantities attributed to the nodes. The input to GRIDGEN is a ***tcl*** script that describes the geometry of the EOS. The computational grid is created transforming (mapping) the physical domain onto an equidistant logical grid in a rectangle. The magnetic field, produced by a system of solenoids, is calculated by the MAGGEN module. Its off-axis components are computed from the field on the axis using paraxial series expansions. A self-consistent solution to the equations (4)–(7) is obtained in the ESRAY module as a result of an iterative procedure in which the solution of the boundary value problem (Poisson equation) and the integration of electron trajectories follow in a succession until the convergence criterion is satisfied. The results of the numerical experiments are post-processed and can be visualized and exported in a form of informative colour plots. The OVIS GUI is based on an own graphical library (an integral part of the ESRAY package), which is programmed in ***Tcl/Tk***.

Figure 4 illustrates the visualization capabilities of the ESRAY package.

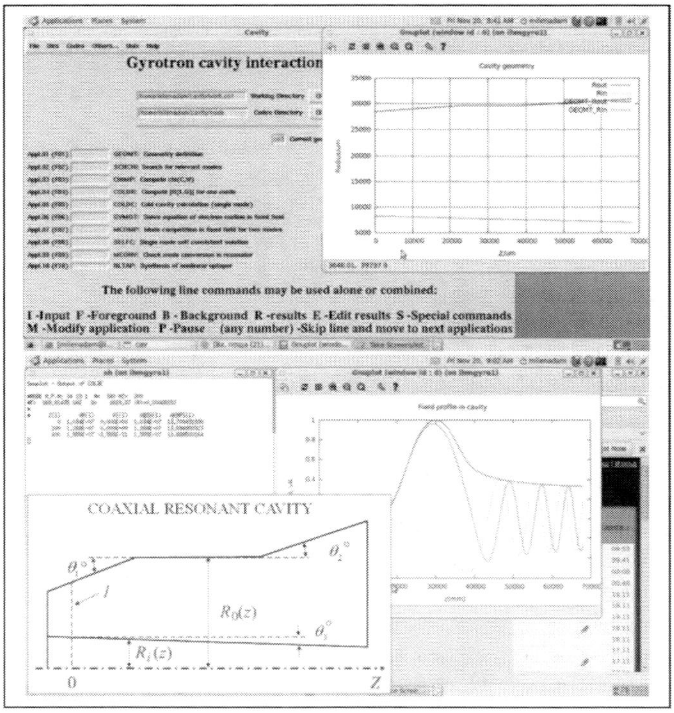

Figure 2. Screenshots, and an example with input and output data of the CAVITY package.

4. GYREOSS package – a work in progress

Recently, the current status of the software tools for CAD of the EOS of gyrotrons has been presented in [4]. A critical examination and benchmarking have revealed not only their advantages and characteristic features but also their limitations. Based on such analysis a concept for further development of

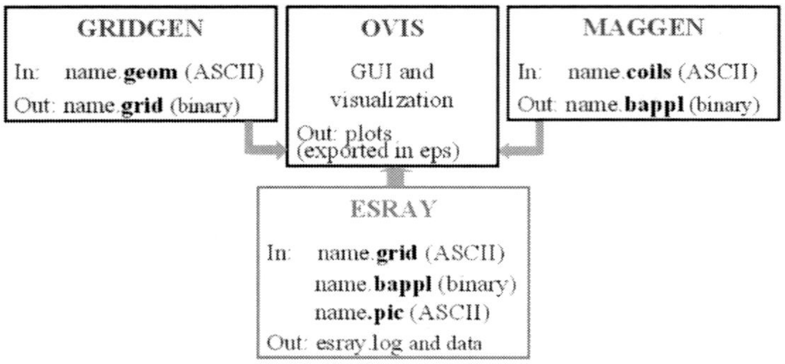

Figure 3. Main modules of the ESRAY package and their input and output files.

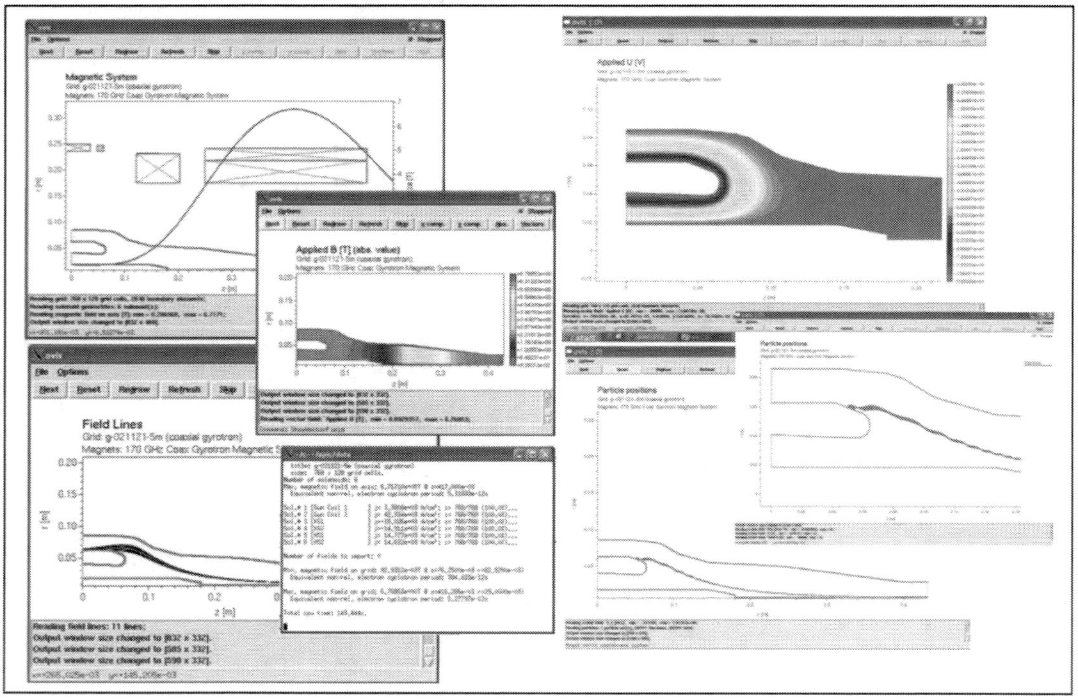

Figure 4. Screenshots with typical results from simulations of a coaxial EOS.

the physical models and computer codes have been developed [4]. The realization of this concept is the main aim of the work on the GYREOSS (which stands for GYRotron EOS Simulation). It has been conceived as a problem oriented software package of computer codes united by a common functional assignment to the simulation of the beam formation in the EOS of gyrotrons. A pivotal goal is to include more physics (i.e., to take into account more physical factors and phenomena) extending the currently used 2D models to more adequate self-consistent physical models formulated in three spatial dimensions. A characteristic example in this respect is the work on the development of a 3D emission model, which takes into account the non-uniformity of the extracted beam current and uses more realistic distributions of the initial velocities on the cathode [6]. Such models make the simulations much more computing intensive and require the utilization of efficient numerical methods, algorithms, programming libraries and integrated development environments. Other two important requirements to the codes under development are their extensibility and portability to different platforms, including high-performance computing (HPC) systems (e.g., multiprocessor workstations, clusters, Grid).

For pre-processing, post-processing and visualization of data from the numerical experiments GYREOSS relies on the 3D finite element mesh generator with a build-in CAD system ***gmsh*** [7]. In the first prototype version of the code the boundary value problem for the electrostatic potential was solved using ***GetDP*** (**G**eneral **E**nvironment for **T**reatment of **D**iscrete **P**roblems), which utilizes ***PETSc*** or ***SPARSKIT*** to solve the corresponding linear systems. It is well known that the two most time consuming procedures in the PIC simulations are the particle pushing and the solution of the field problem. Moreover, the data structure provided by the latter influences strongly the way in which the particles are localized and eventually determines the efficiency of the former procedure. It was found, however, that the algorithms realized using the ***GetDP*** are far from optimal. This motivated us to look for an alternative to this solver. Currently we are experimenting with a number of different short-listed libraries. Among them are ***libmesh***, ***Elmer***, ***GetFEM***, ***FELIB***, ***OpenFOAM***, ***FreeFEM++***, just to name a few. Therefore in its current status GYREOSS is mostly a test bed for experimentation with different solvers, programming libraries and algorithms in a pursuit to find the most appropriate building blocks for the novel package modules under development. More information about the current status of the package is available for the reader at the collaborative site of the project (look at http://gyreoss.wikidot.com).

5. Conclusions

Well-validated and efficient simulation packages based on adequate 2D physical models are used for CAD and optimization of high power gyrotrons for ITER. In parallel with their maintenance, constant upgrade and improvement a work on a novel generation of software tools is in progress now. Its goal is the development of computational modules for realization of more informative, adequate and self-consistent physical models formulated in three spatial dimensions.

Acknowledgements

This work was carried out in the framework of Task 2.1.2 of the scientific programme of the EURATOM-INRNE Association.

References

[1] Thumm M 2010 State-of-the-Art of High Power Gyro-Devices and Free Electron Masers (Update 2009), *KIT Scientific Reports* 7540 1–132

[2] Kern S 1996 Numerische Simulation der Gyrotron-Wechselwirkung in koaxialen Resonatoren, *Wissenschaftliche Berichte* FZKA 5837 1–245

[3] Sabchevski S, Idehara T, Saito T, Ogawa I, Mitsudo S, Tatematsu Y 2010 Physical Models and Computer Codes of the GYROSIM (GYROtron SIMulation) Software Package, FIR Center Report FIR FU-99 (http://fir.u-fukui.ac.jp/FIR_FU99S.pdf)

[4] Sabchevski S, Zhelyazkov I, Thumm M, Piosczyk B, Tran T M, Hogge J-Ph, Pagonakis J Gr 2007 Recent Evolution of the Simulation Tools for Computer Aided Design of Electron-Optical Systems for Powerful Gyrotrons, *Computer Modelling in Engineering & Sciences* **20** 203–220

[5] Illy S 1997 Untersuchungen von Strahlinstabilitaten in der Kompressionzone von Gyrotron-Oscillatoren mit Hilfe der kinetishen Theorie und zeitabhangiger Particle-in-Cell-Simulationen, *Wissenschaftliche Berichte* FZKA 6037 1–145

[6] Sabchevski S, Illy S, Piosczyk B, Borie E, Zhelyazkov I 2008 Towards the Formulation of a Realistic 3D Model for Simulation of Magnetron Injection Guns for Gyrotrons (A Preliminary Study), *Wissenschaftliche Berichte* FZKA 7409 1–34

[7] Damyanova M, Sabchevski S, Zhelyazkov I 2010 Pre- and Post-processing of data for Simulation of Gyrotrons by the GYREOSS Software Package, *J. Phys.: Conference Series* **207** 012032 (6 pages)

On the current layer in the run-down phase of the plasma focus discharge

S Zapryanov, V Yordanov and A Blagoev

Department of Optics and spectroscopy, Faculty of Physics, Sofia University 5 J. Bourchier blvd., 1164, Sofia, Bulgaria

E-mail: blagoev@phys.uni-sofia.bg

Abstract. In the experiments with the 3 kJ plasma focus device in the Sofia University are measured the basic characteristics, namely the discharge current, the current derivative, the soft X-ray and the hard X-ray emission from the plasma. By a set of magnetic probes the velocity of the current sheath during the run-down axial phase is determined. In the last two thirds of the axial phase the current layer velocity is a constant.

1. Introduction

Despite the considerable period of study of the plasma focus (PF) discharge, there is still an interest towards this phenomenon due to the unresolved problems of the discharge dynamics and the resources for different applications of the PF machines.

In the Faculty of Physics at the University of Sofia a Mather's type dense plasma focus device (DPF) is in regular operation for about one and a half years.

The aim of this work is to present the results of the measurements of basic discharge characteristics of this discharge with an emphasis of the current layer during the axial phase.

It is well known that the evolution of the PF discharge consists of several phases: breakdown, axial run-down (in the Mather type PF systems), radial compression, pinch phase followed by expanding and a decay of the hot plasma column. After the breakdown and current lift-off from the insulator, a run-down phase takes place. The current sheath observed in this stage is a strong ionizing shock wave, driven by $(\vec{j} \times \vec{B}_\vartheta)$ force (a magnetic piston). Here as usual $\vec{j}$ is the current density and $\vec{B}_\vartheta$ is the induced magnetic field. This current sheath propagates into the neutral gas, scooping up a part of the mass it encounters and ionizing the particles.

One of the conditions for an efficient high energy density pinch is that the plasma layer should be thin, axially symmetrical and uniform, moving with high snow-plow efficiency. Therefore efforts to find the best conditions during this stage are necessary both for an efficient work of the existing devises and for construction of new ones. The variable parameters are geometry of the PF system, type and pressure of the filling gas, charging voltage, etc.

There are several experimental and theoretical works, considering the axial run-down phase [1 - 4]. Although that in principle the existing models describe satisfactory the effects observed, still there are some questions. For example in the work [5] it was shown that the current sheath structure is not homogeneous. It consists of two zones: a first layer with a steep density gradient and a second one associated with the plasma-magnetic field interface. It is not yet clear whether this division exists only in the beginning of the axial phase.

2. Experimental device

The condenser bank has capacitance of 20 μF, with the maximal voltage of 40 kV. The insulator is a quartz tube (2.6 mm in diameter, 30 mm length) the main switch is a vacuum spark gap. The anode is a copper tube with 20 mm diameter and 145 mm length, the cathode consists of 6 copper rods (8 mm diameter, 160 mm length) mounted in the massive cathode bottom on a circle with a radius 35 mm.

A procedure for optimizing the discharge conditions was carried out, changing the charging voltage and the gas pressure. So far the experiments were conducted in air and the nominal pressure is in the range 1.0 ÷ 1.7 mBar. The voltage operating range is chosen to be 15 ÷ 18 kV. We are monitoring the discharge current, (Rogowski belt) the current derivative (pick-up coil), the soft X-ray (PIN diodes) and the hard X-ray emission (scintillator probe) from the plasma. Thermo-luminescent dosimeters (TLD) were used to measure the full X-ray dose in the stainless steel chamber of our DPF and to control the radiation outside the device.

A set of magnetic probes inserted between the anode and one of the cathode rods allow us to determine the velocity of the current sheath, during the run-down phase.

Figure 1 shows schematically the PF electrodes and the set of magnetic probes. Each probe is formed by a twenty-turn, 2 mm diameter coil. The distance between the probes is 1 cm. The probes shown by one color are connected in series. So we have 4 "channels" of probes each one with 4 cm distance between the separate coils. A Pyrex tube with 5 mm external diameter serves as a container of the probes.

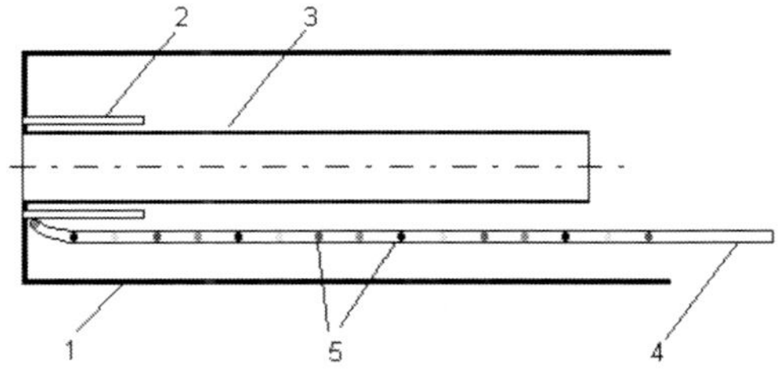

Figure 1. Scheme of the PF electrodes and the Pyrex tube with the set of magnetic probes. 1 – cathode, 2 – insulator, 3 – anode, 4 – Pyrex tube, 5 – probe coils

The data of all of the detectors mentioned above have been recorded by two 4 channel oscilloscopes (TDS 3034C, TDS 2004B).

3. Results and discussions

3.1. Basic measurements

The data, received by the diagnostic tools in the first several hundred shots fired so far, reveal the typical peculiarities known from the literature. The oscilloscope pictures give the moment of the initial breakdown of the gas, followed in a few microseconds by the occurrence of a pinch. A correlation between the observed one or more peaks of soft and hard X-rays and the peculiarities of the dI/dt signal is established, the latter corresponding to the same number of pinches with the same mutual distance in time. Figure 2 presents oscilloscope traces of discharge current, soft and hard X-ray signals.

As it was stated above the optimization of the run-down phase of a specific plasma focus system is necessary in order to obtain high yield of radiation in the pinch phase. Thus the experimental study of this period is required. For this purpose monitoring of the light of the moving layer or magnetic probe measurements are used. An advanced method is visualizing the layer by the Schlieren technique [5].

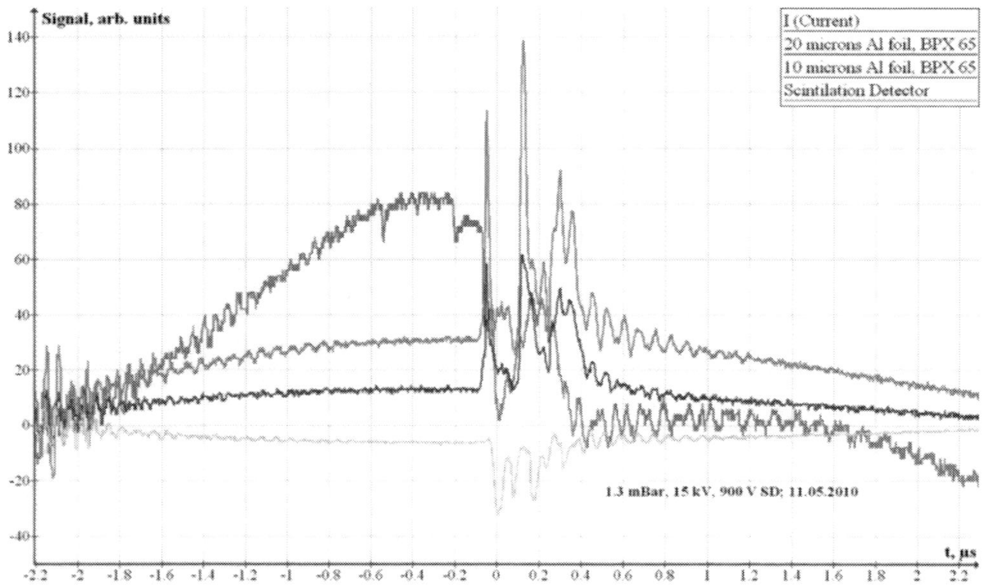

Figure 2. Basic measurements

3.2. Magnetic probe measurements

As it was already mentioned in this work we use a set of magnetic probes shown on figure 1. As output of the set are 4 channels, consisting of connected probes with a distance of 4 cm between the coils.

Figure 3 presents the experimental results derived with this system of probes. Using such system of probes we can determine the velocity and the acceleration of the current sheet as well as it's evolution during the run-down phase in a single shot. It is seen that in the frame of one probe "channel" the intensity of the peaks increases (see for example the behaviour of the red curve). This is due to the current density distribution of the layer [7].

The sharp dips appearing simultaneously for all four channels near 2.8^{th} and 3.2^{nd} microsecond time moments correspond to the electromagnetic signals from the pinch. It is seen also that there are signals from the probes in the second half period of the discharge current too although that no pinch in this case is observed.

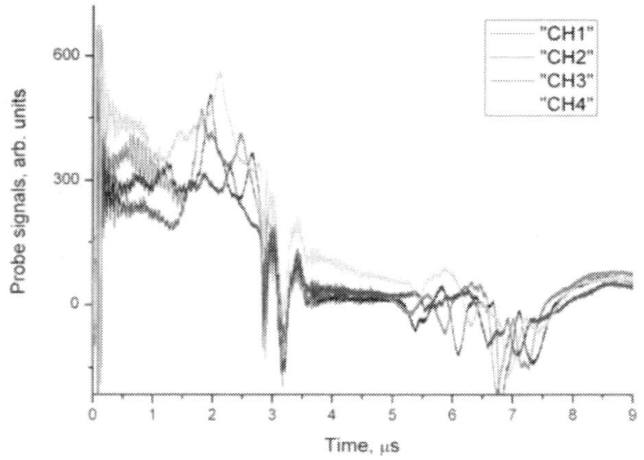

Figure 3. Signals from the magnetic probe system for single shot.

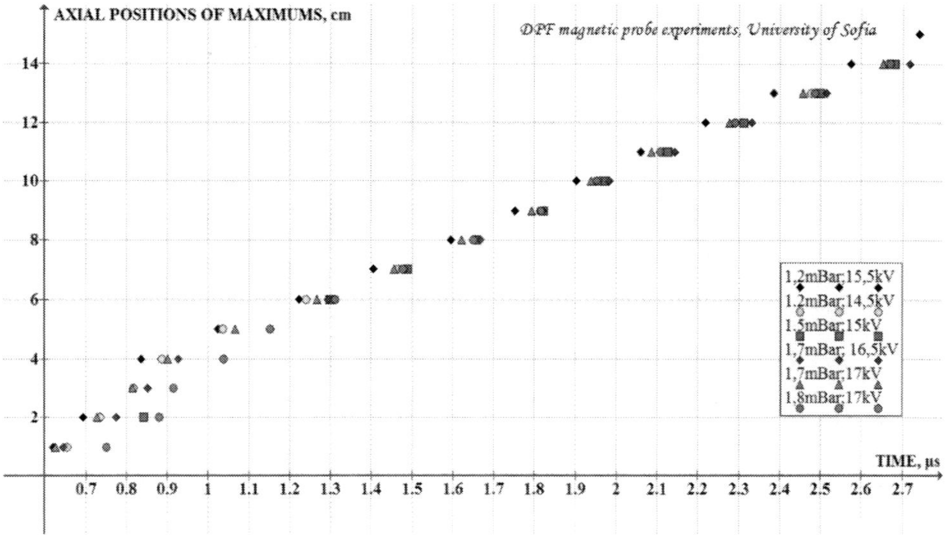

Figure 4. Axial positions of the maximums of the signal of magnetic probe set.

Figure 4 shows the axial positions of the probe peak maximums versus the time scale of the discharge evolution. The points at the (start) beginning of the acceleration phase are determined with a considerable error. Nevertheless after about 1 microsecond from the beginning of the discharge equilibrium of the forces, controlling the current layer is established. It seems that the driving magnetic piston is balanced by the resistance of the compressed gas. That leads to constant velocity of the current layer. I our case between the 1st and 2.8^{th} µs the layer velocity is constant. The values of the velocity of the current sheath were derived by processing of the data of a number of shots fired at different conditions. Before this time interval according to figure 3 from 700 ns to 1 µs the layer is moving faster. Quite interesting is the period before 700 ns but it is impossible to say something for this time interval with the present probe set. For the purpose a radial set of probes has to be used [8,9]. Subtle details in the behavior of the current sheath can be obtained by a model, however this model should be complete enough in order to describe adequately the operation of the specific device (see for example [3]). Naturally constructing such a model is not a trivial task. In the study [9] Bruzzone et al propose a simple model for the same lift-off stage that is intermediate between the breakdown and the axial phase the PF discharge.

4. Summary

Besides the data for discharge current, current derivative, Soft and hard X-ray, we present values of the velocity of the current sheath in the axial run-down phase. The data, derived from a series of shots at conditions: gas pressure 1 ÷ 2 mbar, charging voltage 14 ÷ 17 kV shows that between 1st and 2.8^{th} µs the current sheet has almost equal constant velocity in range of 5 ÷ 6.7 cm/µs.

Acknowledgments: This work was supported by the contract DO-02-136/2008 of the Bulgarian Scientific Research Fund, and by the research contract No 14897/R2 of IAEA.

References

[1] J Gonzales, A Clausse, H Bruzzone and P Florido, "A lumped parameter model of plasma focus", IEEE Trans. on Plasma Science **32** N 3 (2004) 826 - 830

[2] F Casanova, C Moreno and A Clausse "Finite-element numerical model of the current -sheet movement and shaping in coaxial discharges" Plasma Physics Control Fusion **47** (2005) 1239 - 1249

[3] W Stepnyewski, "MHD numerical modelling of the plasma focus phenomena", Vacuum **76** (2004) 51-55
[4] M Favre, P Silva, H Chuaqui, E Wyndam and P Choi "Current sheath studies in a small plasma focus operating in hydrogen-argon mixture" Astrophysics and Space Science **256** (1998) 485 -190
[5] K Kweg, T Tou and S Lee, "Current sheath structure of the plasma focus in the run-down phase" IEEE Trans. on Plasma Science **18** N 5 (1990) 826 - 830
[6] W Bostik, W Prior and E Farber, Phys. Fluids **8** (1965) 745-
[7] H Bruzzone and D Grondona, "Magnetic probe measurements of the initial phase in a plasma focus device", Plasma Phys. Control. Fusion **39** (1997) 1315 - 1326
[8] F Aghamir and R Behbahani, "Study of the sheath formation in a low energy plasma focus" J. Plasma Fusion Res. SERIES **8** (2009) 1256 - 1268
[9] H Bruzzone, H Acuna and A Clausse, "The lift-off stage of the plasma focus discharge", Plasma Physics and Controlled Fusion **49** (2007) 105 - 112

Comparison the Results of Numerical Simulation And Experimental Results for Amirkabir Plasma Focus Facility

Shervin Goudarzi[1], R Amrollahi[2], M Niknam Sharak[2]

[1]North Kargar Ave., Atomic Energy Organization of Iran, Nuclear Science and Technology Research Institute, Plasma Physics & Nuclear Fusion Research School, Tehran, Iran.
[2]Hafez Ave, Amirkabir University of Technology, Nuclear Engineering and Physics Department, Tehran, Iran

E-mail: shgoudarzi@yahoo.com

Abstract In this paper the results of the numerical simulation for Amirkabir Mather-type Plasma Focus Facility (16 kV, 36μF and 115 nH) in several experiments with Argon as working gas at different working conditions (different discharge voltages and gas pressures) have been presented and compared with the experimental results. Two different models have been used for simulation: five-phase model of Lee and lumped parameter model of Gonzalez. It is seen that the results (optimum pressures and current signals) of the Lee model at different working conditions show better agreement than lumped parameter model with experimental values.

1. Introduction

The Plasma Focus devices have been developed independently in 1960's in two different models by Filippov in Soviet Union and Mather in United States [1, 2]. Today many samples of both models in the range of tens of joules to some Megajoules are working in different research centers all over the world [1–7], and the technical problems about design and construction of them are solved completely, but the theoretical description of these devices is very complicated and there is not any proper model for explanation of all of their working phases [2, 7, 8].

In general, the different phases of discharge in Plasma Focus devices can be divided to five different stages [2, 7, 9]: 1 – The initial breakdown and surface discharge, very little is known about the physics of this phase [10], 2 – rundown phase that does not exist in Filippov-type devices, 3 – radial compression, 4 – pinch phase, 5 – instability happening and pinch destruction.

For explanation of rundown and radial compression phases there are no difficulty but there is not any complete and reliable model for description of pinch phase and after that [2, 7–9]. For rundown and radial compression phases some different models are presented that the most acceptable of them are 2 models that the first one is presented by S. Lee that uses the equivalent circuit of the Plasma Focus with a snowplow model for rundown phase and a slug model for radial compression phase [2, 8] and the second model is presented recently by J. H. Gonzalez on the base of lumped parameters [7, 10].

Journal of Physics: Conference Series **516** (2014) 012030 doi:10.1088/1742-6596/516/1/012030

In this paper, The results of these models (discharge current (I) and current derivative (dI/dt) signals, I_{max} and t_{max}, pinch time (t_p) and I_{t_p}) for Amirkabir Mather-type Plasma Focus device "APF"(16 kV, 36μF and 115 nH) for different values of gas sweeping efficiency factor (f_m) and current efficiency factor in run down phase (f_c) and gas sweeping efficiency factor (f_{mr}) and current efficiency factor (f_{cr}) in radial compression phase in the case of using Ar as working gas at different working conditions (different discharge voltages and different initial pressures) have been compared with experimental results. It has been observed that the best agreement between the results of simulation and experimental data would achieve by using the Lee model.

2. Experimental Set–up

The description of "APF" has been reported elsewhere (Table 1) [11]. The diagnostic system consists of: Two 100 MHz and 400 MHz PC-based oscilloscopes, two Rogowski coil for measuring the total discharge current & current derivative signals. NaI scintillator to register hard X-ray yield, plastic scintillator + photomultiplier to register hard X-ray & neutron signals, a pinhole camera to study soft X-ray and hard X-ray emission zones, magnetic probes to study the dynamics of current sheath in the axial phase, and faraday cup to investigate ion emissions.

Table 1. Parameters of Amirkabir PF

Symbol	Parameter	Value
R_i	Anode Radius	1.39 cm
R_e	Cathode Radius	4.47 cm
Z_0	Anode Length	14.8 cm
C_0	Capacity	36 μF
L_0	External Inductance	115 nH
R_0	Stray Resistance	0 mΩ

3. Results and Discussion

The equations of Lee & Gonzales models have been completely described in a lot of references [2, 8–11], here we only present the numerical results of using them for Amirkabir Plasma Focus (APF). For numerically solving the equations of these models for this device, we have used a step-by-step method with MATLAB 6.5 code. Here we have compared some results of predictions of these models (discharge current (I) signal, the pinch time (t_p) at different initial pressures and discharge voltages and optimum pressure at different discharge voltages) for "APF" with Ar as working gas for different values of f_c, f_{cr}, f_m and f_{mr} with experimental results. It is observed from the comparison of the experimental results by the results of two models for several groups of the values of f_c, f_{cr}, f_m and f_{mr} that the results of two model are very close together but the Lee model is a little more compatible with experimental results than lumped parameter model (figures 1,2), to calculation the proper factors for model, several experimental results in different working conditions have been compared with the results of the simulation with different factors and finally it is observed that the results of the model for $f_m = 0.02$, $f_{mr} = 0.15$, $f_c = 0.75$ and $f_{cr} = 0.7$ have the best agreement with experimental results.

In tables 3–5 the predictions of this model for these optimum values of f_c, f_{cr}, f_m and have been compared with experimental data. One can observe that the results of the model for these values have very good agreement with experiment.

Table 2. The best factors with Argon in Amirkabir PF

Symbol	Parameter	Value
f_m	Axial Sweep Efficiency	0.02
f_c	Axial Current Efficiency	0.75
f_{mr}	Radial Sweep Efficiency	0.15
f_{cr}	Radial Current Efficiency	0.7

Table 3. Optimum pressure vs. charge voltage

Charge Voltage [kV]	Optimum Pressure [Torr]			Charge Voltage [kV]	Optimum Pressure [Torr]		
	Measured	Computed	E_{rel} %		Measured	Computed	E_{rel} %
10.5	0.952	0.89	6.56	13.0	1.359	1.36	0.09
11.0	1.048	0.98	6.52	13.5	1.506	1.47	2.39
11.5	1.210	1.07	11.59	14.0	1.558	1.58	1.41
12.0	1.255	1.16	7.56	14.5	1.617	1.70	5.12
12.5	1.344	1.26	6.22	15.0	1.662	1.82	9.51

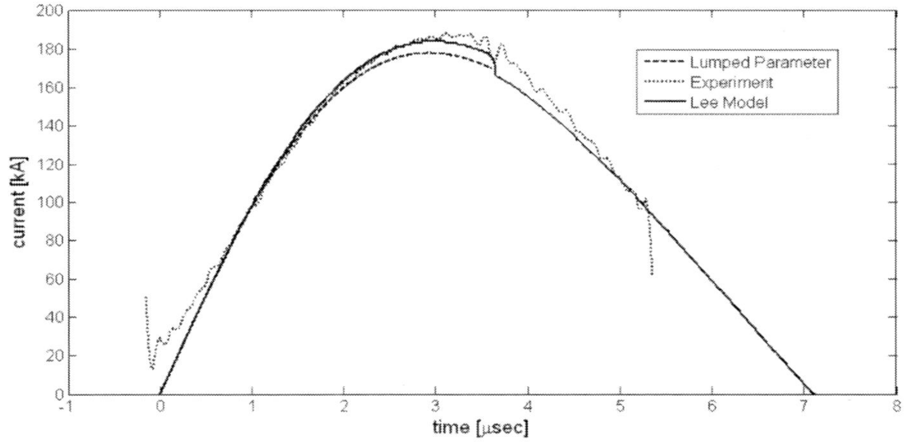

Figure 1. Total current for P_0=1.8 torr and V_0= 12 kV

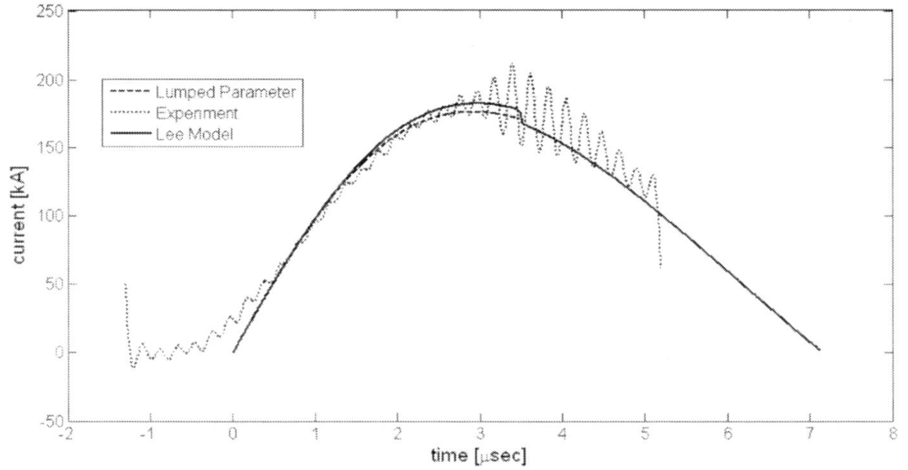

Figure 2. Total current for P_0=1.6 torr and V_0= 12 kV

Table 4. Pinch current vs. charge voltage in optimum pressure

Charge Voltage [kV]	Pinch Current [kA]			Charge Voltage [kV]	Pinch Current [kA]		
	Measured	Computed	E_{rel} %		Measured	Computed	E_{rel} %
10.5	102.574	104.194	1.58	13.0	127.985	128.893	0.71
11.0	107.505	109.244	1.62	13.5	132.911	133.872	0.72
11.5	109.759	114.171	4.02	14.0	138.611	138.673	0.04
12.0	116.979	119.162	1.87	14.5	140.482	143.626	2.24
12.5	126.115	124.153	1.56	15.0	145.793	148.229	1.67

Table 5. Pinch time vs. operation pressure

Charge Voltage V_0 = 11.5 kV				Charge Voltage V_0 = 13.5 kV			
Operation Pressure [Torr]	Pinch Time [μsec]			Operation Pressure [Torr]	Pinch Time [μsec]		
	Measured	Computed	E_{rel} %		Measured	Computed	E_{rel} %
0.8	3.111	2.942	5.43	0.8	3.042	2.684	11.77
1.0	3.209	3.133	2.38	1.0	3.101	2.858	7.85
1.2	3.329	3.314	0.44	1.2	3.210	3.016	6.01
1.4	3.412	3.473	1.78	1.4	3.280	3.154	3.85
1.6	3.484	3.613	3.71	1.6	3.330	3.281	1.44
1.8	3.550	3.753	5.75	1.8	3.459	3.398	1.76

4. Conclusions

Two models on the base of lumped parameters for description of the operation of Mather-type Plasma Focus devices from the end of breakdown phase until the pinch time are presented, Lee model & the lumped parameter model (Gonzalez model). It is seen that the predictions of the first model for "APF" Plasma Focus device with Ar as working gas at different discharge voltages and different initial pressures have shown better agreement than the second model with experimental results and the best values of gas sweeping efficiency factors in rundown & radial compression phase (f_m, f_{mr}) and current efficiency factors (f_c, f_{cr}) in several working conditions have been found. It must be noted that these values are calculated only for Argon as working gas. The main reasons of the differences between the experimental results are that the real values of f_c, f_{cr}, f_m and f_{mr} are not constant in different experiments and the surface discharge and breakdown phase has been neglected in it.

References

[1] Bernard A, Cloth P, Conrads H, Coudeville H, Gourlan G, Joals A, Maisonnier CH and Rager J P 1977 *Nucl. Instrum. Methods* **145** 191–218
[2] Lee S 1989 *Proceedings of the Spring College on Plasma Physics* ICTP Trieste 113–169
[3] Castillo F , Milanese M, Moroso R and Pouzo J 2002 Brazil. J. Phys. 32 3–12
[4] Silva Patricio, Soto L, Kies W and Moreno J 2004 *Plasma Sources Sci & Tech.* 13 329–332
[5] Hussain S, Ahmad S, Khan M Z, Zakaullah M and Waheed A 2004 *J. Fusion Energy* **22** 195–200
[6] Rawat R S, Zhang T, Lim G J, Tan H, Ng S J, Patran A and Hassan S 2004 *J. Fusion Energy* **23** 49-53
[7] Soto L 2005 *Plasma Phys. Control. Fusion* **47** A361–A381
[8] Lee S and Serban A 1996 IEEE Trans. Plasma Sci. 24 101–1105
[9] Mathuthu Manny, Zengeni Teddy G. and Gholap Ashok V 1997 *IEEE Trans. Plasma Sci.* **25** 1382–1388
[10] Gonzalez Jose H 2004 *IEEE Trans. Plasma Sci.* **32** 1383-1391
[11] Habibi M and Amrollahi R 2008 *Plasma Devices and Operation* **16** 163–169

Experimental results of breakdown in "Dena" plasma focus device

Shervin Goudarzi*, S M Hoseinian, A Raeisdana

North Kargar Ave., Atomic Energy Organization of Iran, Nuclear Science and Technology Research Institute, Plasma Physics & Nuclear Fusion Research School, Tehran, Iran

E-mail: shgoudarzi@yahoo.com

Abstract. In spite of the intense research activities on Plasma Focus devices, the physics of the initial breakdown and surface discharge phase has not been realized completely. In this paper we have analyzed the surface discharge and initial breakdown phase in Filippov-type Plasma Focus Facility "Dena" (90 kJ, 25 kV) on the base of the current and current derivative measured signals by using Argon, Neon and Krypton as working gases at different discharge voltages and gas pressures, and the effects of working conditions (atomic weight, discharge voltage and gas pressure) on the breakdown and surface discharge phase have expressed. Also, on the base of these results, we have investigated about the relation of this phase with final pinch phase.

1. Introduction

From the development of the Plasma Focus devices in 1960s in two different models by Filippov in Soviet Union and Mather in United States [1, 2]. A lot of samples of them in the range of less than one Joule to some Megajouls have been constructed in different research centers all over the world [1–8]. In spite of the intense research activities on these devices, some theoretical problems in them are not resolved completely and there are not proper models for explanation of them [8, 9–11]. In general, the different phases of discharge in Plasma Focus devices can be divided to five different stages [1, 2, 3, 8, 12]: 1 – The initial breakdown and surface discharge , very little is known about the physics of this phase [13], this phase has a great influence on the formation of final pinch , 2 – rundown phase that does not exist in Filippov-type devices, 3 – radial compression , 4 – pinch phase , 5 – instability happening and pinch destruction. In this paper we would present and analyze the results of initial breakdown & surface discharge phase in "Dena" Filippov-type Plasma Focus Facility (90 kJ, 25 kV) in different working conditions.

2. Experimental set-up

Our experiments have been performed in Filippov-type Plasma Focus "Dena" (90 kJ, 25 kV). A schematic diagram of "Dena" with all of its diagnostics is shown in Figure1. The description of this device has been reported completely elsewhere [13]. The diagnostic system consists of a 4-channel PC-based data acquisition system including two GPIB compatible oscilloscopes (50MHz) and two fast (500MHz) digital storage Tektronix oscilloscopes. In these experiments, several discharges have been made at different working conditions (discharge voltage, initial pressure, gas composition, insert

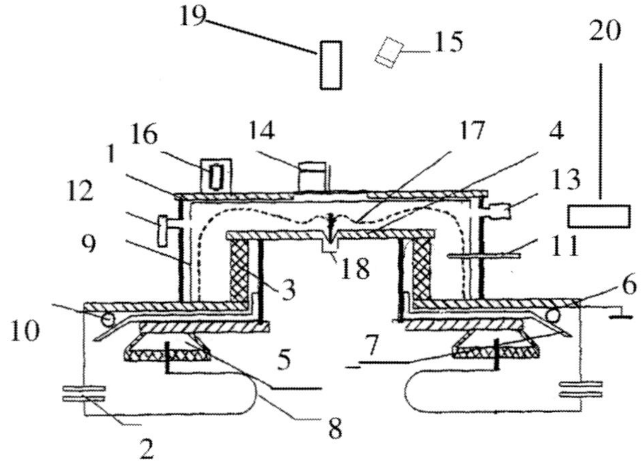

Fig.1. Schematic diagram of the Dena plasma focus and its diagnostics.

1. Upper flange of the vacuum chamber
2. Capacitor energy storage
3. Porcelain insulator
4. Anode
5. Ring-like spark gap switch
6. Collector plate
7. Insulation
8. Matching inductance
9. Liner
10. Rogowski coil
11. Magnetic probe
12. Pinhole camera
13. SXR detector (SPPD-type)
14. Detector-monitor of integral HXR(PhEC)
15. Scintillation detector with multiplier
16. Detector of integral neutron yield (G-M)
17. Plasma-current sheath
18. conic insert anode
19. End–on G-M counter
20. Side-on G-M counter

anode) to study the variations of total discharge current and current derivative signals to investigate the correlation between the breakdown phenomena and working conditions. The total discharge current and current derivative signals are measured by a Rogowski coil and a magnetic probe, respectively.

3. Results and discussion

In these experiments Hydrogen, Neon and Argon have been used as working gases. We have done 3 series of experiments, 1- Using Neon and Argon as working gases at constant initial pressures and different discharge voltages, 2- Using Hydrogen, Neon and Argon as working gases at fixed discharge voltages and different initial pressures and 3- Using Hydrogen, Neon and Argon as working gases at fixed initial pressures and discharge voltages. From the investigation of the results of these experiments (discharge current, current derivative), we have suggested about the breakdown phenomena in "Dena".

In the first series of our shots we have observed that as could be predicted at constant pressures (for example, 0.5 torr) for both gases, by increasing the discharge voltage (for example, from 14 k V to 16 kV) and therefore, increasing the pinch intensity and tightness and decreasing the pinch time (tp). The breakdown phenomenon occurs earlier.

By changing the pressure in the experiments with Hydrogen, Neon and Argon at constant voltages, it is seen that for each gas when the pressure approaches to the optimum pressure of x-ray emission (therefore, increasing the pinch intensity and its tightness and decreasing the pinch time) the breakdown happens earlier.

Also, it is seen from the results of the experiments with Hydrogen, Neon and Argon at constant pressures and discharge voltages that the breakdown in Argon happens earlier than two other cases and in Neon earlier than Hydrogen, we can conclude that commonly the breakdown in heavier gases happens earlier than lighter gases. This is acceptable because the heavier atoms can release the electrons of their outer orbits easier than lighter atoms. These results show again the same relation as two other modes between breakdown time and the pinch intensity and tightness and the pinch time

4. Conclusions

The presented results in this paper show that the working conditions (initial pressure, discharge voltage and gas type) have high influences on discharge and breakdown phase. At constant discharge voltage and gas pressure by using heavier gases, surface discharge and breakdown phase happens earlier. In a definite gas with a constant pressure, by increasing the discharge voltage, the breakdown happens earlier and pinch time will decrease. Also, in a definite gas and constant discharge voltage, when the pressure approaches to the optimum pressure, the time of starting the breakdown phase decreases .These results have shown clearly the strong relation between pinch intensity and its tightness and pinch time (t_p) with breakdown time. Commonly, in the discharges with tighter and stronger pinches the breakdown & pinch times are less than the weak pinches

References

[1] Bernard A, Cloth P, Conrads H, Coudeville H, Gourlan G, Joals A, Maisonnier CH and Rager J P 1977 *Nucl. Instrum. Methods* **145** 191-218
[2] Lee S 1989 *Proceedings of the Spring College on Plasma Physics* ICTP Trieste 113-169
[3] Castillo F, Milanese M, Moroso R and Pouzo J 2002 *Brazil. J. Phys.* **32** 3-12
[4] Silva Patricio, Soto L, Kies W and Moreno J 2004 *lasma Sources Sci & Tech.* **13** 329-332
[5] Hussain S, Ahmad S, Khan M Z, Zakaullah M and Waheed A 2004 *J. Fusion Energy* **22** 195-200
[6] Rawat R S, Zhang T, Lim G J, Tan H, Ng S J, Patran A and Hassan S 2004 *J. Fusion Energy* **23** 49-53
[7] Shafiq M, Hussain S, Sharif M and Zakaullah M 2001 *J. Fusion Energy* **20** 113-115
[8] Soto L 2005 *Plasma Phys. Control. Fusion*, 47, A361-A381
[9] Gribkov V A 2000 *Nukleonika* **45** 149-153
[10] Zakaullah M, Akhtar I, Waheed A, Alamgir K, Shah Anwar Z and Murtaza G 1998 *Plasma Sources Sci.Technol.* **7** 206-218
[11] Castillo F 2000 *J. Phys. D: Appl. Phys.* **33** 141-147
[12] Mathuthu Manny, Zengeni Teddy G and Gholap Ashok V 1997 *IEEE Trans. Plasma Sci.* **25** 1382-1388
[13] Goudarzi S, Amrollahi R, Sadat Kiai S M, Morshedian N and Nasiri N 2005 *Czech. J. phys.* **55** 45-53

Journal of Physics: Conference Series **516** (2014) 012032 doi:10.1088/1742-6596/516/1/012032

Measurement and analysis of the radiation losses in DAMAVAND Tokamak

Shervin Goudarzi, Amir Raeisdana

North Kargar Ave., Atomic Energy Organization of Iran, Nuclear Science and Technology Research Institute, Plasma Physics and Nuclear Fusion Research School, Tehran, Iran

E-mail: shgoudarzi@yahoo.com

Abstract. Radiation losses play an important role on reaching to break-even conditions in Tokamaks. In this paper the results of measurement by a bolometer in Damavand Tokamak have been presented & analyzed. Meanwhile, we have explained our future research program on the base of last modifications in the control system of the DAMAVAND.

1. Introduction

Between candidates of the future Fusion reactors, Tokamak is the most important one and in 4 last decades very big projects have been done on it and in the 2006 the building of ITER have started in France. In Tokamaks fusion reactions happen in very high temperature plasmas (tens of million Kelvin) [1–5]. In the research with them, multiple parameters of the plasma like plasma temperature, plasma density, plasma current, radiation losses and so on must be measured [1, 6]. In a power plant the output power must be considerably more than the input power, thus the losses have high importance in energy balance in Tokamaks and must be known carefully. A lot of power losses in Tokamaks are related to the line radiation of impurities and radiation of neutral particles, especially in the VUV and SXR range [5,7].

It must be noted that a bolometer only obtains a spatial canal of information and the recorded signals by it related to radiated power from a finite part of the Tokamak plasma. With respect to that in our experiments the determination and control of the plasma position was impossible, we could not certainly conclude about the amount of radiation losses and only we could investigate the variation of it in general form.

By a new control system that installed 6 months ago on DAMAVAND, we can control and change the position of the Plasma with high accuracy, thus it is possible to bring the dense Plasma region in front of the window of bolometer and record the main part of the radiation losses, it is estimated that the accuracy of the results would be improved about 10 times.

2. DAMAVAND Tokamak

DAMAVAND is a Tokamak with a noncircular cross section (D-shape). For research activities in this device, several diagnostics have been used like: rogowskii coils for measuring the current, 40 magnetic probes for study of the magnetic field position, 2 saddle loops: one in the z- direction for study the vertical movement and another in the r-direction for study the horizontal movement, Langmuir probes, photoelectron spectrometer, microwave interferometer, charge-exchange analyzer, hard x-ray detector, bolometers for measuring the radiation losses, and so on. The main parameters of this device are noted in Table 1.

Table 1. The main parameters of the DAMAVAND

Parameter	Value
Major radius	36 cm
Minor radius	7 cm
Aspect ratio	5.1
Elongation of cross section	1.2 – 1.4
Toroidal magnetic field	1.2 Tesla
Maximum density of the plasma	3×10^{19} m^{-3}
Ion maximum temperature	150 eV
Electron maximum temperature	300 eV
Maximum plasma current	35–40 kA
Volume of the vacuum vessel	0.91 m^3
Maximum confinement time of the plasma	21 ms

3. Results and discussion

In DAMAVAND a bolometer with a pyroelectric detector (PD) is used for measurement of heat losses from the plasma by radiation and particle fluxes. It is made from a pellet of the ceramics made of lead circonate-titanate, this pellet is 10 mm in diameter and 400-500 μm thick. The PD sensivity is 5mV cm^2/W. The relation between amplitude of the output signal of the bolometer and the radiated power is approximately P = 1.18 V, that P and V are the power at watt and voltage at volt, respectively. The PD has high registration efficiency in the wave length range 0.5–300 A and in the range of the charge exchange neutrals with the energies 0.1-10 keV. The bolometer is placed upon a mobile coaxial lead-in and by its rotation moves back and forth.

In figures 1–4 four samples of the best results of measured signals by the bolometer in DAMAVAND in the experimental conditions noted in below with Hydrogen as working gas. P = 4×10^{-5} torr. Voltages of the capacitor banks:

TFB= 1.8 kV EFB$_1$= 1.7 kV EFB$_2$= 1.2 kV
FIB= 4kV SIB= 2.6 kV SFB= 2.6 kV

This is the optimum mode of the performance of DAMAVAND that in it the best results are I_{pmax}> 30 kA and Δt_{plasma} = 21 ms have been achieved; only in this mode the results of the experiments are repeatable and controllable.

In these figures horizontal axis shows time in milliseconds and the vertical axis shows the output signal of bolometer in volts that as noted before, its conversion factor is P = 1.18 V. It is observed from these figures that in all of these cases the maximum of the radiation dissipation power at any time is less than 6 watts and the plasma life time is about 18 ms or less. One can observe that the output signal of bolometer in this time interval changes rapidly and unorderly, that it can happen for 2 reasons. First, because the little distance of the bolometer from Tokamak, from its point of view the plasma seems as a great source of radiation and by this diagnostic we can only investigate a part of it and because that in these experiments the position of plasma is not controllable, with the movement of the plasma the zone of it that is situated opposite to the bolometer changes and therefore the input radiation flow of the bolometer will change according to the features (plasma density, density and type of impurities and so on) of this zone. Second, because the life time of the plasma of the DAMAVAND is low, the number of sampling that we can do in this interval is low, therefore, it is possible that in different samplings the number and energy of arriving photons to bolometer change randomly that it may make the differing of its output signal.

Also it is seen that in these cases the maximum values of signals and their time width are close together, the general forms of them are similar to each other, too. These results show that generally the values of the dissipated radiation power in all of these experiments are very close together.

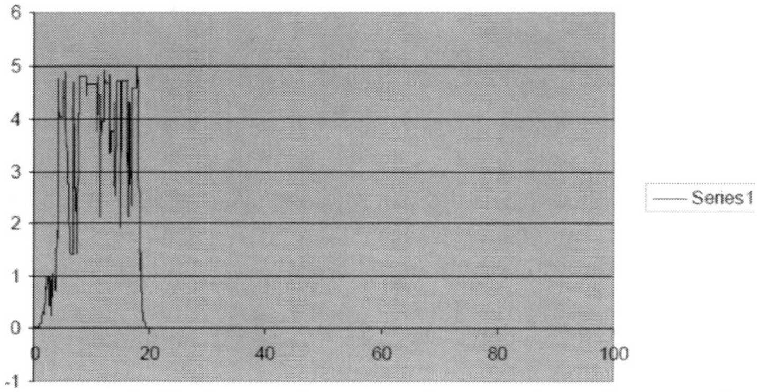

Figure 1. The output signal of bolometer in an experiment at the optimum mode of operation

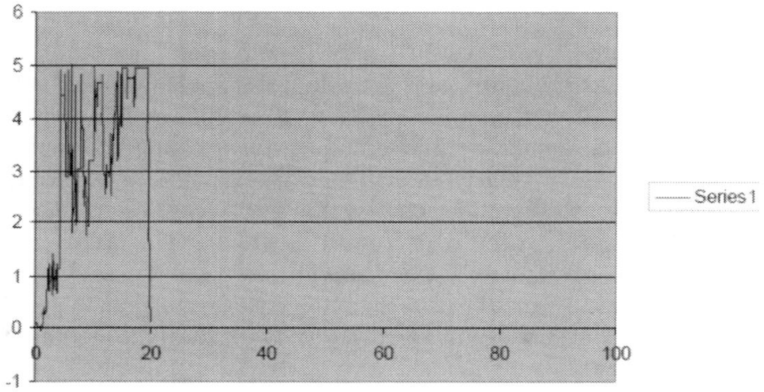

Figure 2. The output signal of bolometer in another experiment at the optimum mode of operation

From the investigation of the 20 experiments that have done in 2 days in DAMAVAND, it has been observed that in all of them the value of the amplitude of the plasma current and life time of the plasma have not a big difference (the maximum about 7–8 percent), the measured signals by bolometer in these experiments change rapidly up and down, but the maximum amplitudes of them and the life times of the plasma and the maximum of the plasma currents in all of experiments are very close together, that it can be a sign of the optimizing working condition in our experiments that causes the maximum value of plasma current and its life time and reproducibility of the experimental results.

It must be noted that a bolometer gives only a spatial channel of data. Therefore, these signals are related to the radiated power from a limited zone of the plasma of DAMAVAND that with respect to that in these experiments we had not the possibility of the determination and control of the position of this plasma, it was not possible to suggest carefully about the amount of the dissipated radiation power and we could only discuss about the form of its variation. By using a new control system that presently installed on DAMAVAND, it is possible to control the position of the plasma very carefully and change it desirable [8] and move the dense and hot zone of plasma opposite the window of bolometer and measure the main part of the dissipated radiation power, it is estimated that the results of the future experiments will be exact more than 10 times in comparison with now.

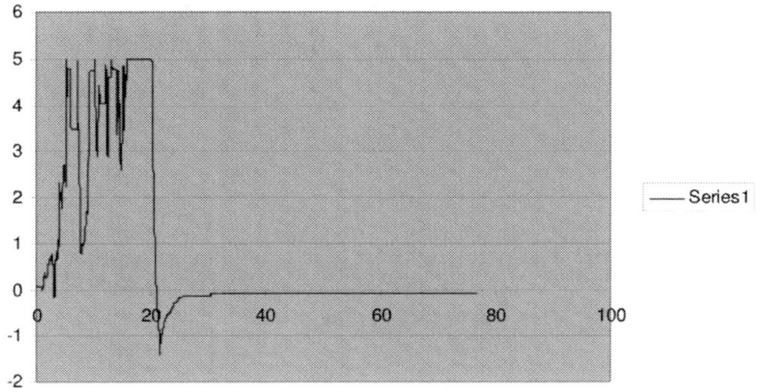

Figure 3. The output signal of bolometer in another experiment at the optimum mode of operation

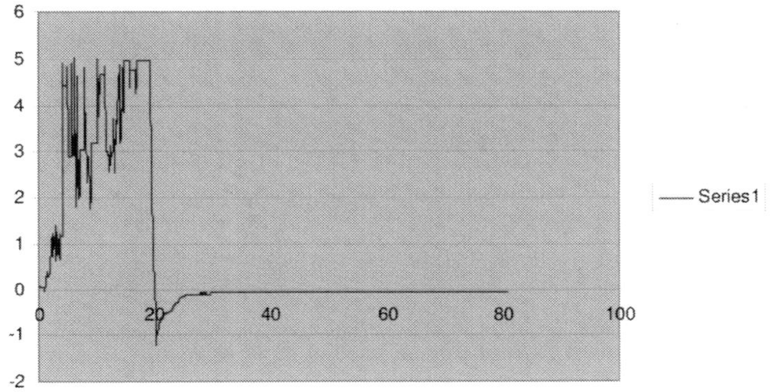

Figure 4. The output signal of bolometer in another experiment at the optimum mode of operation

4. Conclusions

The measured signals by bolometer in DAMAVAND Tokamak change very rapidly and abnormally. From investigation and analysis of these signals it seems that value of the radiation power dissipation in DAMAVAND in lifetime of the plasma is highly variable but there are not notable difference between the maximum of its amplitude and the total amount of it in multiple accomplished experiments at optimum mode. But for accurate estimation about the radiation dissipation we need to a proper control system that by it we can bring plasma to proper position and keep there as possible, such system is constructed for DAMAVAND and installed on it now and it is predicted that in future experiments the accuracy of the results will increase notably. Also, for accurate estimation of the dissipated power, one bolometer is not enough and an array that made of several bolometers must be used and by finding the spatial profile of the radiation power do an accurate estimation and it is one of the future programs in our research lab.

References

[1] Tatarov J W 1987 Tokamak *The Oxford Engineering Science Series*, (Oxford: Clarendon Press)
[2] Mc Cracken G and Stott P 2005 *Fusion : The Energy of Universe*
[3] Strelkov V S 2001 *Plasma Physics Reports* **27** 867–872
[4] Pandya S and Govindarajan J *Application of Thermal Imaging to Tokamak Fusion Plasma*
[5] Gray D S, Luckhardt S C, Chousal L and Gunner G 2004 *Rev. Sci. Instrum.* **75** 376–381
[6] Hutchinson H 1994 *Principles of Plasma Diagnostics* **4** (Cambridge: Cambridge Uiversity Press)
[7] Kundu M 2004 Time-Dependent Study of Hall Thruster Plasma *Plasma India* **19**
[8] Rasouli H, Fatehi A, Rasouli CH, Hefzolsehe A, Habibi H 2008 The control of vertical position of the plasma in DAMAVAND tokamak *Nuclear Conference of Iran*, Yazd

Avoided crossings of modes of a finite target with adiabat shaping in inertial fusion implosions

Ángel De Andrea González
Departamento de Física. Escuela Politécnica Superior.
Universidad Carlos III de Madrid. Av. Universidad, 30. 28911 Leganés. Spain.

E-mail: aandrea@fis.uc3m.es

Abstract. In Inertial Confinement Fusion (ICF), a cold target material is accelerated by a hot low density plasma. In this work, the small amplitude disturbances in ablative Rayleigh-Taylor instability in the presence of entropy gradients is studied using a sharp boundary approximation, being the inverse entropy-gradient scale length the quantity $K_o = \partial LnS_o / \partial y$. For simplicity, the target considered here is a finite incompressible slab. Due to ablation, the phenomenon of the interactions of two types of modes (Rayleigh mode and entropic mode) is shown. This manifest itself in the occurrence of rapid shifts of rate growth (bumping, avoided crossings). The entire situation is strongly reminiscent of the well-know "avoided crossings" of modes of two coupled oscillators. The growth rate of one mode (Rayleigh mode) approaches that of another one (entropic mode) which is "bumped" to a quite different rate growth, while the "bumping" mode settles at roughly at the original rate growth of the bumped mode.

1. Introduction

Anderson et al. [2], obtained a RT (Rayleigh-Taylor) dispersion for a semi-infinite incompressible slab relation, using a sharp boundary approximation, which is valid for kL_s ~kd >> 1 where d is the target thickness, and L_s is the entropy gradient scale length. However, this model does not take account the thickness slab; therefore, don't show the disappearance of the cut-off wave number due to the resonant interaction of the RT mode with the most unstable entropic mode.

The model presented here overcomes this obstacle: a more accurate prediction about the RT mode is obtained by assuming a slab with finite thickness. Our model is more complete than the Anderson´s model because, determinates all the entropic modes of the slab, showing the disappearance of the cut-off wave number. For simplicity, we are considering a planar target of thickness d (small compared with the ablation radius R of the spherical shell), which is moving with a constant acceleration $\vec{g}$ due the ablation pressure, generated by the heat flux coming from corona. Moreover, an entropy gradient $\vec{\nabla} S$, of the same direction and sense that acceleration, exist in the slab. This slab is continuously ablating with a mass ablation rate.

For simplicity, the ambient flow speed in the slab is very low, and it is generally useful to neglect it taking account the conservation of mass in the ablation front. Thus, the dynamical pressure inside the target and in the region of the ablating plasma close to the ablation front is small compared with the thermal pressure (the ablating plasma is very subsonic close to the ablation front). Also, the effect of the gravity on the ablating fluid can be neglected. In order to solve the initial value problem, we

Journal of Physics: Conference Series **516** (2014) 012033 doi:10.1088/1742-6596/516/1/012033

consider single interface (ablation front) which separates an impressible fluid (the slab region) from a non-uniform ablating plasma. In ablative RT [1], the growth rate is characterized by the Froude number, $F_r = v_a^2/(gL_o)$, where v_a is the ablation velocity, $v_a = \dot{m}/\rho_a$, $\dot{m}$ is the mass ablation rate, and ρ_a is the ablation density, being L_o the characteristic thickness of the ablation front.

2. Linear stability analysis and the initial problem value

Let $w_{sk}(y,s)$ denote the Laplace transform with respect to time, and the Fourier transform with respect to y of the flow velocity. So, for the y component of the perturbed flow we obtain the equation:

$$\frac{d^2 w_{sk}(y,s)}{dy^2} + \frac{K_o}{\gamma}\frac{d\, w_{sk}(y,s)}{dy} - k^2\left(1 - \frac{K_o g}{\gamma s^2}\right) w_{sk}(y,s) = Q(y,s) \tag{1}$$

Where γ is the ratio of specifics heats and g the slab acceleration, and:

$$Q(y,s) = -\frac{k^2 H(y,0)}{s^2} + \frac{1}{s^2}\frac{d^2 H(y,0)}{dy^2} + \frac{K_o}{\gamma}\frac{1}{s^2}\frac{d\, H(y,0)}{dy} \tag{2}$$

$$H(y,0) = s\, w\ (y,0) + \frac{d\, w\ (y,0)}{dt} \tag{3}$$

The boundaries conditions on each side are:

$$\left.\frac{s^2}{k^2 g}\left(\frac{K_o}{\gamma} w_{sk}(y,s) + \frac{d\, w_{sk}(y,s)}{dy}\right) - \frac{K_o H(y,0)}{\gamma k^2 g} - \frac{1}{k^2 g}\frac{dH(y,0)}{dy} - w_{sk}(y,s)\right|_{y=-d} = 0 \tag{4}$$

$$\left.\frac{s^2}{k^2 g}\left(\frac{K_o}{\gamma} w_{sk}(y,s) + \frac{d\, w_{sk}(y,s)}{dy}\right) - \frac{K_o H(y,0)}{\gamma k^2 g} - \frac{1}{k^2 g}\frac{dH(y,0)}{dy} - w_{sk}(y,s) + \frac{\beta}{r_d} w_{sk}(y,s)\right|_{y=0} = 0 \tag{5}$$

As shown in (Piriz et al.) [3] taking the blowoff density as a density in the blowoff region calculated at a distance of the order of the perturbation wave length from front the ablation front, the density jump r_d, for Fr>>1, yields $r_d = [4\beta/(5F_r)]^{2/5}$, being $\beta = kv_a^2/g$ the normalized wave number . Moreover, $\varepsilon = K_o d/\gamma$ is the normalized inverse entropy-gradient scale length; kd the normalized thickness; $s\, v_a/g$ the normalized growth rate.

2.1. *Results and discussion*

These transforms can be inverted explicitly to express the fluid variables as integrals of Green´s functions multiplied by initial data. For more simplicity, taking $\varepsilon = 0.05$ and $\mathrm{Fr} = 6$, we obtain numerically the following dispersion relations which are poles of the Green's function (see figure 1). It is explicitly demonstrated that when the growth rates of different modes are almost identical, strong mode coupling occurs (resonant interaction). The existence of coupling between branches shifts their point of intersection into the complex region: an avoided crossing is produced.

The obtained figure shows, in finer detail, the avoided crossings between the Rayleigh Taylor mode and entropic modes. It easy to notice that the most dangerous mode is the Rayleigh mode. Moreover, the eigenmodes have an accumulation point at 0.

So, when the normalized wave number diminishes, the phenomenon of mode "bumping" [4-5] can be observed: the growth rate of a certain mode approaches that another one which is "bumped" to a quite different growth rate, while the "bumping" mode settles at roughly the original growth rate of the bumped mode. In this work, when the normalized wave number increases, the first entropic mode appears to be the continuation of the Rayleigh mode; the second entropic mode appears to be the continuation of the first entropic mode...

It is interesting to obtain the analytical relation dispersion for entropic modes without ablation, which is in good agreement with the numerical result shown in figure 1 (except at avoided crossings):

$$\frac{sv_a}{g} = \frac{kd\sqrt{\frac{K_o d}{\gamma}\frac{kv_a^2}{g}}}{\sqrt{(m\pi)^2 + (kd)^2 + \frac{(kd)^2}{4}\left(\frac{K_o d}{\gamma}\right)^2}}; \quad m=1,2,3... \tag{6}$$

The entropic modes form an infinity discrete set. As for entropic modes, these are quantized. In practice, one is usually interested in the first two or three fastest growing modes.

On the other hand, as $kd \to \infty$ the separation of the entropic eigenvalues goes to zero. Thus, we approach a continuous spectrum of entropic eigenvalues.

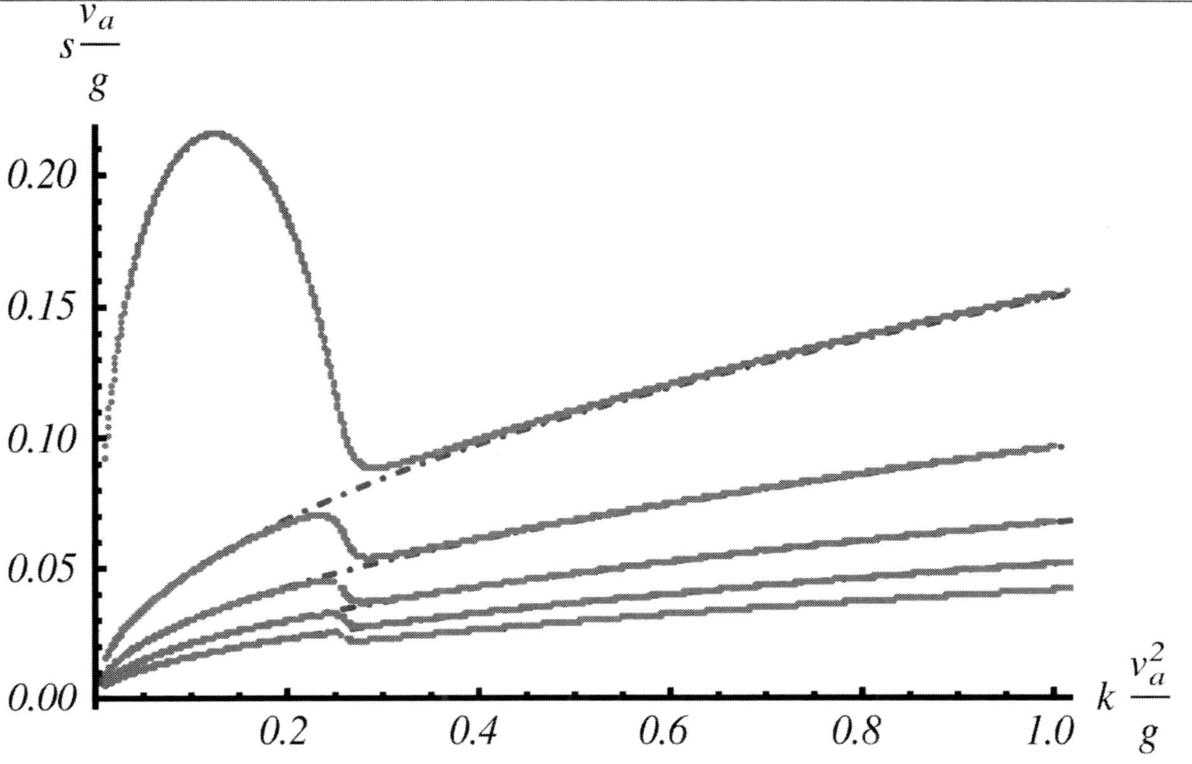

Figure 1. Dimensionless normalized growth rate versus dimensionless normalized wave number for $\varepsilon = 0.05$, $\mathrm{Fr} = 6$, and the normalized thickness kd=3. Dot-Dashed line corresponds to the analytical results; The solid lines corresponds to our model . Rapid shifts of rate growth (bumping, avoided crossings) are shown.

3. Conclusion

The effect of avoided crossings in ablative Rayleigh-Taylor instability in the presence of entropy gradients is analyzed. So, we observe the mode "bumping" which explain the disappearance of the cut-off wave number in a slab with finite thickness.

4. References

[1] Gardner J H, Bodner S E and Dahlburg J P 1991 *Physics of Fluids B: Plasma Physics* **3** 1070-74
[2] Anderson K and Betti R 2003 *Physics of Plasmas* **10** 4448-62
[3] Piriz, A R, Sanz J and Ibañez L F 1997 *Physics of Plasmas* **4** 1117-26
[4] Osaki Y 1975 *Astronomical Society of Japan* **27** 237-58
[5] Christensen-Dalsgaard J 1999 *Mon. Not. R. Astr. Soc.* **190** 765-91

Journal of Physics: Conference Series **516** (2014) 012034 doi:10.1088/1742-6596/516/1/012034

Positron Momentum Calculation of Defects in α-Fe and Fe-Cr Containing Hydrogen and Helium

E Popov, P Staykov, K Berovski, S Peneva, N Djourelov and T Troev

Institute for Nuclear Research and Nuclear Energy, Bulgarian Academy of Sciences, 72, Tzarigradsko Chaussee Blvd., 1784 Sofia, Bulgaria

E-mail: evpetropov@abv.bg

Abstract. Lattice defects that contain hydrogen and helium in Fe and Fe-Cr have been studied by positron model calculations using Density Functional Theory (DFT) in its Local Density Approximation (LDA) implementation. It was found that the value of positron lifetime for perfect Fe lattice is 101.6 ps. and for mono vacancy is 176.6 ps. which a close to the experimental values. The calculated values of Positron Life-Time (PLT) for Fe-2.5% wt Cr is 104.7 ps., for Fe-5% wt Cr is 127.6 ps., for Fe-8% wt Cr is 128.9 ps. and for Fe-9% Cr is 128.2 ps. A correlation between the positron lifetime and H and He atom concentration has been established. The model calculations for PLT and Coincidence-Doppler-Broadening (CDB) for Fe-Cr containing vacations in the presence of different numbers of hydrogen and helium atoms will be presented.

1. Introduction

Positron quantum-mechanical calculations were carried out for investigation of defects in Fe, Cr and Fe-Cr alloy containing hydrogen and helium atoms. Recently the convergence of electronic structure calculations for super sell size have been studied for mono-vacancy, bi-vacancy and vacancies contacting hydrogen and helium atoms [1,2]. The electron wave function in iron, chromium and Fe-Cr alloy has been obtained in LDA [3] to the DFT. On the base of the DFT and its two-component extension (TCDFT) [4] determination of electron and positron states in solid has been done. The Positron Life-Time (PLT) and Coincidence Doppler Broadening (CDB) have been solved a set of one-particle Kohn-Sham equation for electrons and positron. The positron model calculations have been performed in bcc iron lattice containing vacancies and various number of hydrogen and helium atoms into them.

On the base of the calculated results, the behaviors of the vacancies and vacancies with different number of hydrogen and helium atoms were discussed. There is a change of the influence of core electrons with the increase of the percentage of chromium in the Fe-Cr alloy. The lifetime of trapped positron is sensitive to the vacancies defect as well as to vacancies containing various ratio of hydrogen and helium atoms [5,6].

2. Methodology

The theory of this presentation is very short. For more exact treatment the reader can be referred to the papers of Schultz, Lynn and Saarinen, Hautojärvi, Corbel [1, 7].

2.1. PLT

The positron annihilation rate is proportional to the electronic density at the positron position and can be calculated from the overlap integral as:

$$\lambda = \frac{1}{\tau} = \pi r_0^2 c \int d\,\mathrm{r}\, n_+(\mathrm{r}) n_-(\mathrm{r}) g(0;n_+;n_-), \tag{1}$$

where r_0 is the classical electron radius, c is the speed of light, n_+ and n_- are the positron and electron densities and $g(0;n_+;n_-)$ the electron-positron pair correlation functional evaluated at the positron [1]. Explanation for the obtaining of the densities and the factor g has been done in the work of Puska and Nieminen [2]. The zero-positron density limit of the pair-correlation function $g(0;n_+;n_-)$ is denoted by γ and called the enhancement factor of the electron density at the positron. The most trivial form is simply $\gamma = 1$, which is called the Independent-Particle Model (IPM) approximation.

2.2. CDB

The momentum distributions are calculated separately for each electron state described φ_j by according to the equation:

$$\rho_i(\mathrm{p}) = \pi r_0^2 c \frac{\lambda_j^{LDA}}{\lambda_j^{IPM}}(0) \left| \int d\,\mathrm{r}\, e^{-i\mathrm{p}\cdot\mathrm{r}} \varphi_+(\mathrm{r}) \varphi_j(\mathrm{r}) \right|^2, \tag{2}$$

where φ_+ is the positron wave function [8]. The λ_j^{LDA} is the annihilation rate within the local density approximation. It is calculation as:

$$\lambda^{LDA} = \pi r_0^2 c \int d\,\mathrm{r}\, \gamma^{LDA} |\varphi_+(\mathrm{r})|^2 |\varphi_-(\mathrm{r})|^2 . \tag{3}$$

The final momentum distribution is then obtained by summing $\rho_i(\mathrm{p})$ in Eq. (2) over all electron states.

A parameterized positron wave function is used in the calculation of momentum distributions,

$$|\varphi_+\rangle \approx a_0 + a_1 \left(erf\left(\frac{r}{a_2} \right) \right)^{a_3} = a_0 + a_1 \left(\frac{1}{\sqrt{2\pi}} \int_{-\infty}^{r} e^{\left(-\frac{r}{\sqrt{2}a_2}\right)^2} dr \right)^{a_3}, \tag{4}$$

where r is the distance from the center of the atomic sphere and a_i are fitted parameters, $i \in \{0,...,3\}$. The parameters have been calculated using the LMTO-ASA method [9].

3. Results

From calculations ware obtained the fallowing results:

3.1. PLT

In the Table 1 are sown a results for PLT that are obtained with LDA method.

Table 1. PLT Fe-Cr alloys - Model calculations (LDA method comparison with [10,11])

wt 2.5%Cr	bulk 104.7 ps.			
	1v 229.1 ps.	1v1H 162.3 ps. 1v1He 141.4 ps.	1v2H 158.7 ps. 1v2He 126.4 ps.	1v4H 129.8 ps. 1v4He 105.3 ps.
	2v 236.0 ps.	2v2H 167.4 ps. 2v2He 146.2 ps.	2v4H 160.7 ps. 2v4He 134,8 ps.	2v8H 131.1 ps. 2v8He 108.9 ps.

wt 5%Cr	bulk 127.6 ps.			
	1v 233.1 ps.	1v1H 165.4 ps. 1v1He 143.8 ps.		
	2v 239.1 ps.			2v8H 133.2 ps. 2v8He 109.7 ps.
wt 8%Cr	bulk 128.9 ps.			
wt 9%Cr	bulk 128.2 ps.			
	1v 233.9 ps.	1v1H 165.9 ps. 1v1He 144.3 ps.		
	2v 239.7 ps.	2v2H 169.7 ps. 2v2He 147.8 ps.		2v8H 133.5 ps. 2v8He 109.9 ps.
Cr – crystal comparison with [9]	perfect lattice 103.4 ps.			
	1v 179.7 ps.			
	2v 202.3 ps.			2v8H 111.2 ps. 2vHe 101.6 ps.
Fe - crystal comparison with [8]	perfect lattice 101.6 ps.			
	1v 176.6 ps.	1v1H 137.6 ps. 1v1He 129.2 ps.	1v2H 122.6 ps. 1v2He 112.4 ps.	1v4H 103.8 ps. 1v4He 99.1 ps.
	2v 198.8 ps.	2v2H 154.3 ps. 2v2He 140.0 ps.	2v4H 134.6 ps. 2v4He 119.1 ps.	2v8H 109.0 ps. 2v8He 99.6 ps.

3.2. CDB

It can be compared an experimental results with model calculation for iron. This is done in Figure 1.

There is a deviation of the model calculations from the experimental results for α-Fe. It is necessary to complete the relocation component in the wave function of the positron. An improvement has been done to the electron-positron correlation-energy functional and to the exchange-correlation energy in [15]. However via this calculation a good quality description can be obtained of the momentum distribution. On the fallowing Figures 2a÷2c has been done study of the momentum distributions for Fe and Fe-Cr alloys with 2.5% and 9% Cr with 2 vacancies containing 8H and 8He.

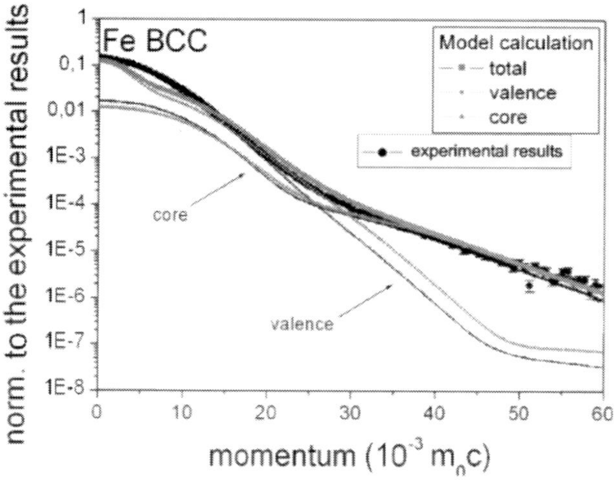

Figure 1. Comparison of the momentum distribution for α-Fe of our model calculations with experimental results [14]

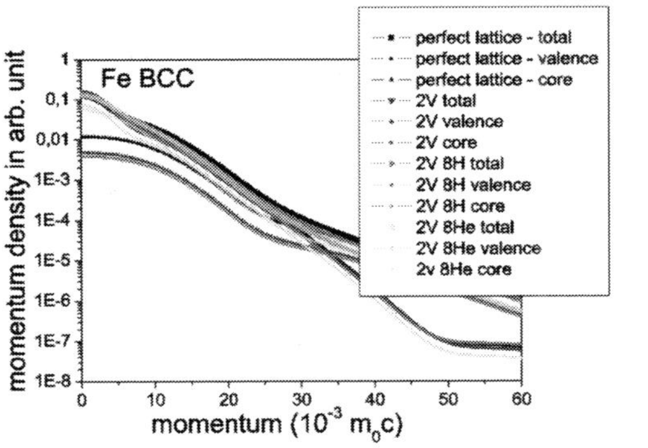

Figure 2a. α-Fe momentum distribution for 2 vacancies containing 8H and 8He

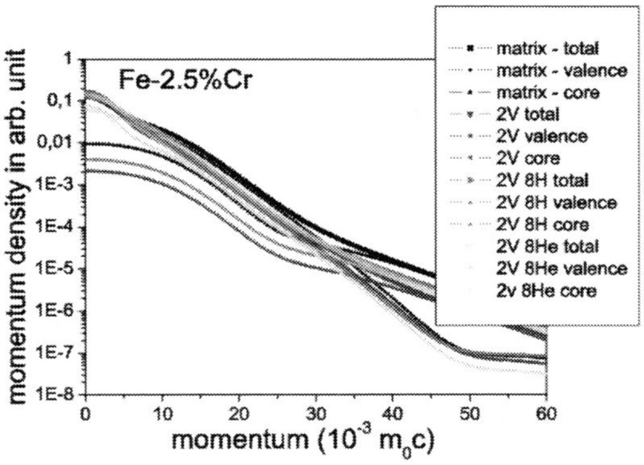

Figure 2b. Fe-2.5%Cr momentum distribution for 2 vacancies containing 8H and 8He

4. Conclusions

Model calculation of the positron annihilation characteristics in Fe and Fe-Cr have been done by LDA method.

The influence of H and He atoms on the momentum distribution of Fe and Fe-Cr are carried out for different concentration of gas atoms.

It is found a correlation between the calculated positron lifetime and the Cr concentration in Fe matrix.

The calculated momentum distribution and positron lifetime are different from those in the bulk as well as in pure vacancies and vacancy-clusters (nano-voids) containing hydrogen and helium atoms.

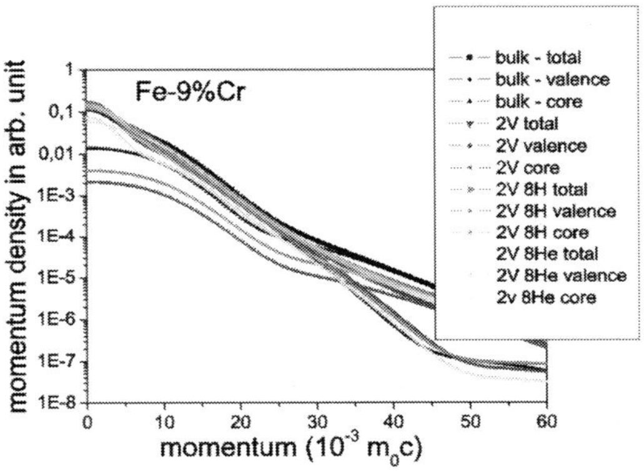

Figure 2c. Fe-9%Cr momentum distribution for 2 vacancies containing 8H and 8He

References

[1] P J Schultz, K G Lynn, Rev. Mod. Phys., 1988, **60**, 701-779
[2] M J Puska, R M Nieminen, Rev. Mod. Phys., 1994, **66**, 841-897
[3] B L Shivachev, T Troev, T Yoshiie, J. Nucl. Mater., 2002, **306**, 101
[4] R M Dreizler, E K U Gross, Density Functional Theory, Springer-Verlag, 1990
[5] Laboratory of Physics Helsinki University of Technology, 2003, Doppler: A program to model positron states and annihilation in solids
[6] T Ishizaki, Q Xu. T Yoshiie, S Nagata, T Troev, J. Nucl. Mater., 2002, **961**, 307-311
[7] K Saarinen, P Hautojärvi, C Corbel, Academic Press, 1997, **A 388**, 434
[8] M Alatalo, B Barbiellini, M Hakala, H Kauppinen, T Korhonen, M J Puska, K Saarinen, P Hautojärvi, and R M Nieminen, Phys. Rev., 1996, **B 54**, 2397
[9] P Valminen, Special Assignment, Helsinki University of Technology, Espoo, 2002
[10] V Kršjak, V Slugeň, M Mikloš, M Petriska and P Ballo, 2008, Ap. Surf. Sc. **255** 153–156
[11] M. Lambrecht, Open Report SCK CEN-BLG-1067, 2009, 48-50
[12] T Troev, A. Markovski, S. Peneva, T Yoshiie, J. Nucl. Mater., 2006, **359,** 93-101
[13] L Petrov, T Troev, N Nankov, E Popov, J. of Phys.:Conference Series, 2010, **207,** 012037
[14] Z Tang, M Hasegawa, Y Nagai, M Saito, 2002, Phys. Rev. B, **65**, 195108
[15] H Stachowiak, E Boroński, 2008, ACTA Phys. Pol. A, **113,** 1524-1532

AUTHOR INDEX